Aerospace Psychology and Human Factors

Aerospace Psychology and Human Factors

Applied Methods and Techniques

Ioana V. Koglbauer and Sonja Biede-Straussberger

(Editors)

Library of Congress of Congress Cataloging in Publication information for the print version of this book is available via the Library of Congress Marc Database under the Library of Congress Control Number 2024940701

Library and Archives Canada Cataloguing in Publication
Title: Aerospace psychology and human factors : applied methods and techniques / Ioana V. Koglbauer and Sonja Biede-Straussberger (eds.).
Names: Koglbauer, Ioana V., editor. | Biede-Straussberger, Sonja, editor.
Description: Includes bibliographical references.
Identifiers: Canadiana (print) 20240403231 | Canadiana (ebook) 20240403274 | ISBN 9780889376472 (softcover) | ISBN 9781616766474 (PDF) | ISBN 9781613346471 (EPUB)
Subjects: LCSH: Aeronautics—Human factors. | LCSH: Aviation psychology.
Classification: LCC TL553.6 .A38 2024 | DDC 629.1301/9—dc23

www.hogrefe.com

PUBLISHING OFFICES

USA:	Hogrefe Publishing Corporation, 44 Merrimac St., Newburyport, MA 01950 Phone 978 255 3700; E-mail customersupport@hogrefe.com
EUROPE:	Hogrefe Publishing GmbH, Merkelstr. 3, 37085 Göttingen, Germany Phone +49 551 99950 0, Fax +49 551 99950 111; E-mail publishing@hogrefe.com

SALES & DISTRIBUTION

USA:	Hogrefe Publishing, Customer Services Department, 30 Amberwood Parkway, Ashland, OH 44805 Phone 800 228 3749, Fax 419 281 6883; E-mail customersupport@hogrefe.com
UK:	Hogrefe Publishing, c/o Marston Book Services Ltd., 160 Eastern Ave., Milton Park, Abingdon, OX14 4SB Phone +44 1235 465577, Fax +44 1235 465556; E-mail direct.orders@marston.co.uk
EUROPE:	Hogrefe Publishing, Merkelstr. 3, 37085 Göttingen, Germany Phone +49 551 99950 0, Fax +49 551 99950 111; E-mail publishing@hogrefe.com

OTHER OFFICES

CANADA:	Hogrefe Publishing Corporation, 82 Laird Drive, East York, Ontario, M4G 3V1
SWITZERLAND:	Hogrefe Publishing, Länggass-Strasse 76, 3012 Bern

Printed and bound in the Czech Republic

ISBN 978-0-88937-647-2 (print) • ISBN 978-1-61676-647-4 (PDF) • ISBN 978-1-61334-647-1 (EPUB)
https://doi.org/10.1027/00647-000

Dedication

To my family, Reinhard, Alina and Dan, for their love and support

Ioana V. Koglbauer

To my daughter, Mina, who inspires with courage and strength

Sonja Biede-Straussberger

Acknowledgments

The editors would like to thank the following organizations and individuals:

For Supporting This Book Project

Gunnar Steinhardt, President of the European Association for Aviation Psychology (EAAP), and the Board of Directors of the EAAP – Renée Pelchen-Medwed, Mickaël Causse, Julia Behrend, Jennifer Eaglestone, Robert Bor, and Jóhann Wíum

For Expertly Reviewing Parts of This Book

René Amalberti, Ciprian Baciu, André Droog, Renée Pelchen-Medwed, and Peter Sandl

Each chapter of this book was reviewed by three independent peer reviewers.

For Their Assistance From Contracting to Publishing and Marketing

The dedicated team at Hogrefe Publishing

Contents

Foreword

Our professional community is dedicated to developing and sharing knowledge beyond our own domains, enriching disciplines and ensuring we anticipate future challenges to get it right from the onset when crafting new designs or procedures. As the aerospace sector encounters new technological and societal challenges affecting operators and passengers, it is essential for practitioners and scientists to collaborate, refining the integration of the human element into the overarching sociotechnical system. This book, co-edited by Ioana Koglbauer and Sonja Biede-Straussberger and featuring contributions from experts in academia, industry, and international agencies, marks a significant step in advancing the human aspect of aerospace. It delves into both present and future methodological trends in aviation psychology and human factors. This volume fosters interdisciplinary learning and collaboration, essential for effective human performance management. It does so by offering discussions on research methods, practical "how to do it" guidance, insights from past experiences, and projections of future trends. The ethos of this book portrays the spirit of the European Association for Aviation Psychology (EAAP), which aims at promoting applied psychology and human factors in aviation, ensuring the dissemination of information and experience. Everyone, whether newcomers or seasoned experts from academia, industry, or government, interested in the human-centric approach in aerospace systems design and operation, will find invaluable insights and guidance in these chapters.

Gunnar Steinhardt
President of the European Association for Aviation Psychology

Preface

How to Put People First in the Design and Operation of Aerospace Systems

Nearly 80 years ago, human factors became an area of interest in aviation. Since then, the world has rapidly evolved: Changes have occurred, knowledge has improved, experience has grown, society has changed, and new technologies have been invented. As people and organisations involved in aviation and space dream bigger and as the technical possibilities develop at a fast pace, focus needs to be maintained on integrating the human element in the system. More than ever, the maintenance of and even the increase in the current level of safety are of utmost importance.

Despite all these advances, human factors and aerospace psychology professionals still need to strive for the integration of the human element throughout business, development and operations. Especially in such a complex system as that of aerospace, we need to ask ourselves whether we are solving the right problems. Once the right problems are identified, the next question is how to solve them in the right way. Which industry standards are suitable and applicable? Where are the gaps? Which scientific methods can help bridge the gap between the status quo and future performance expectations? Is the human element appropriately addressed in each stage of a system's life cycle? Are interdisciplinary perspectives convergent and harmonised?

Thus, as psychologists and human factors specialists who drive and enable these innovations, we are often confronted with new questions that cannot be answered by conventional means. Sometimes we need to adapt or develop new methods or tools to address them. In this book experts working in the industry and in academia share methods and techniques of aerospace psychology and human factors that are currently used in research and development. Thus, our intention with this new book is to provide a wide range of methods, techniques and tools for promoting the application of aerospace psychology and human factors. All of this serves to build better products for operators. These operators want to be efficient in their tasks. Their objective is to deliver safe and efficient operations.

Several chapters of this book try to grasp the role of human factors from a more global perspective, such as describing the current practice in specific organisations, whereas others zoom in on addressing specific problems, such as how to capture human performance. At the beginning, design methods are addressed. The first chapter provides a systems-theoretic perspective and a method for modelling emergent system properties in existing systems or in those that are in development.

Professionals looking for a powerful and efficient tool to identify the right problems in a system and to address them will find it here. In the next chapter, the use of a cockpit philosophy is highlighted to support the transition from initial concepts to detailed designs that justify certification requirements. A special chapter is dedicated to human factors challenges in cabin design for commercial aircraft. New questions on how to design assistance are addressed in a different chapter, a topic which has rapidly spread in multiple industries over recent years.

The design environments have a number of different aerospace psychology and human factors topics in common that are addressed in the following sections of the book. The introduction of new artificial intelligence technology poses new challenges and requires new solutions on how a system can explain information to operators. Combined measures of workload and situation awareness are integrated in a model to support the assessment of performance from a team's perspective. Operators' attention and awareness are addressed in the context of pilot monitoring, and the benefits and drawbacks of the current eye-tracking technology are analysed for both design and training. Furthermore, techniques to improve fatigue risk management systems by adding additional parameters to identify and monitor risks are presented. These are expanded with a chapter on hazards related to human space flight and methods to analyse them. Another chapter is dedicated to the development of a free flying virtual companion for an astronaut and methods to implement various humanlike features in the area of tension between the machine and the "uncanny valley."

Another section of the book is dedicated to the use of virtual, augmented, or mixed reality technologies that found their way into daily aviation business. They are studied in depth to investigate how they may better support operations and training, in application fields from cockpit to air traffic control or even maintenance. A number of chapters cover human factors methods and techniques related to this issue. A special chapter is dedicated to virtual reality applications for developing and testing the Argonaut Lunar Lander. An additional chapter addresses applications of extended reality for studying human behaviour in immersive conditions, manipulating mental workload, prototyping, and evaluating complex interfaces. In addition, challenges of virtual reality and techniques to overcome them are presented in the context of pilot training. This section is rounded up by a chapter on methods to prioritise and implement augmented reality-based innovations for pilot training in a sustainable manner.

The final section of the book includes methods and techniques that provide a broader view of how to systematically learn from past research and to plan future developments. Thus, the method for conducting a meta-analysis is explained, an approach that will be more frequently used to gain knowledge by aggregating results of a large number of studies. In a different chapter, a method for integrating the assessment of human readiness level in the Single European Sky Air Traffic Management Research (SESAR) is presented. The final chapter takes a look at where a major aircraft manufacturer stands in the process of integrating human and organisational factors throughout the organisation along key principles to be taken into account (e.g., competencies) and anticipates the impact of new technologies and a changing society.

The sociotechnical aerospace system has rapidly evolved and continues to change, as we see in the current sociopolitical context. New challenges will emerge that are far from being anticipated today. Whatever those challenges will be, our strongly connected and interdisciplinary community of professionals will strive to put the human at the centre and do their best for society.

Sonja Biede-Straussberger & Ioana V. Koglbauer

Chapter 1

Integrating Human Factors Into the System Design Process

Brittany Bishop, Pauline Harrington, Nancy Leveson, and Rodrigo Rose

Abstract

Hazard analysis is the basis of engineering for safety. However, in such analyses, human factors are often oversimplified as simply "human failure," disregarding the systemic issues that lead to flawed decisions. A new, more powerful hazard analysis technique, called "system-theoretic process analysis" (STPA), combines sophisticated human factors, hardware design, software design, and even social systems in one integrated model and analysis. STPA can be used to identify conditions and events that can lead to an accident or mission loss so that designs can prevent or minimize losses. Safety assurance is typically carried out separately from system design and in later stages of development. By the time these assurance processes are used, it is often too late to effectively modify a system to address any safety issues that are found. STPA assists in overcoming these problems when used by an integrated team of engineering specialists, including human factors experts, to identify potential scenarios leading to unsafe behavior starting from the beginning of the design process.

Keywords

aviation psychology, human factors in system engineering, system safety engineering, STPA

The Goal

Hazard analysis is the foundation of engineering for safety. It is used to identify the hazards, which are defined as system states or sets of conditions that, together with a particular worst-case environment, will lead to a loss (Leveson, 2012). Once identified, this information can be used in system development and operations to eliminate these hazards or, if that is not possible, to reduce their likelihood or to minimize their potential impact. Unfortunately, the complex software-intensive systems being built today cannot be fully analyzed using traditional hazard analysis techniques. In addition, human contributions to risk have traditionally been oversimplified by engineers in the hazard analysis process, thus limiting the usefulness of the hazard analysis process in reducing overall system risk.

The role of humans is changing as our systems become increasingly automated. Rather than directly controlling a potentially dangerous system, operators today are more often supervising automation and taking over in the cases where automation is not able to cope. It is no longer useful to only look at simple human mistakes in reading a dial or operating controls. The cognitively complex activities in which operators are now engaged do not lend themselves to simple failure analyses.

At the same time, some systems are designed such that a human error is inevitable, and then the loss is blamed on the human rather than on the system design (Leveson, 2019). Hazards may result from automation design that induces erroneous or dangerous operator behavior. Sometimes interface changes can alleviate these human errors, but often interface design fixes alone are not enough.

Human–machine *interactions* are greatly affected by the design of both the software and the hardware in concert with the design of the activities and functions provided by the operator. Changing the software, hardware, and human activities is the most direct and effective way to eliminate interaction problems as opposed to simply changing the interface between the human operator and the rest of the system. To reduce risk most effectively, the design or redesign of the functionality of the software and hardware and of the activities assigned to the operator and to the automation is needed rather than merely the design or redesign of the displays and controls.

In addition, today's complex, highly automated systems argue for the need for integrated system analyses and design processes. In the analysis and design of complex systems, it is not enough to separate the efforts in hardware design, software design, and human factors. Successful system design can only be achieved by engineers, human factors experts, and application experts working together. Obstacles to this type of collaboration stem from limitations in training and education, the lack of common languages and models among different specialties, or an overly narrow view of one's responsibilities. These obstacles need to be overcome to successfully build safer systems. This chapter presents an approach involving new modeling and analysis tools that will allow all the engineering specialties to use common tools and work more effectively together.

An overriding assumption in this chapter is the systems theory principle that human behavior is impacted by the design of the system in which it occurs. If we want to change operator behavior, we have to change the design of the system in which the operator is working. For example, if the design of the system is confusing the operators, (1) we can try to train the operators not to be confused, which will be of limited usefulness, (2) we can try to fix the problem by providing more or better information through the interface, or (3) we can redesign the system to be less confusing. The third approach will be the most effective.

Simply telling operators to follow detailed procedures that may turn out to be wrong in special circumstances or relying on training to ensure they do what they "should" do – when that may only be apparent in hindsight – will simply guarantee that unnecessary accidents will occur. The alternative is to ask how we can design to reduce operator errors or, conversely, identify what design features induce human error. In other words, we must design to support the operator.

A New Foundation for Integrated System Analysis

Achieving this goal will require new modeling and analysis tools. Traditional hazard modeling and analysis techniques do not have the power to handle complex systems today. They are based on a very simple model of causality that assumes accidents are caused by component failures. A new model of accident causality, called the "system-theoretic accident model and process" (STAMP), comprises more complex types of causal factors, including interactions among system components and including the operators (Leveson, 2012). In this enhanced model of causality, accidents may result from unsafe interactions among components that may not have "failed." In other words, each system component satisfies the specified requirements but the overall system design is unsafe. For example, the software and hardware satisfy their specified requirements and the operators correctly implement the procedures they were taught to use.

As an example, consider the crash of a Red Wings Airlines Tupolev (Tu-204) aircraft that was landing in Moscow in 2012. A soft touchdown made runway contact a little later than usual. There was also a crosswind, which meant that the weight-on-wheels switches did not activate. Because the software did not think that the aircraft was on the ground and because it was programmed to protect against activation of the thrust reversers while in the air (which is hazardous), the command of the pilot to activate the thrust reversers was ignored by the software. At the same time, the pilots assumed that the thrust reversers would deploy as they always do, and quickly engaged high engine power to stop sooner. Instead, the pilot command accelerated the aircraft forward, eventually colliding with a highway embankment (Leveson & Thomas, 2018).

Note that nothing failed in this accident. The software satisfied its requirements and behaved exactly the way the programmers were told it should. The pilots had no way of knowing that the thrust reversers would not activate. There were no hardware failures. The software performed exactly as it was designed to do. The humans acted reasonably. In complex systems, human and technical considerations cannot be isolated.

These types of accidents are enabled by the inability of designers and operators to completely predict and understand all the potential interactions in today's tightly coupled and complex systems. That is, the error is in the overall system design and how the system components interact and not in the individual components. These types of accidents, which are increasingly occurring in today's complex systems, cannot be handled with the traditional linear causality model and hazard analysis techniques.

STAMP, by contrast, treats safety as a control problem rather than a failure problem. In other words, accidents result when the system design does not control hazardous system states. Those hazardous states may result from component failures, but they may also arise from overall system design flaws.

In this chapter, we describe and illustrate a new hazard analysis technique, called "system-theoretic process analysis" (STPA), which is built on STAMP and is more powerful than the traditional hazard analysis techniques. STPA is a structured step-by-step process for identifying the ways that hazards can occur in a system. It integrates hardware, software, and human factors into one modeling and analysis process and enhances the design process by allowing for shared modeling and analysis efforts (Leveson, 2012; Leveson & Thomas, 2018).

The Concept of Control in Safety

As noted, STAMP treats safety as a control problem. The system design must control both the component failures and the unsafe interactions among the components.

STPA uses a simple model of control in the form of feedback control loops. Such a loop is illustrated in Figure 1.1. At the top of the figure is the operator, who provides commands to automation as well as, in some cases, directly to the controlled process. For example, the driver of the vehicle may issue acceleration and braking commands. The operator gets feedback about the state of the controlled process directly (e.g., by feeling or seeing the vehicle slow down or accelerate) or through electronic displays. Even with highly automated systems, human operators often get feedback in addition to that provided by the displays, such as from sound, vibration, etc., which cannot easily be communicated through an electronic interface.

Figure 1.1 shows two components within the operator box. A human mental model contains the information the operator uses to make control or monitoring decisions. The mental model contains what the operator *thinks* is the current state of the automation (e.g., the brakes are being activated), the controlled process (the aircraft is slowing or accelerating), and relevant parts of the environment. The mental model is updated by various means, but primarily from feedback. Other information operators may use to update this mental model include beliefs they have about how the process can change and inferences about the effect of previous commands the operator issued to the automation – and assumes were executed

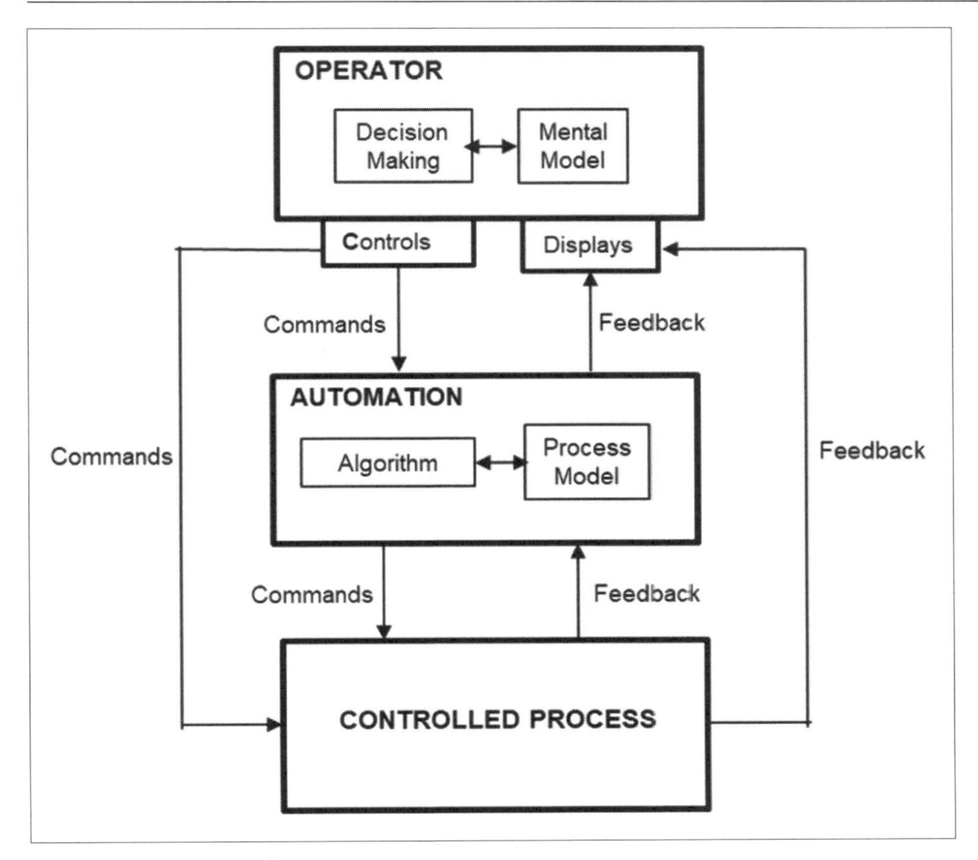

Figure 1.1. A basic model of system control.

correctly. An example of the latter is the belief that the thrust reversers would activate on the Red Wings aircraft mentioned earlier because the operator had commanded the software to activate them.

Note that computer automation also has a model of the state of the process. This model is usually much simpler than human mental models and may simply be represented as a few variables in the memory of the computer or in the software algorithm.

The automated controllers update their process models through direct feedback from sensors in the system and through human controller input. For example, an altimeter tells the automated controller the altitude of an aircraft, and a pilot may tell the automated controller what the desired altitude is.

Human controllers update their mental models of the controlled process, the environment, and the automation through direct feedback they receive from the system (i.e., displays, alerts, observed system behavior, etc.). Updates to the mental model of the environment can also occur through direct stimuli from visual, auditory, or vestibular systems, such as a pilot seeing clouds through the windshield. Any of these models can also be updated by information from another human controller in the system, such as the copilot in an aircraft. In the control structure, these relationships will be modeled as arrows flowing into the human and automated controllers.

In basic feedback control loops like the one shown in Figure 1.1, feedback from the controlled process is used by the controller of that process to adjust the system's behavior to achieve the system goals and avoid hazards. In this way, the feedback received by the controller is used to guide decision-making for future control actions.

In a feedback control loop, the actions available to controllers to manage the process are termed "control actions" (commands) and are represented by downward arrows (see Figure 1.1). *Feedback*, which is used to inform the decisions about these actions, is represented by upward arrows. For example, an automated cruise control system on a car might have "accelerate" and "brake" as control actions. The car determines which action to take based on feedback from sensors about the car's current speed and from commands by the human operators about the desired speed.

Each controller uses their process model to make decisions about the changes they need to enact on the controlled process. To ensure that each controller's decision-making process is adequately informed, the process model(s) of the human controller needs to match the process model(s) of the automation, both of which need to match the reality of the system and environment. If these models do not match, the control actions coming from any of these controllers may become unsafe. If the pilots think the aircraft is not in a stall, they will not behave properly regarding the stall.

The safety of control actions depends on the context in which the actions occur, namely, the state of the overall system and its environment. Mismatched process models between controllers or misunderstanding of the context for a particular action can lead to unsafe control actions. Therefore, the human controller should understand what control actions are safe or unsafe in each context. This understanding may come from the human controller's prior experience, training, or any additional resources that they can consult, such as manuals.

Accidents often happen when the operator's mental model or the automation's process model become inconsistent with the real state of the controlled process and the environment. For example, the driver or the vehicle automation thinks that the lane to the left is clear, when it is not, and moves into that lane. Another example is that the human operator or the automation thinks that the helicopter state is fine when, in fact, some equipment is overheating and a control action is required to prevent an accident.

This chapter uses the example of mode confusion to illustrate the new STAMP-based design tools. In mode confusion, the human controller's process model about the mode of the automation and/or the controlled process does not match the actual mode. Two potential examples are:

- The human operator believes the system is in mode A, when the system actually is in mode B.
- The human operator knows the system is in mode A, but does not know the implications of mode A on the state of the system.

Frequently, accidents are related to such mode confusion, that is, well-trained controllers believe that they are making the right decision to maintain safe operations because they are confused about the current mode of the aircraft or automation. One reason such confusion may occur is that the automation changes the aircraft mode without any inputs to do so by the pilots. As an example of such indirect mode changes, an A320 crashed while landing at Bangalore, India, in 1990. The pilot selected a lower altitude while the automation was in the *altitude acquisition* mode. This command resulted in the activation of the *open descent* mode, where speed is controlled only by the pitch of the aircraft and the throttles go to idle. In that mode, the automation ignores any preprogrammed altitude constraints. To maintain the pilot-selected speed without power, the automation had to use an excessive rate of descent, which led to the aircraft crashing short of the runway (Sarter & Woods, 1995).

How could this happen? There are several different ways to activate *open descent* mode without the pilot directly commanding it. The investigators suspected that the inaccurate pilot mental model resulted from the automation design that activates *open descent* mode when a pilot selected a lower altitude while in *altitude acquisition* mode. The pilot must not have been aware the aircraft was within 200 ft of the previously entered target altitude, which triggers *altitude acquisition* mode and thereafter *open descent* mode. He therefore may not have expected selection of a lower altitude at that time to result in a mode transition and did not closely monitor his mode annunciations during this high workload time. He discovered what happened 10 s before impact, but that was too late to recover with the engines at idle (Sarter & Woods, 1995).

Accident investigators often blame operators in such cases for poor decision-making or blame a loss of situational awareness (Leveson, 2012). However, neither of these bring us closer to preventing future incidents. Redesign of the interface is also often not the right solution. Instead, redesign of the automation may be more effective.

In the STAMP model terminology, mode confusion occurs when one or more controllers have different models of system status and behavior (Leveson & Palmer, 1997). This is occurring more often as operators move from active control roles to monitoring (Leveson et al., 1998), as is common in new systems with primarily automated controls. Previous studies by Sarter and Woods (1995), Leveson et al. (1998), and Bredereke and Lankenau (2005) have examined and described the cognitive processes that underlie mode confusion. This chapter shows how the STPA process enables the analyst to identify sources of mode confusion in a specific system design and to generate recommendations for improvement.

The rest of this chapter presents an example of the use of STPA in the identification and prevention of potential mode confusion in the autopilot design of a Boeing 777 aircraft. This example is adapted from Bishop et al. (2023).

STPA and an Example of Its Use

STPA consists of four basic steps to identify why a particular system might behave in a hazardous manner and what requirements should be implemented to prevent losses (see Figure 1.2). Leveson and Thomas's *STPA Handbook* provides a detailed guide on how to properly follow these steps (Leveson & Thomas, 2018). A brief overview will be provided here to explain the basic process and illustrate its use with respect to mode confusion.

The first step in STPA is to define the purpose of the analysis.

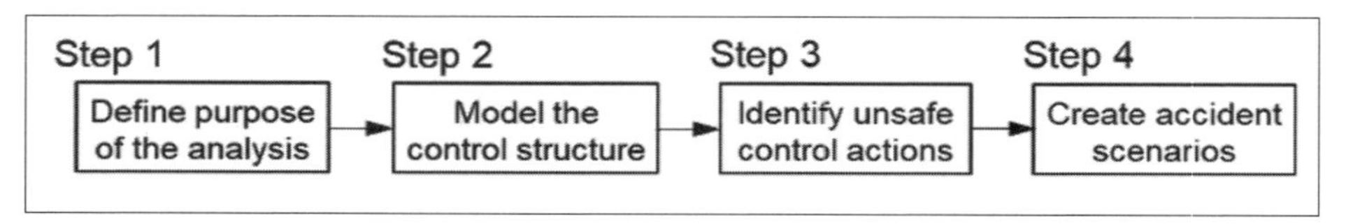

Figure 1.2. The four steps in STPA.

Step 1: Identifying the Goals of the Analysis

The first step in any engineering activity is to identify the goals or purpose. This step involves identifying the system and its boundaries, the potential system-level losses and hazards, and the necessary constraints of system behavior to avoid those hazards.

A *system* is the set of components that work together to accomplish specific objectives. The system and its boundaries must be defined to clearly understand which design aspects can be controlled to prevent hazards. The *boundary* separates the environment, which is *not* under the control of the designers, from the entities within the system, which *are* under the control of the designers (Leveson, 2012). For instance, an aircraft designer might define their aircraft as the system but consider airport infrastructure as part of the environment and thus not under the control of the designer.

In safety, the goal is to prevent losses. A *loss* involves anything of value to stakeholders. Examples of losses include loss of life or injury to people, loss of or damage to the system, loss of or damage to objects outside of the system, loss of mission, and even loss of reputation.

The Boeing 777 features an autopilot (A/P) with various pitch, roll, and thrust modes to manage the speed and direction of flight. Selected pitch and roll modes may impact the set thrust mode and vice versa. Thus, it is imperative for pilots to know the current mode of the autopilot and the consequences of changing that mode. The pitch modes are managed by the autopilot flight director system (AFDS). Additionally, pilots can engage autothrottle (A/T) to have the autopilot manage thrust. The current modes for the AFDS and A/T are displayed in the flight mode annunciator (FMA), a rectangle at the top of the primary flight display.

For commercial aircraft, the highest priority losses to be prevented commonly include *loss of life involving passengers or crew, destruction or damage to the aircraft,* and *loss of mission.*

After defining the unacceptable losses, the system-level hazards are identified. The *system hazards*, as defined previously, are the system states that will lead to a loss given a particular set of worst-case environmental conditions. Hazards refer to the overall system and not to individual components. Hazards identified by an aircraft designer could include the aircraft coming too close to terrain or losing controllability. To narrow the example to one that can be included in this chapter, we select the hazard as *H-1: Loss of control of the aircraft.*

After generating losses and hazards, the safety constraints for the system are defined. Safety constraints are simply statements of what the system should not do. Traceability remains a key component throughout as constraints are linked to hazards that are connected to losses. The safety constraint here is simply that the aircraft must always be controllable.

Step 2: Creating a Model of the Control Structure

Figure 1.3 shows a simplified control structure for the Boeing 777 autopilot system. In this case the human controller is the pilot, the automated controller is the autopilot, and the controlled process/system is the aircraft itself. The control actions and feedback lines identified in this figure are not meant to be exhaustive but are sufficient to generate the UCAs and scenarios in the example shown in the following sections. Note that this model does not contain design details and could be constructed early in the system design process. That would make it possible to generate a safe design from the start without having to undo earlier design decisions.

Steps 3 and 4: Identifying Unsafe Control Actions and Scenarios

In the third step of STPA, users identify *unsafe control actions* (UCAs). A UCA is a control action that will lead to a hazard given specific worst-case conditions.

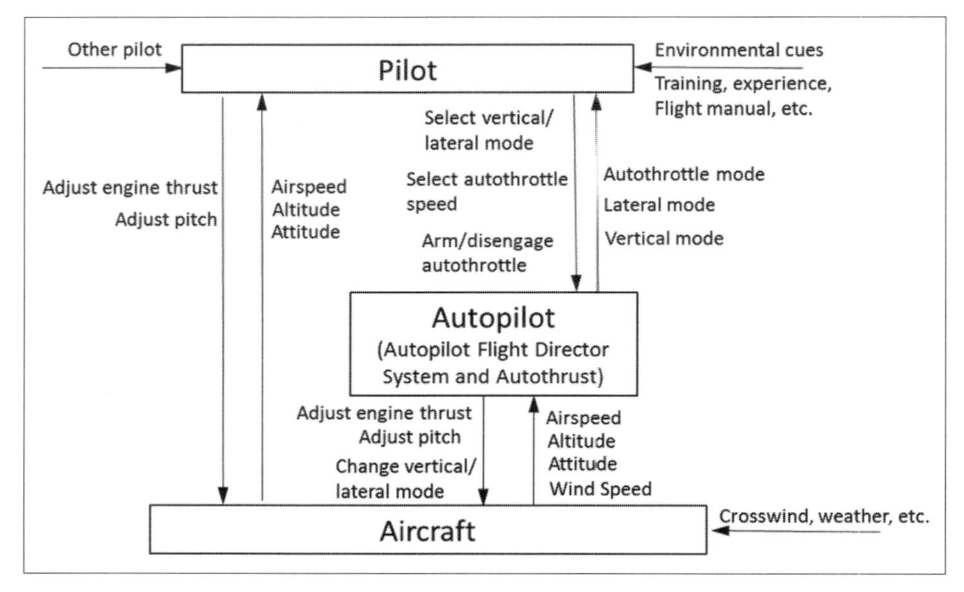

Figure 1.3. Simplified control structure for the Boeing 777 autoflight system.

There are four ways in which a UCA can occur: (1) not providing the control action leads to a hazard; (2) providing the control action leads to a hazard; (3) providing a control action too early, too late, or in the wrong order leads to a hazard; and (4) the control action lasts too long or is stopped too soon, which leads to a hazard. For example, a control action might be applying the brakes in a car. A driver could: (1) not apply the brakes when an obstacle is in front of the car; (2) apply the brakes when there are cars close behind; (3) apply the brakes too late to fully stop; or (4) apply the brakes for too short of a time to decelerate to a safe speed.

The fourth and final step of STPA is to identify potential loss scenarios by analyzing the causal factors that would lead to UCAs. In other words, identify the reasons that a UCA might be taken. Among other things, this step involves asking why a controller would reasonably take a UCA. One possible reason (involving mode confusion) is that they misunderstand the true mode of the controlled process or automated controller and issue a UCA as a result.

Within the STPA framework, controllers' choices of control actions are understood through their process/mental models. Inadequate mental/process models may occur when controllers receive incorrect feedback; they receive feedback but interpret it incorrectly; they do not receive feedback when needed; or the necessary feedback does not exist (Leveson & Thomas, 2018). For example, in a plane, if the altimeter sensor is broken, the pilot will get incorrect feedback from the display. If the altimeter is in a different mode than expected, the pilot may misinterpret the altitude. These unsafe control actions involving feedback can be captured when identifying loss scenarios.

By providing a systematic method to identify hazards and potential loss scenarios, STPA allows users to efficiently analyze the system architecture, generate effective requirements for safety and reliability, and ultimately identify gaps where changes need to be implemented. When applied in a particular way, STPA can be leveraged to effectively identify sources of mode confusion and generate recommendations to improve system design in that regard.

UCAs are identified when using STPA by applying a rigorous process described by Leveson and Thomas (2018), which requires more space to describe than is possible in this chapter. Two examples and some potential scenarios that could lead to UCAs are shown instead.

The first example involves the control action of a Boeing 777 pilot to "engage autothrottle THR REF mode." In THR REF mode, thrust is set to the reference thrust limit displayed on the engine indication and crew alerting system (EICAS; National Transportation Safety Board [NTSB], 2014). An example UCA for the pilot related to this control action is that:

UCA-1: The pilot does not engage autothrottle TRH REF mode when the pilot intended to do so.

The causal scenarios by which this unsafe control action could develop are diverse, and to conduct an exhaustive search the analyst should consider all contexts by which the THR REF mode would not engage despite a pilot's intent to engage it. One potential scenario is the following:

- **Scenario 1**

  The pilots do not engage autothrottle THR REF mode when they had the intent to engage it because they press the incorrect button and do not verify engagement of the mode by checking the FMA. The pilot may not verify the engagement of the mode due to task saturation or expectation that the mode will engage when a button is pressed because their prior experience has always been that the mode engages at the press of a button.

This scenario explains the unsafe control action in terms of a non-update to the pilot's process model of the automation. Recommendations stemming from scenarios like this will relate to ensuring *appropriate* feedback, rather than including *new* feedback, because the feedback was not perceived by the pilots despite it being available to them.

A second potential scenario is the following:

- **Scenario 2**

  Pilots do not engage autothrottle THR REF mode when they had the intent to engage it because a single press of the takeoff go around (TO/GA) switch will not engage THR REF mode if the aircraft is in a landing configuration with go-around mode armed. Engaging THR REF mode during a go-around requires a double push of the TO/GA switch (Air Accident Investigation Sector [AAIS], 2020). The pilot may not verify THR REF engagement in the FMA after the first click due to task saturation or expectation that the mode will engage when a single button is pressed. Or the pilots may not be aware that TO/GA is engaged.

This scenario also explains the unsafe control action in terms of a non-update to the pilot's process model of the automation, but also involves a misunderstanding of the implications of a control action during a particular mode. Both of these scenarios could ultimately lead to pilot mode confusion, as they all involve inadvertent activation (or non-activation) of modes coupled with an opposing belief.

The generation of UCAs for automated controllers is aided by a thorough understanding of mode transition logic and criteria, but it is also necessary that each potential mode transition be analyzed for unsafe interactions with different states of the system and environment, rather than in isolation. Specifically, in the Boeing 777 example, an available control action to the autopilot is "change vertical mode." There are various vertical modes, two of which are TO/GA mode and ALT mode. In TO/GA mode, the autopilot acquires and maintains a takeoff speed reference after liftoff, or a go-around speed reference after initial go-around rotation (NTSB, 2014). In ALT mode, the autopilot adjusts the pitch of the aircraft to stay on a target altitude (Federal Aviation Administration [FAA], 2022). An example UCA for this mode transition is:

UCA-2: A/P changes default TO/GA mode to ALT mode too early when the aircraft is still on ground without adequate feedback to flight crew.

This action involves a specific mode transition coupled with a context in which this mode transition becomes unsafe. Causal scenario identification for this UCA should consider how the mode transition logic would allow the transition to happen in this unsafe context. One potential scenario for this is:

- **Scenario 1 for UCA-2**

  The autopilot changes the pitch mode from default TO/GA mode to ALT mode because a realignment of the air data inertial reference system was initiated when the flight director was ON and the MCP selected altitude was within 20 ft of the barometric altitude (FAA, 2022). The pilot may not perceive this transition because they are expecting the default TO/GA mode. Thus, they become mode confused when they take off and upon liftoff, A/P commands nose-down pitch to obtain the set altitude for ALT mode (sea level).

This scenario explains the unsafe control action in terms of an update to the automation's process model by an action (realignment of the air data inertial reference system) but also involves a miscommunication of that process model update to the flight crew. Once again, this could lead to pilot mode confusion by the inadvertent activation of a mode that a pilot would not expect in a particular context.

Once the potential scenarios are identified, recommendations can be derived from them to design the system to eliminate or mitigate them.

Summary and Outlook

As modern systems grow increasingly complex, it becomes impossible to simply "train out" pilot and automaton behavior that can lead to hazards. Instead, the sources of such unsafe behavior should be identified and designed out of the system. STPA provides a methodology to do this by abstracting systems in terms of functional control feedback loops. The results of the STPA analysis can be used by designers to identify effective design requirements and reduce hazardous behavior.

In addition, the use of a modeling and analysis technique, such as STPA, makes it easier for hardware, software, and human factors engineers to work together to create safer design and operational procedures.

References

Air Accident Investigation Sector. (2020). *Runway impact during attempted go-around. Aircraft accident final report AIFN/0008/2016*. https://reports.aviation-safety.net/2016/20160803-0_B773_A6-EMW.pdf

Bishop, B., Harrington, P., Rose, R., & Leveson, N. (2023). System theoretic process analysis for identification of sources of mode confusion. *Proceedings of the Human Factors and Ergonomics Society Annual Meeting, 67*(1), 2397–2403. https://doi.org/10.1177/21695067231192457

Bredereke, J., & Lankenau, A. (2005). Safety-relevant mode confusions – modelling and reducing them. *Reliability Engineering & System Safety, 88*(3), 229–245. https://doi.org/10.1016/j.ress.2004.07.020

Federal Aviation Administration. (2022). *Special airworthiness information bulletin: Autopilot flight director system: ALT HOLD engaged on takeoff on Boeing Model 777/787 Common Fleets*.https://ad.easa.europa.eu/blob/AIR-22-09R1.pdf/SIB_AIR-22-09R1_1

Leveson, N. G. (2012). *Engineering a safer world: Systems thinking applied to safety*. The MIT Press. https://doi.org/10.7551/mitpress/8179.001.0001

Leveson, N. G. (2019). *CAST handbook: How to learn more from incidents and accidents*. http://sunnyday.mit.edu/CAST-Handbook.pdf

Leveson, N. G., & Palmer, E. (1997). Designing automation to reduce operator errors. *Computational Cybernetics and Simulation 1997 IEEE International Conference on Systems, Man, and Cybernetics, 2*, 1144–1150. https://doi.org/10.1109/ICSMC.1997.638104

Leveson, N. G., Pinnel, L. D., Sandys, S. D., Koga, S., & Reese, J. D. (1998). Analyzing software specifications for mode confusion potential. *Safety Analysis of FMS/CTAS Interactions During Aircraft Arrivals*. https://ntrs.nasa.gov/citations/19990063822

Leveson, N. G., & Thomas, J. P. (2018). *STPA handbook*. http://psas.scripts.mit.edu/home/get_file.php?name=STPA_handbook.pdf

National Transportation Safety Board (NTSB). (2014). *Descent below visual glidepath and impact with seawall Asiana Airlines flight 214 Boeing 777-200ER, HL7742 San Francisco, California July 6, 2013* [Aircraft accident report NTSB/AAR-14/01, PB2014-105984]. https://www.ntsb.gov/investigations/accidentreports/reports/aar1401.pdf

Sarter, N., & Woods, D. (1995). How in the world did we ever get into that mode? Mode error and awareness in supervisory control. *Human Factors, 37*(1), 5–19. https://doi.org/10.1518/001872095779049516

Chapter 2

From Requirements to Cockpits – Considerations on the Design Process

Christoph Vernaleken and Daniel Dreyer

Abstract

When designing new aircraft cockpits from scratch or substantially upgrading existing ones, the initial steps of human factors requirements elicitation are relatively clear from a methodological perspective, and the corresponding tools – such as hierarchical task analysis – are very well documented. Likewise, the applicable airworthiness certification requirements are, in spite of occasional evolutions, common knowledge and accessible to everyone. By contrast, the creative process of transforming these requirements into a viable cockpit or flight deck design is much less obvious. One of the key challenges is that the airworthiness requirements to be considered from a human factors perspective typically cover a wide range from the relatively generic guidance provided in certification specification (CS) 25.1302 and associated explanatory material, down to very detailed and often implicit requirements regarding the precise shape and location of cockpit controls, or detailed prescriptions regarding the use of certain colours. This chapter describes a pragmatic approach in which the requirements baseline is initially used to derive a consistent cockpit philosophy that discusses the main design trade-offs that have been made. On the basis of the cockpit philosophy, and the design office data detailing the physical space available for the cockpit, an initial cockpit concept encompassing both geometry and layout, as well as initial ideas for display formats and modes, can be created. This is the foundation for detailed cockpit design, again based on well-known and excellently documented human factors criteria for functional allocation as well as on heuristics such as the Gestalt laws or Wickens' 13 principles.

Keywords

human factors engineering, cockpit philosophy, cockpit design, display formats

Introduction

When facing the task of designing a new aircraft cockpit from scratch or performing a substantial upgrade of an existing flight deck, the initial steps of the requirements elicitation process are typically rather straightforward, as they are defined by the airworthiness certification baseline on one hand and complemented by some form of mission and task analysis, for example a hierarchical task analysis, on the other. Of course, there might be additional customer or programme requirements that could have a significant impact on the cockpit. It should also be noted that the initial airworthiness requirements for large aeroplanes laid down in certification specifications (CS/Part 25 (European Union Aviation Safety Agency [EASA], 2021; Airworthiness Standards, 2023)[1] are by no means exhaustive when it comes to flight deck design, and that important requirements regarding equipment to be fitted, with potentially a substantial impact on flight deck displays and controls, can be found elsewhere, depending on the type of operation intended, for example Part 121 and Part 135 for the Federal Aviation Administration (FAA; National Archives, Code of Federal Regulations, 2023a, 2023b). Similarly, in Europe, regulations such as the CS-AWO (EASA, 2022a) address all-weather operations or, in the case of CS-ACNS, access to certain airspaces (EASA, 2022b). As an example, CS-25 does not contain any mandate to equip transport-category aircraft with an autopilot, although there are several specific requirements pertaining to controls, indications and alerts for engaging/disengaging the autopilot if fitted (see CS 25.1329; EASA, 2021), which reflects the implicit assumption that most transport-category aircraft will be equipped accordingly. By contrast, CS-AWO will typically mandate the installation of an autopilot for certain types of instrument landing system approaches, as detailed in CS AWO.B.CATIII.106 (EASA, 2022a).

Irrespective of whether the envisaged aircraft mission is a basic commercial air transport operation (passenger, cargo) or some other potentially more complex civilian or military mission, certain basic pilot task templates can virtually always be used, in the following order of descending priority (after Appendix D to CS-25, in EASA, 2021):

- **Aviate:** maintain aircraft flight path control in terms of attitude, altitude, speed and course, and prevent collisions with other traffic
- **Navigate:** know own position in relation to the planned flight path, terrain and obstacles in the vicinity of the envisaged route, as well as the destination (final, intermediate or alternative)

1 For simplicity, we only refer to CS-25 in the following, given the high degree of commonality between FAA Part 25 and EASA CS-25. For reasons of traceability and legibility, we also directly reference the applicable CS or AMC paragraph in the following, without repeated reference to the overall document.

- **Communicate:** communication within the crew (procedures or checklists, situation assessment, intentions) and with air traffic control (intentions, and, especially in an emergency, type of support requested)
- **Manage systems:** operate and monitor aircraft engines, fuel, and systems
- **Command decisions:** strategic decision-making – can the mission be performed as envisaged, or have conditions arisen that require a change of plan, such as a diversion to another airport?

Of course, a full mission and task analysis must consider pilot tasks in the context of the respective flight or mission phases, and also adequately assess the priority of these tasks in normal operation, compared to abnormal/emergency situations. Typically, military missions are substantially more complex, as the flight profile might include air-to-air refuelling, both as a tanker or receiver, weapons delivery, aerial delivery, search-and-rescue missions, maritime patrol, and many more. Usually, such missions are performed in an environment that requires additional considerations regarding the defence against enemy forces in the air and on the ground throughout the mission. Considering this wide range of possible aircraft mission profiles, the smallest common denominator is the basic flight operation from an Origin A to a Destination B. Therefore, this chapter will focus on the CS-25 (EASA, 2021) regulations, which contain the requirements for large transport-category aircraft.

The Requirements Conundrum

One of the key challenges is that the airworthiness requirements to be considered from a human factors perspective for flight deck design typically cover a wide range of granularity and detail. Some of the requirements are very generic, such as CS 25.777 (a), which mandates that cockpit controls "must be located to provide convenient operation and to prevent confusion and inadvertent operation" (EASA, 2021, p. 423). Likewise, according to CS 25.1302 (a), "flight deck controls must be installed to allow accomplishment of [...] tasks [associated with their intended function], and information necessary to accomplish these tasks must be provided" (EASA, 2021, p. 718).

In many cases, nevertheless, CS-25 provides an exhaustive overview of the information to be presented to the flight crew for accomplishing the *aviate* and *navigate* task. CS 25.1303 lists the required flight and navigation instruments, while CS 25.1305 specifies the powerplant-related indications and alerts that are necessary. For the flight and navigation instruments, CS 25.1321 (b) additionally prescribes the required arrangement in the well-known "basic T" layout.

By contrast, other CS-25 requirements already provide detailed design solutions. As an example, there are very detailed design requirements for landing gear and flap controls with respect to both location (CS 25.777) and control knob shape

(CS 25.781). According to CS 25.777 (e), "wing-flap controls and other auxiliary lift device controls must be located on top of the pedestal, aft of the throttles, centrally or to the right of the pedestal centre line, and not less than 25 cm (10 inches) aft of the landing gear control" which in turn "must be located forward of the throttles", as required by CS 25.777 (f) (EASA, 2021, p. 423). It is important to note that CS 25.777 (e) also contains the implicit assumption that there is a pedestal, and that it provides additional details concerning the location of the throttles that go beyond the requirements on engine controls in CS 25.777 (d), which only mandates "identical powerplant controls for each engine" that "must be located to prevent confusion as to the engines they control" (EASA, 2021, p. 423). Similarly, CS 25.1143 merely requires "separate power or thrust control for each engine" permitting both "separate control of each engine" and "simultaneous control of all engines" (EASA, 2021, p. 692). Interestingly, the word "pedestal" only appears in the context of CS 25.777 in the entire CS-25 document! The example above also shows a certain degree of redundancy and fragmentation in airworthiness requirements.

As a helpful analogy to better understand the underlying challenges, one might think of the airworthiness requirements in CS-25 as something equivalent to a partially completed old master's oil painting, in which the artist has almost fully completed some areas, whereas other parts of the artwork only exist as sketches of varying levels of detail, but such that the overall subject of the painting is already unambiguously recognisable. Thus, while there could be some completely blank areas on the canvas, there would nevertheless be no doubt as to whether one was looking, for example at a Nativity Scene or the Marriage Feast at Cana. Tasked with the completion of this painting, the workshop of the old master would have to paint in a style that is in harmony with the already completed parts, but at the same time true to the overall subject and intention of the painting. Likewise, flight decks must be designed in such a way that they are consistent with the highly detailed requirements discussed above, while at the same time conforming to the overall intentions behind the high-level requirements. What is required, therefore, is a guideline that provides information on how to fill in the blanks and to complete the sketches. The next section proposes a *cockpit philosophy* as a viable, pragmatic means of achieving this.

A Potential Solution: Cockpit Philosophy

Defining the cockpit philosophy is a multi-disciplinary effort, which – like any flight deck design activity – must of course involve pilots and operators from the very beginning. In order to document the overall intentions behind the flight deck design, one of the first steps in defining a cockpit philosophy consists of stating and, where necessary, detailing the fundamental underlying design principles. In this context, it also seems worthwhile to document the main design trade-offs that have been made. Of course, the cockpit philosophy can and should make reference to existing airworthiness and industry standards. As an example, one would typically reaffirm a human-centred design approach, or design against human error, and might then reference the corresponding paragraphs of CS 25.1302. The following further aspects should be covered by the fundamental design principles:

- **Flight Crew Philosophy**
  Of how many pilots and potentially other operators is the flight crew composed of, and what are their envisaged roles and responsibilities, including the envisaged task sharing? How are they intended to be kept in the loop on the status of the aircraft's systems?
- **General Philosophy of Interaction and Information Presentation**
  Is a "quiet, dark and silent" flight deck philosophy used, or another approach (see AMC 25-11; EASA, 2021)? Will there only be head-down displays, or are solutions for presenting information overlaying the outside view, such as head-up displays, available (see SAE ARP 5056; SAE International, 2013)? What is the general approach for flight crew procedures and checklists? Will electronic checklists be used, and if so, what is the envisaged level of integration with information from the aircraft systems, that is will some checklist items automatically be "ticked" if the desired system status has already been achieved? Is a paperless cockpit envisaged, and if so, how is this realised, via electronic flight bag type devices, or as part of the main avionics display suite?
- **Automation Philosophy**
  What is the role of automation on the flight deck, and for which tasks and purposes is it predominantly used? Which factors and criteria determine whether a task will be automated? How is it ensured that crew members are adequately kept in the loop to preclude automation surprises and other human factors issues related to automation? At least the following aspects should be addressed:
  - *Flight Crew Role:* What is the role and authority of the flight crew when using automation? It is essential in this context to define whether the flight crew partially or generally serves as a redundant back-up in case automated functions fail, or whether the design of the automation is redundant and fail-safe.
  - *Task Allocation:* Are tasks statically allocated to be performed either by the flight crew or by automated systems, or is task allocation dynamic, that is do pilots have a choice of delegating a task to the automation? A very basic example of dynamic task allocation would be the availability of an auto-thrust function giving pilots the option to either manage thrust manually or automatically.
  - *Automation Awareness and Feedback:* How can pilots understand, supervise and monitor what the automation is doing? For complex automated functions, how are their modes and particularly automatic mode changes enunciated? Do automatic trim, automatic pilot and automatic thrust physically move the corresponding physical controls in the cockpit when active, or not?

- *Adaptive Automation:* Does the behaviour or level of automation change automatically in certain situations, for example triggered by external events? If yes, how does the flight crew maintain automation mode awareness?

- **Approach Versus Novel Technologies**
What are the criteria that will be applied when it comes to the introduction of new technologies not previously encountered in cockpits? Evidently, the use of novel technologies might constitute a certification risk, since neither airworthiness regulations nor industry standards addressing the use of these new technologies on the flight deck might already be available. At the same time, innovative technologies might, for example support new operational capabilities, enhance safety or increase efficiency, thus a considerate trade-off including a risk-benefit analysis is required. Consequently, solely using new technologies for the sake of novelty is not appropriate. Furthermore, it is important to note that, for airworthiness certification, the degree of novelty does not only consider whether a technology is new on the market, but also considers whether a manufacturer newly applies an existing technology (see AMC 25.1302; EASA, 2021).

The above list is, of course, by no means exhaustive. For further inspiration, Table 2.1 gives an overview of the high-level cockpit philosophies of Airbus and Boeing, which gives a good impression of similarities and differences. For the first point in both philosophies, the similarity is not surprising, since this constitutes an almost verbatim reference to the role and responsibility of the pilot-in-command as defined in ICAO Annex 2, "Rules of the Air" (International Civil Aviation Organization [ICAO], 2005). Furthermore, it is interesting to see that both philosophies explicitly refer to human factors engineering (5) and outline general principles and conditions for the use of novel technologies (6). Particularly for the items where the philosophies are virtually the same regarding intent (grey background), the wording is partially identical (bold font).

Table 2.1. Comparison of Airbus and Boeing high-level flight deck philosophies

Airbus	Boeing
1. **The pilot is** ultimately responsible **for the** safe **operation of the aircraft.** The pilot has **final authority** with adequate information and means to exercise this authority.	**The pilot is** the **final authority** for **the operation of the airplane**.
2. The design of a cockpit is dictated by **safety, passenger comfort, and efficiency** in that order of priority.	Flight crew tasks, in order of priority, are: **safety, passenger comfort, and efficiency**.

Table 2.1. continued

Airbus	Boeing
3. The **design** of a cockpit accommodates for a wide range of **pilot** skill levels and **experience** acquired on previous aircraft.	**Design** for crew operations based on **pilots'** past training and operational **experience**
4. The automation is considered as a complement available to the pilot, who can decide when to delegate and what level of assistance is desirable, according to the situation.	Apply automation as a tool to aid, not replace, the pilot.
5. The human–machine interfaces are designed considering system features, together with the pilot's strengths and weaknesses	Address fundamental human strengths, limitations, and individual differences – for both normal and non-normal operations.
6. The **use** of **new technologies** and implementation of new functionalities are dictated by: • Significant safety benefits • Obvious **operational advantages** • A clear response to a pilot's needs	**Use new technologies** and functional capabilities only when: • They result in clear and distinct **operational** or efficiency **advantages**, and • There is no adverse effect to the human–machine interface.
7. State-of-the-art human factors considerations are applied in the system design process to manage the potential pilot errors.	Design systems to be error tolerant.
8. The overall cockpit design favours crew communication.	Both crew members are ultimately responsible for the safe conduct of the flight.
9. The cockpit design aims at simplifying the crew's task by enhancing situational and aircraft status awareness.	The hierarchy of design alternatives is: simplicity, redundancy, and automation.
10. The full authority, when required, is obtained with simple intuitive actions, while aiming at eliminating the risks of over-stress or over-control.	

Note. Adapted from SAE ARP 5056 (SAE International, 2013, p. 28)

Of course, while doubtlessly very helpful as guidelines on the main intentions behind the flight deck design, these fundamental design principles are nevertheless not yet specific enough to aid in transforming requirements into a concrete conceptual design. Therefore, any viable cockpit philosophy must, at a minimum, detail the approach to be taken regarding control, interaction and information presentation. In some cases, it might make sense to split the information presentation philosophy into dedicated sections for electronic display system formats, flight deck alerting and an overarching colour philosophy.

One of the key objectives of a cockpit philosophy is, of course, to ensure *consistency* across the flight deck. Accordingly, information should be presented in a way consistent with the flight deck design philosophy in terms of symbology, location, control, behaviour, size, shape, colour, labels, dynamics and alerts. This evidently applies to multiple displays on the same flight deck. Likewise, acronyms and labels should be used consistently, and messages/enunciations should contain text in a consistent way, also with related labels located elsewhere in the flight deck (see AMC 25-11; EASA, 2021).

The *information presentation philosophy* needs to establish the relation between information that is either permanently presented on the flight deck displays, or automatically brought up when certain conditions are fulfilled, and which information is generally available, but must actively be accessed by the pilots, for example by calling up a certain system synoptics page. This is, of course, strongly linked to aspects of automation, flight deck alerting and electronic checklist.

Regarding the *philosophy for electronic display systems*, the criteria for using digital readouts versus analogous indications should be clearly laid down. Likewise, it seems very useful to limit the number of scale and dial types to the necessary minimum to ensure consistency. Generally, the orientation of the information presented and the flight crew's frame of reference should be correlated. Accordingly, indications for systems unambiguously located on the left side of the aircraft (e.g. engines, fuel tanks) should be shown on the left side (see AMC 25-11; EASA, 2021). This document also cautions that, to avoid visual clutter, graphical representations should only be included if they reduce flight crew access or interpretation time, or decrease the probability of interpretation error.

A further very important aspect in this context is the *colour philosophy* to be used, particularly for information presentation on electronic flight deck displays. It is common human factors knowledge that, if colour is used for coding, only six or fewer different colours should be used (see AMC 25-11; EASA, 2021). The underlying rationale is that humans are limited to five to seven distinct values in the scope of their absolute judgement (Wickens et al., 2004). The particular challenge in a flight deck environment, however, is that the colours red and amber are already reserved for flight deck alerting purposes to present cautions and warnings (CS 25.1322; EASA, 2021), or operationally significant limitations of parameters (as, e.g. in CS 25.1305; EASA, 2021) to the pilots. Likewise, white is typically used for scales and dials, whereas brown is reserved for presenting the ground on the artificial horizon, which also requires a blue shade for visualising the sky. Given that the simultaneous use of amber/yellow is discouraged (see, e.g. AMC 25-11; EASA, 2021), the task of defining a colour philosophy is therefore commonly reduced to the definition of easily memorable and consistent meanings for the use of the colours green, blue (usually a cyan tone), and magenta. What is the reason for typically using these colours? Another less obvious requirement regarding the colours used for coding information is that they should be sufficiently close to elementary colours which can be unambiguously identified and named by the pilots, in order to preclude confusion in flight deck communication. Consequently, a colour philosophy relying on colours named, for example "dark mauve" or "light apricot" must be discouraged even if this choice of colours might fulfil the requirement that colours should have sufficient chrominance separation to be identifiable and distinguishable in all foreseeable lighting conditions to be encountered in operations. In other words, the "keep it simple" principle also applies to colour names and, consequently, colours. In this setting, using different shades of grey can be useful to prevent an all-white representation of current values, labels, and units, or to provide intuitive background information, which might be shaded terrain on a navigation display, or a depiction of a fuselage as a backdrop for a system synoptic page presenting the status of aircraft doors. In view of the potentially difficult lighting conditions that might be encountered in operations, it is also generally recommended not to use colour as the sole means of coding, but rather to additionally use at least one other distinctive coding parameter, such as shape, location, or size etc. (e.g. AMC 25-11; EASA, 2021).

Last but not least, the *control philosophy* should be laid down. This includes primary and secondary flight controls (e.g. the decision on whether to use a conventional yoke or a side stick), but is even more important when it comes to the various other controls on the flight deck. One of the fundamental decisions to be made is which forms of software control will be used, and what the enabling hardware controls are. In particular, it must be outlined whether line-select keys on the display bezel, cursor control devices, or touch screens will be used, and in which combination if applicable. Irrespective of this, the number and types of classic "single-purpose" hardware controls should be defined. When will, for example classic rocker switches be applied, what is the concept for pushbuttons, and which are the criteria for installing rotary selectors? Are there any non-standard variants of these controls, or special-purpose hardware controls present?

Control feedback is a further essential aspect of the control philosophy. Selection feedback provides information as to whether a control input or selection has actually been accomplished, irrespective of the system response. Accordingly, all hardware and software controls should provide immediate feedback for any pilot input. For hardware controls, with the exception of some momentary action pushbuttons, there is virtually always some tactile feedback that provides at least sufficient selection feedback. In many cases, however, the selector or switch position is not deemed sufficient to provide adequate feedback regarding system status, in particular for systems of high complexity or high criticality. For software controls, selection feedback is more complex, due to the typically missing tactile feedback, but in part also because software controls (such as data entry fields) allow for more complex and sophisticated inputs. Consequently, especially for software controls, there should be a clear and unambiguous feedback when pilot input is not accepted or followed by the system. According to AMC 25.1302 (EASA, 2021), this feedback should be primarily realised by visual indications and/or audio.

From Philosophy to Concept

In combination with design office data defining the physical space available, the cockpit philosophy paves the way towards the first cockpit concept encompassing both geometry and layout, as well as initial ideas for display formats and their management and moding. The key step in turning a cockpit philosophy into a concrete cockpit concept is the first complete functional allocation. One of the central principles for the functional allocation on the flight deck is that *controls and information must be accessible and usable by the flight crew in a manner consistent with the urgency, frequency of use, and duration of their tasks* (see AMC 25.1302; EASA, 2021).

- The aspect of *urgency* chiefly addresses controls or information used in abnormal or emergency operations. As an example, controls and indications for the engine fire extinguishing system must be readily and immediately accessible, even in conditions of dense cockpit smoke.
- *Frequency of use* is self-explanatory – the underlying principle is that, based on the task analysis and/or in-service experience, the most frequently used controls and indications must ideally be placed where they can conveniently be reached and in the primary field of view. By contrast, less frequently used controls and information can be placed elsewhere. For electronic display formats, less frequently used information can also be integrated in display (sub-)formats that the flight crew has to explicitly select. Evidently, there may be a competition between urgency and frequency of use – for the fire extinguishing system, the expected frequency of use is ideally very low, whereas the urgency is comparatively high irrespective of this. Consequently, part of the cockpit real estate must be reserved for controls with urgently required access.
- *Duration of use* is an additional aspect – when the interaction with a control or indication is of longer duration, or even virtually continuous, it must of course be ensured that this can be achieved with a convenient posture. The reference is always the "normally seated position" of the respective flight crew members (see CS 25.1302; EASA, 2021).

Nevertheless, these three principles should not be applied in isolation, but must be complemented by additional considerations and boundary conditions to

prevent a fragmentation of the indications and controls required for a certain task, and to retain an operationally meaningful grouping of these. Some exemplary boundary conditions can be found, as follows:
- Ideally, the controls and indications required for a certain task should be distributed over a very limited number of locations in the cockpit (functional or task grouping).
- Except for flight controls, if controls for a certain task cannot be integrated at a single location, the allocation to different locations should be based on either, for example useful sub-functions, or a delineation between normal and abnormal/emergency operations.
- Location cueing should be applied such that the relative location of cockpit controls and indications reflects the actual location of the respective system on the aircraft.

In this context, it is additionally very useful to define distinct *visibility and accessibility zones*. In the primary field of view, information can be perceived without head and eye movement, whereas the larger immediate field of view already requires eye movement. To fully cover the peripheral field of view, pilots must turn their heads. Anything beyond this also requires body movement. For controls, useful distinctions are whether they can be reached immediately without body movement (e.g. leaning forward), and whether they can be used by both hands, or only by one hand.

It also seems appropriate at this stage to take a closer look at *design against human error.* Regulators acknowledge that even well-trained and proficient pilots will make errors. However, both the likelihood of occurrence of some of these errors and the flight crew's capability to detect and to mitigate errors can strongly be influenced by the flight deck design. In an attempt to prevent design-induced human error as far as reasonably possible, the following principles from AMC 25.1302 (EASA, 2021) should be considered:
- Avoid indications and controls which are complex and inconsistent, either with one another or other systems on the flight deck.
- Procedures must be consistent with the design of the equipment.
- Inadvertent operation of systems which might subsequently result in a hazard must be discouraged and prevented by guarded switches or other efficient means.
- The effects of any error on the functions and capabilities of the aircraft must be evident to the flight crew.
- Minimise the impact of human error by system logic or redundant, robust or fault-tolerant system design.
- Enable positive transfer of learning from other designs where possible, and ensure there is no negative transfer, that is that the same pilot action results in different system responses.

Again, this list is not exhaustive, but touches upon some of the most important points, and may serve as a checklist for scrutinising the cockpit concept at various stages of maturity. In the following, this chapter will focus on the visual aspects of cockpit design considerations in order to illustrate the engineering process with more details.

Detailed Cockpit Design Considerations

Once the overall concept regarding cockpit geometry and layout has reached a sufficient maturity, work on the detailed cockpit design follows suit. In some cases, there are specific industry standards defining the appearance of certain indications required by airworthiness regulations. Most prominently, Appendix A to SAE ARP 4102-7 (SAE International, 2007) provides a detailed definition of the primary flight display (PFD), which contains all the indications required for the aviate task as per CS 25.1303. Similarly, Appendix B of the aforementioned document standardised the appearance of the navigation display (ND), which not only provides the indications for the navigate task required by CS 25.1303, but also displays information regarding radio navigation as well as the flight plans defined in the flight management system (FMS). Last but not least, a way of presenting the engine indications required by CS 25.1305 is defined in Appendix C to SAE ARP 4102-7 (SAE International, 2007). It is important to understand that, in contrast to the airworthiness requirements, the aforementioned industry standards are not mandatory in application. Nevertheless, as they are explicitly referenced by airworthiness guidance material in AMC 25-11 (EASA, 2021), most commercially available avionics display suites on the market consequently exhibit a high degree of adherence to these industry standards. This serves as an explanation of why the PFDs and NDs exhibit a very similar appearance even on aircraft by different manufacturers, because the certification of already standardised solutions is much faster and poses a substantially lower risk in terms of evidence, documentation and consequently cost.

Generally, however, the detailed design of cockpit controls and indications should again be based on well-known and excellently documented human factors criteria for functional allocation as well as on heuristics such as Gestalt psychology (Wertheimer, 1923) or the Wickens 13 principles (Wickens et al., 2004). This particularly applies to both the design of controls and indications for the various aircraft systems.

The example of the Heinkel He 177 fuel indication and management panel in Figure 2.1 can be used to illustrate some of the most important Gestalt psychology laws (Wertheimer, 1923) in an aviation context. Evidently, the panel uses electromechanical indications, but the insights discussed in the following can of course be directly transferred to electronic displays as well.

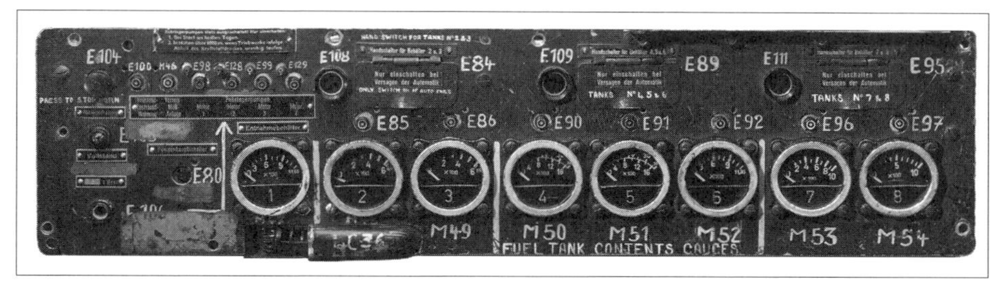

Figure 2.1. Heinkel He 177 fuel indication and management panel. Photo by Dane Penland, Smithsonian National Air and Space Museum (NASM2014-06924). Reprinted with permission.

- **Law of Proximity (Wertheimer, 1923)**
  Elements which are spatially close to one another are grouped by human perception, and consequently perceived as belonging together. This apparently holds true for the eight fuel gauges numbered "1" to "8" in Figure 2.1, which are installed in very close proximity, with an inter-gauge spacing of between 7 mm and 10 mm for instruments with a diameter of 61 mm. By contrast, elements which are farther apart will be perceived as separate and independent. This leads to the highly important conclusion that deliberately adding empty spaces between elements can visually code the important information that they are separate or belong to different groups. Adding space is therefore not necessarily a waste of space!
- **Law of Similarity (Wertheimer, 1923)**
  Elements which are similar are also grouped by human perception. Conversely, items that differ in important characteristics are perceived as separate or independent of one another. In the Heinkel He 177 example in Figure 2.1, the fuel gauges are similar because they have identical form factors (shape and size), layout and colour coding. By contrast, the differences in the gauge scales reflecting the different fuel tank capacities are much less salient at first glance.
- **Law of Continuation (Continuous Line) (Wertheimer, 1923)**
  Our perception also groups elements arranged on a continuous line or curve. This is particularly evident for the single row of eight fuel gauges in Figure 2.1.
- **Law of Closure (Wertheimer, 1923)**
  Items with a closed outline or elements that are surrounded by a line are grouped by our perception, that is understood as belonging together. By contrast, elements separated by lines do not seem to belong together. Figure 2.2 shows the main instrument panel of the three-engine Junkers Ju 352 aircraft, a less-than-successful successor of the famous Junkers Ju 52. The required "identical powerplant instruments for the engines" (see CS 25.1321 (c)(1); EASA, 2021, p. 811), for the three BMW 323 radial engines are visually separated by giving the instruments for each engine a closed white outline. This constitutes an early ap-

Figure 2.2. Junkers Ju 352 main instrument panel. © Deutsches Museum, München, Archiv CD77704. Reprinted with permission.

plication of the law of closure. Unfortunately, it could not be established whether this particular design choice was made intuitively or with explicit knowledge of Wertheimer's publications. As a more contemporary application, the engine indication and crew alerting system (EICAS) format on the Dornier Do 328 aircraft (Figure 2.3) uses vertical and horizontal lines to visually delineate areas for engine indications (left) and alert messages (upper right), as well as cabin pressurisation and flight control system information (lower right), respectively.

There are two highly important conclusions that can be drawn from the above Gestalt laws. First, *consistency* (which is incidentally also one of the Wickens 13 principles) in cockpit design is always the result of intentional *similarities*, which offer the possibility of visually connecting elements even over larger distances. Second, the Gestalt laws are rarely applied in isolation. Eventually, the fuel gauges in Figure 2.1 are perceived as belonging together because of their proximity, similarity and arrangement on a continuous line. Apparently, when applied in combination, the Gestalt laws visually reinforce one another.

Ironically, however, the resulting arrangement of the eight fuel gauges on the Heinkel He 177 panel is so much perceived as a unit due to the aforementioned Gestalt laws that the originally intended functional grouping of the fuel tanks had to be restored by a somewhat "makeshift" application of the law of closure, as can be seen from the three vertical lines separating the four functional groups of fuel tanks in Figure 2.1. The functional groups are as follows: the leftmost fuel gauge (1) shows the fuel in the feeder tank, which is then followed by two gauges (2, 3)

Figure 2.3. Dornier Do 328 EICAS format. © Dr. C. Vernaleken. Reprinted with permission.

for the fuel tanks in the left wing, three gauges for the fuselage tanks (4, 5, 6), and two more fuel tanks in the right wing (7, 8). Evidently, increasing the inter-instrument spacing from 7 mm to 10 mm to delineate these functional groups did not result in sufficient visual saliency due to the dominance of the law of proximity and, partially, the law of continuity. As a consequence, the three vertical white lines had to be applied in addition to the slightly increased spacing to restore the correct perception of the functional groups.

This example nicely illustrates the interrelations between the Gestalt laws, and that it must be carefully validated with pilots whether the resulting effect of their combined application is in line with intentions – a purely analytical prediction of their combined effect is virtually impossible to achieve.

Summary and Outlook

Neither airworthiness regulations nor industry standards provide a sufficiently homogeneous level of detail that would enable a comprehensive single-step, clean-sheet flight deck design. Rather, for those aspects for which only more generic requirements are available, a creative process is necessary to complement the partially very detailed airworthiness requirements in other areas in such a manner that a consistent overall flight deck design results. It has been shown that a cockpit philosophy documenting the most important design principles, trade-offs and objectives is a practicable initial step toward achieving this consistency. Subsequently, however, this must be supplemented by general criteria for functional allocation in combination with heuristics such as the Gestalt laws (Wertheimer, 1923) or the Wickens 13 (Wickens et al., 2004) principles to arrive at a viable detailed design.

To ensure that these heuristics have the desired effect, and that the overall flight deck concept is viable, the design must undergo iterative validation cycles involving, at a minimum, flight crews, other operators, where applicable, and human factors specialists, as early as possible. While some human factors issues might already be apparent in static renditions of the envisaged display formats created with, for example a vector graphics program, it is typically the dynamic environment of a flight simulator that enables a full assessment of all human factors aspects.

Future cockpits developments will face new human factors engineering design challenges such as single pilot operation, mixed remotely piloted aircraft systems, the introduction of artificial intelligence on the flight deck, and new display and interaction technologies. Usually, technology evolves faster than the underlying airworthiness regulations, so that aircraft manufacturers and authorities will have to spend a considerable amount of effort harmonising these two worlds in order to keep aviation the safe, efficient and reliable transportation method it is today.

References

Airworthiness Standards. (2023). *Transport category airplanes, 14 C.F.R. § 25.* https://www.ecfr.gov/current/title-14/chapter-I/subchapter-C/part-25

European Union Aviation Safety Agency. (2021). *Certification specifications and acceptable means of compliance for large aeroplanes* (CS-25, Amendment 27). https://www.easa.europa.eu/en/document-library/certification-specifications/cs-25-amendment-27

European Union Aviation Safety Agency. (2022a). *Certification specifications and acceptable means of compliance for all-weather operations* (CS-AWO, Issue 2). https://www.easa.europa.eu/en/document-library/easy-access-rules/easy-access-rules-all-weather-operations-cs-awo

European Union Aviation Safety Agency. (2022b). *Certification specifications and acceptable means of compliance for airborne communications, navigation and surveillance* (CS-ACNS, Issue 4). https://www.easa.europa.eu/en/document-library/certification-specifications/cs-acns-issue-4

International Civil Aviation Organization. (2005). *Annex 2 to the Convention on International Civil Aviation. Rules of the air* (10th ed.). https://www.icao.int/Meetings/anconf12/Document%20Archive/an02_cons%5B1%5D.pdf

National Archives, Code of Federal Regulations. (2023a). *Operating requirements. Part 121 – domestic, flag, and supplemental operations, 14 C.F.R. § 121.* https://www.ecfr.gov/current/title-14/chapter-I/subchapter-G/part-121

National Archives, Code of Federal Regulations. (2023b). *Operating requirements: Commuter and on demand operations and rules governing persons on board such aircraft, 14 C.F.R. § 135.* https://www.ecfr.gov/current/title-14/chapter-I/subchapter-G/part-135

SAE International. (2007). *Electronic displays (ARP4102/7).* https://doi.org/10.4271/ARP4102/7

SAE International. (2013). *Flight crew interface considerations in the flight deck design process for Part 25 aircraft. SAE ARP 5056.* https://doi.org/10.4271/ARP5056

Wertheimer, M. (1923). Untersuchungen zur Lehre von der Gestalt [Considerations on the theory of Gestalt]. II. *Psychologische Forschung, 4*(1), 301–350. https://doi.org/10.1007/BF00410640

Wickens, C. D., Lee, J., Gordon, S. E., & Liu, Y. (2004). *An introduction to human factors engineering* (2nd ed.). Pearson Prentice Hall.

Chapter 3

Applied Human Factors in Aircraft Cabin Design

Thomas Müller and Hans-Gerhard Giesa

Abstract

The aircraft cabin provides facilities for working, sleeping, eating, personal hygiene, and other activities in a limited space. This includes safe and efficient working conditions for the cabin crew and a reasonable level of comfort for the wide variety of passengers (age, physical and mental abilities, cultural background, stature, and corpulence). Therefore, specific human factors requirements for individual areas of the cabin such as passenger compartment and seating, galleys, lavatories, stowage, technical devices, and tools need to be considered.

Keywords

aircraft cabin, ergonomics, human factors methods, passengers, cabin crew

Design Scope and Design Process

The cabins of today's commercial aircraft provide living space for many hours of work, leisure, and recreation for humans and serve as protection against a hostile environment. The needs of a high diversity of end users in the cabin have to be considered for the cabin design, both for professionals (such as well-trained cabin crew, cleaning staff, security staff, catering staff) and for passengers (including those with special needs such as older people, children, and persons with disabilities). There is a wide scope in anthropometrics, cultural backgrounds, and special needs to be covered. One of the main challenges is that due to technical, economic, and environmental restrictions, space is a very limited resource in aircraft cabins.

There are several clusters of human factors (HF) aspects to be considered in parallel in all stages of cabin design:

- Environment, health, and comfort (e.g., pressurization, air conditioning, temperature and humidity, noise, vibration, lighting)
- Physical ergonomics and anthropometry (e.g., reachability of stowage, head clearance, actuation forces, visibility and legibility of signs and labels, lavatories for wheelchair users)

- Working conditions to enable efficient operations in terms of organization, crew communication, and cooperation as well as ergonomic workplace design for normal cabin operations (e.g., cabin boarding or food and drink service), as well as emergency operations (e.g., firefighting or emergency evacuations)
- Well-being and user experience (e.g., seat comfort, adjustments in the personal space such as reading lights and air outlets, inflight-entertainment systems, and social areas)

As it is not feasible to cover all these aspects comprehensively in this chapter, the topics are presented in the form of selected case studies that give insights into basic knowledge about HF issues and the related application of methods. The selected case studies address typical modules offered inside the cabin of passenger aircraft. The focus of this chapter is on physical ergonomics, workplace design, and seating comfort.

Human Factors Methods in Cabin Design

Basically, the entire range of HF methods (e.g., Stanton et al., 2005) is applied in aircraft cabin design. This includes analytical methods (e.g., calculations based on ergonomic tables and standards, biomechanical models, analysis of airline procedures and failure reports), simulation (e.g., modeling of cabin operations, boarding sequences, computer manikins in 3D cabin models), laboratory and mock-up trials, observation and questioning, technical measurements (e.g., lighting, noise, operation forces), and performance measurements.

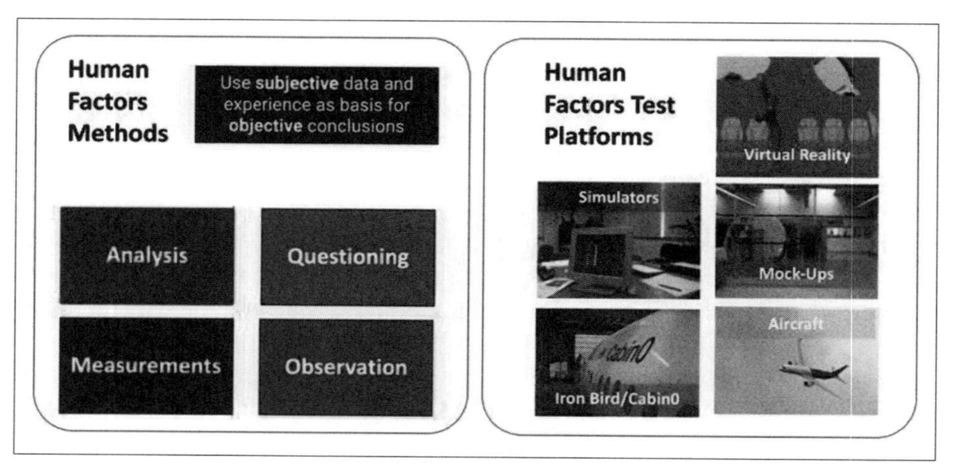

Figure 3.1. Human factors methods and test platforms. © Airbus Operations GmbH. Reprinted with permission.

Generally, the selection of tools and methods depends on the maturity of the design. In early stages analytical techniques and numerical simulations are predominantly used. As soon as a proper 3D model is available, virtual reality techniques under the use of virtual manikins are used, in later stages practical trials in mock-ups (preferably with skilled test persons, e.g., professional cabin crew for galley operations) and finally in-flight tests and observations (Figure 3.1).

Case Studies

Galley Design

Galleys in aircraft are highly specialized areas for stowage and food processing including service preparation. Thus, galleys are equipped with various types of devices such as ovens, coffee makers, and hot water boilers. To enable loading of food as well as disposal of waste quickly at differently equipped airports, specialized standard container formats are in use, the so-called standard units (or standard containers) and service trolleys. The standard units (SUs) are usually aluminum boxes, which can have a gross weight of up to 20 kg. SUs are typically stored in specialized shelves and are equipped with a door on the front side enabling withdrawal of items without taking the entire SU from the shelf. However, for some operations, cabin crew need to take the units from the storage place. This is particularly true when SUs are stowed double deep (i.e., one behind the other) for space efficiency. The service trolleys are another type of storage device, and they are used for stowage as well as for meal and beverage service in the passenger area. They are available in two sizes as a full-size type (FST) and a half-size type (HST). The gross weight is up to 100 kg for the FST and 60 kg for the HST. They are equipped with four castors each for moving on the floor and brakes for safe parking. Typically, the service trolleys are stowed in trolley bays underneath the work-surfaces, thus define the height of the latter in the range of 1100 mm above the floor.

HF activities related to galley equipment and galley operations cover aspects of safety and health, feasibility, acceptability, and efficiency. Assessments typically include checks of feasibility, efficiency, and acceptability (reachability, handling space), measurement of actuation forces, safety procedures (e.g., in the case of turbulence), and risk of injuries (e.g., finger pinching, burns). Regarding safety and health, the main topics to be considered from an HF point of view are handling of hot items, manual handling of bulky and heavy items, slips and falling, and turbulence-related injuries. Various devices may have hot surfaces during operation, which may cause a risk of burns. These risks can be assessed on the basis of research data on the contact with hot surfaces, for example, as compiled in ISO 13732-1 (International Organization for Standardization [ISO], 2006). In particular in galleys for economy class, a large number of meals need to be prepared. Hot meal components are heated in casseroles placed on trays in the ovens. The hot oven trays and the casseroles need to be handled manually to be placed on the individual meal trays for passengers. In addition to heat protection gloves, sufficient space for placing the hot items in front of or near the ovens is necessary to reduce the risk of burns.

The restricted space is a general challenge in galley design. The upward reach of shorter cabin crew defines a limit of feasibility, which can be extended by provision of foldable footrests in the galley monument plus handholds to stabilize posture. For handling of the SUs above shoulder height, more restrictive weight limitations are applied that are compatible with relevant local occupational safety and health regulations even if not explicitly applicable in the aviation context. Tools for assessment of lifting tasks can be applied, for example, the National Institute for Occupational Safety and Health (NIOSH) lifting equation (Waters et al., 1994/2021), Liberty Mutual Insurance tables (Liberty Mutual Insurance, 2012) and related equations (Potvin et al., 2021), or ISO 11228-1 (ISO, 2021). Depending on the size and the operational concept, a galley needs to provide enough space for one or more cabin crew accessing stowage, operating equipment, and performing manual work (e.g., preparation of meals). The minimal footprint is typically defined by the space required for maneuvering trolleys in the galley, for example, extracting from the trolley bay, loading with meals and beverages for service, plus some additional clearance. Generally, galley design needs to be usable for cabin crew as well as other parties (caterer, cleaning stuff) to enable efficient operations. Therefore, assessments with end users, especially professional cabin crew, are usually conducted in early stages (operative mock-up stage).

Stowage for Cabin Luggage

A large part of the passengers' luggage will be checked in for cargo transport. Nevertheless, a significant part of the luggage is carried as hand luggage in the cabin. The most important stowage spaces for this luggage in the cabin are the overhead stowage compartments installed above the seat rows. This is particularly true for economy class. We find two basic types of such compartments in today's aircraft cabins: (1) static installations with an upward swinging door opening to the aisle ("fixed bin" or "shelf bin") and (2) installations where the entire bin moves downward to provide access to the interior ("moveable bins"; depending on kinematic, also denominated as "pivot bin" and "translating bin"). Both types have advantages and disadvantages from an ergonomic point of view:

Fixed bins are easy to open since doors are typically spring-loaded. Generally, there is a trade-off between easy reach with view into the bin for shorter users and comfortable access to seats with sufficient head clearance for taller passengers when sitting under the bin. The opening of movable bins is basically supported by gravity. Spring systems or power-drives are sometimes installed for support of the closing of the bin since not only the bin but also the content needs to be lifted. In an open position, view and reach into the bin should be comfortable also for shorter users. However, the open (i.e., lowered) bins may reduce the head clearance for the access to the seats. Following the trend to bring more and larger luggage into the cabin, newer developments of overhead bins provide a higher loading capacity. An ergonomic design of overhead bins should focus on the view into the bin, the reach into the bin, and the operation forces:

- View
  Generally, a comfortable view into the bin is desirable for all users, for example, to see smaller items in the rear of the bin, find and read placards within the bin (e.g., maximal loading), and to stow luggage safely. However, it is essential for cabin crew to check proper stowage and perform pre-flight security checks. Lines of sight can be estimated in 2D models, with anthropometric manikins in 3D cabin virtual mock-ups as well as in physical mock-ups with appropriate test persons. Shoes are to be considered as well as the aids provided, for example, steps and fixed or movable mirrors.
- Reach
  Regarding reach into the bin, basically the same considerations apply as above for view into the bin. Again, assessments can be performed in virtual and in physical mock-ups to identify the need for steps or other aids.
  For opening and closing of the bin, latches and other contact points need to be reachable for the user population in general and the cabin crew in particular.
- Forces
  For typical fixed bins with a spring-loaded flap opening upward, the actuation forces are not an issue under normal conditions. However, forces for latching and unlatching of overloaded bins need to be considered.
  The operation of movable bins can be optimized by positioning the pivot point close to the center of gravity or can be supported with a power drive since the bin and the content need to be lifted. Passive spring systems may reduce the closing force but may increase effort to open an empty bin.
  Depending on the geometry, height above floor of the bin, and accessibility, the actual force application regarding force application point and force direction can differ considerably from theoretical technical assumptions. Tools for assessment of lifting tasks can be used as a guideline for assessments, for example, NIOSH lifting equation (Waters et al., 1994/2021), Liberty Mutual Insurance (2012) tables, ISO 11228-1 (ISO, 2021) as well as local safety and health regulations.

Lavatories

Lavatories in aircraft are used in addition to the basic toilet functionality for various other purposes including changing clothes, nursing of babies, dental care, and self-medication (e.g., insulin injection). Typically, lavatories are designed to be used by a single person or an adult plus a (minor) child. Persons with reduced sensory and motor capabilities (referred to as "persons with reduced mobility" [PRM]) need special consideration.

This includes:

- Good visibility and detectability of controls (toilet flush button, water faucet, door-locking mechanism, cabin crew call button) for persons of various stature and visual abilities. Generally, high-contrast designs are preferable.
- Reachability of the touch points relevant for operation by persons of various stature and possibly reduced agility has to be ensured.
- The operation of controls should be designed for adequate actuation forces and simple movements/kinematics. For reasons of hygiene, touchless controls are under discussion.

In smaller aircraft and within high-density layouts (economy class), lavatories tend to have smaller footprints, which can be challenging for corpulent persons.

The requirements to be applied for accessible lavatories are not regulated homogeneously worldwide. The main references are the ICAO manual on access to air transport by persons with disabilities (International Civil Aviation Organization [ICAO], 2013), the U.S. DOT part 382 (Nondiscrimination on the Basis of Disability in Air Travel, 2008), and the regulations and guidance provided by the European Community (European Civil Aviation Conference, 2021; European Parliament, Council of the European Union, 2006).

Accessible lavatories usable with an on-board wheelchair have higher space requirements. This is particularly true if the wheelchair user needs assistance from another person. The required space can be achieved by temporarily merging two smaller adjacent lavatories that are prepared to get the dividing wall removed/opened. The functions and devices intended to be operated without assistance have to be tested carefully in close-to-reality mock-ups. The sample should also include persons with special needs such as PRMs (e.g., wheelchair users, visually impaired persons) or seniors. Tests should cover the visibility of relevant items, comprehensibility of placards or signs, reachability of controls and handles (e.g., for support to stand up from the toilet seat, stabilizing posture), actuation forces, and feasibility of procedures for transfer from and to the wheelchair with and without assistance.

Seating Comfort

Seating comfort for passengers is an important aspect of commercial aviation. Recently, there are some challenging trends:

- Passengers tend to spend longer times on aircraft seats, for example, on ultra-long-range flights.

- There is a tendency toward reducing the space per seat to deal with market requests. This results in lateral densification (adding a seat per row) as well as densification along the longitudinal axis of cabin (reduced seat pitch).
- Today's passengers tend to be taller and more corpulent than in the past.

In parallel, we also have improvements of seats. Newer seat models have thinner backrest paddings providing the same comfort (e.g., due to improved foam materials), but more leg space (or the same leg space at reduced seat pitch).

Generally, aspects of seating comfort can be divided into three groups:

- Anthropometry-related aspects of passengers' fitting into the seat, that is, the relation of body dimensions to seat dimensions
- Seat design-related comfort parameters: cushion material (e.g., stiffness), cushion profile (e.g., lumbar support), adjustment facilities (e.g., backrest recline), avoidance of sharp edges and corners.
- Nonphysical (i.e., context related) comfort parameters including passengers' expectations and non-sitting in-flight options, for example, possibilities to walk around or areas designated for standing (e.g., social areas)

Anthropometry-related issues can be investigated with analytical methods as a first step. With a geometric comfort criterion (e.g., x mm clearance to the sidewall, armrests, passengers sitting in the next seat or front seat), simulation algorithms can calculate the portion of passengers sitting comfortably (in terms of the criterion); recommended techniques are compiled in EN 4730 (European Committee for Standardization, 2018). Regarding anthropometric raw data (i.e., a dataset for each participant of an anthropometric survey), persons with various body dimensions can be randomly placed on the seats of a set row (e.g., by means of Monte Carlo simulation). Anthropometry-related techniques can be used in the very early stages of development, for example, based on a hypothetical cabin layout. Sufficient space does not guarantee comfort, but insufficient space contradicts comfort. Generally, anthropometry-related issues are most dominant in economy class seating.

Physical measurements include measurements of the stiffness of cushions, thermal conduction of the seat, and similar parameters. Most prominent are seat pressure maps (i.e., the spatial distribution of pressure between the occupant's body and the seat surface). Although there are many approaches with which to establish pressure mapping as an objective technique for the prediction of discomfort (Mergl, 2006; Lanzotti et al., 2011; Shen 1994; Zenk, 2008), validity is not assured (Hartung, 2006). However, pressure mapping can support the identification of sources of discomfort.

The group of nonphysical comfort parameters is more diverse and strongly based on individual experience and preferences. Instead of analytical methods, observation and questioning are more appropriate. Regarding passenger comfort in aircraft, it was found that the in-flight comfort experience is impacted not only by properties of the aircraft and passengers' environment (e.g., seat properties, temperature, odors, or other passengers), but also by pre-flight experiences such as the check-in procedure or delays (Vink & Brauer, 2011). Therefore, the attribution of comfort or discomfort to certain pieces of equipment (e.g., seats) may include methodological issues. Unless the situational aspects are of primary interest, seating comfort surveys should be conducted in a neutral environment with a "standardized" context. Tests should not be performed in-service, but under a realistic occupation of neighboring seats (e.g., a triple seat occupied with three test participants selected from a representative sample) and of the proximal environment (e.g., the forward seat row and a side wall with realistic curvature) are required as a minimum if tests are not performed in a real cabin or in full cabin mock-up. Since comfort statements (e.g., on numerical scales) are hard to interpret by themselves, comparative tests with a well-known reference seat are recommended.

Observation methods typically focus on seat postures during various activities and sleep (Fenety et al., 2000; Tan et al., 2009). Basically, the spatial possibility to change seat postures is a benefit. However, frequent minor changes can indicate discomfort. Pressure mapping can support the identification of such discomfort-triggered changes. A larger number of comfort questionnaires for product design are available (e.g., Anjani et al., 2021) including thermal comfort or postural comfort. Ratings of perceived (dis)comfort may address individual activities (e.g., access, sleeping, resting, eating, or working). For seat design, the localization of issues may be helpful. This may address body parts such as the neck and shoulders (Tan et al., 2013) or seat regions (e.g., headrest, forward edge of seat pan, armrests). Typically, 10 to 20 regions are rated individually. Ratings may be requested at several intervals during a test (e.g., once per hour). Further indicators of seating comfort/discomfort are the maximal acceptable travel time on the seat, sleep duration, and quality of sleep (observation, questioning, physiological measurements). The duration of a test session on a particular seat per participant may range from a few minutes (e.g., assessment of electrical adjustment of seat) up to 8–16 hr for sleep tests or full-scale tests for ultra-long flights.

Information and Communication Devices

Although information and communication tasks in the cabin area are typically not very complex, the use of displays, controls, signs, and placards needs careful ergonomic assessment. General test cases are:

- Visibility: The easiest way to check visibility is by means of computer manikins (considering the range of body dimensions) in a 3D model of the cabin when looking at the scene "through the manikin's eyes."
- Perceptibility: In cases where the change of a sign should be noticed even if not looking at the sign (e.g., activation of "fasten seatbelt" sign), it should be checked whether the sign is in the range of peripheral vision.

- Legibility: This addresses primarily the size of the letters or pictograms as well as the colors and contrast for certain viewing distances.
- Comprehensibility: In an international environment, textual messages are critical. The typical bilingual placards (local language of the airline plus English) may be insufficient due to limited knowledge of languages or illiteracy. Pictorial presentations (icons) may perform better.

Generally, the assessment needs to consider the user population, in particular cabin crew being trained with devices and procedures versus passengers, who should be assumed to be naïve users with unknown background knowledge, knowledge of languages, and literacy.

The following aspects need to be considered for crew information, intra-crew communication, and control of cabin systems:

- The flight attendant panel (FAP), usually wall-mounted at the crew station in the front door area of the cabin, serves as the main interface for control and status display of many cabin system parameters (e.g., temperature, lighting). Due to the fixed position on the wall, legibility and reachability of control elements (e.g., software buttons or sliders, or further control hardware) for a range of users need special consideration in the design phase. Comprehensibility of the intended workflow and understanding of interactions can be tested with eye-tracking and complementary questioning techniques. As far as applicable, relevant human–machine interface (HMI) standards and recommendations (e.g., for character sizes, color contrasts, or response times) should be considered (ISO 9241 series; International Organization for Standardization, 2024; ARINC Report 837, 2011).
- Smaller versions of control panels ("mini FAPs") with less functionality are installed at other crew stations (areas next to the doors and galleys where cabin attendant seats – jump seats – are installed).
- For control of in-flight entertainment (IFE) systems, a control center is typically installed in a closet next to a galley.
- The interphone is a telephone-like voice communication net between the crew stations and cockpit. The handsets are typically wall mounted next to cabin attendant seats and equipped with a numerical keypad and potential keys for special destinations (e.g., cockpit). Furthermore, the handset is used as a microphone for voice messages in the cabin. Since it needs to be accessible also in critical situations, the reachability has to be ensured and is verified (e.g., in a mock-up).
- Other indication panels are installed at the door areas/crew stations. They include a display giving information on communication (e.g., cockpit call) or emergencies (e.g., smoke in the lavatory). An ergonomic criterion is that the information shall be readable from distant positions (e.g., when working in the galley).

- Panels consisting of a set of small light indicators in different colors called "area call panels" (ACP) are installed in the cabin ceiling. The light signs give a visual indication of the origin and nature of passenger or crew calls.
- "Fasten seatbelt" signs need to be visible for the crew when sitting in cabin attendant seats as well as during galley operations or service in the passenger compartment. Therefore, installation locations need to be checked.
- Placards are the means for displaying static information such as door operation or maximum loading of galley stowage. Placards may be in written form or pictorial. For the latter, comprehensibility needs to be proved. A methodology for the comprehensibility testing of pictogram-based aircraft signs and placards is described in SAE ARP 8996 (SAE International, 2020). The approach is based on test methods for comprehensibility described in ISO 9186-1:2014 (ISO, 2014a), ISO 9186-2:2008 (ISO, 2008), and ISO 9186-3:2014 (ISO, 2014b).

The following aspects have to be considered for information and communication devices for passengers:

- Passenger service units (PSUs) are typically installed above the seats and include a display for passenger signs ("fasten seatbelt," "no smoking" etc.), a loudspeaker to hear voice messages, and reading lights including a switch, air nozzles, and a cabin crew call button. Reachability, visibility, and legibility are important criteria that should be ensured for the range of eye-height positions. Peripheral vision is relevant in cases where the change of a sign should be noticed even if not looking at the sign (e.g., activation of "fasten seatbelt" sign).
- IFE systems are available in two basic versions:
  - Centrally/cabin crew-controlled systems with ceiling-mounted displays that can be used for presentation of safety videos after boarding and for cinema shows. Sound is provided via cabin speakers.
  - Individual devices – typically integrated into the backrests of forward seats – which can be controlled individually by passengers (selecting and watching movies, listening to music or travel information from an aircraft entertainment server). Sound is provided via headphones. Such devices can also be operated in a centralized mode overrunning individual selections, for presenting safety videos or other general information.
- Exit signs (installed on the ceiling and next to exit doors) and – in the case of emergency evacuations – additional floor-integrated light strips guide passengers to the exits. Exit signs need to be visible for persons of any stature standing in the aisle.
- Information signs may give additional information for passengers' convenience, for example, "lavatories occupied."
- Placards provide static information regarding safety equipment (e.g., stowage of life vests), correct usage of equipment (e.g., maximum loading of overhead bins), and orientation in the cabin (seat row numbering).

Cabin Features for Emergency Evacuation

In the case of an emergency situation, a rapid evacuation of the passengers might be required (e.g., after an emergency landing or due to a fire on-board). Emergency exits, emergency egress assist means, escape routes, markings, lighting, and the cabin layout need to be compliant with the requirements provided by the certification authorities. For the type certification of a commercial passenger aircraft for more than 44 passengers, a full-scale evacuation demonstration has to be conducted to show that the maximum number of passengers can be evacuated in less than 90 s through 50% of the available exits (Airworthiness Standards, 2023; European Union Aviation Safety Agency, Certification Specifications and Acceptable Means of Compliance for Large Aeroplanes, 2021) Such a full-scale demonstration is conducted only once in the lifetime of an aircraft model unless there are major changes in cabin layout or passenger count.

However, smaller engineering evacuation tests are more frequent, for example, for the extension of a type certification due to an increase in passenger count. Tests typically focus on a particular element (e.g., showing that a new evacuation slide has the same or better performance than the old one). Even if the number of evacuees is typically much smaller than for a full-scale test, their age and gender distribution follow the same rules (Federal Aviation Administration, 2016). Engineering tests are typically comparative (e.g., different design solutions) and differences are analyzed through appropriate statistical tests. Due to the risk of personal injury and the costs involved in performing large-scale evacuation tests, simulation tools will be more important in the future (Galea et al., 2002; Gobbin et al., 2021).

Summary and Outlook

Generally, the field of HF needs to pursue technical trends and innovations in the field of cabin engineering, but also, sociodemographic trends that have an impact on cabin design:

- The changing ethnic composition of the worldwide flying population, since more people from emerging market economies participate in air travel. This includes variation in anthropometric dimensions, cultural specificity, and expectations.
- Anthropometric trends. This is (1) the trend toward higher statures (NCD Risk Factor Collaboration [NCD-RisC], 2017a) as well as (2) the trend toward more corpulence (NCD-RisC, 2017b), sometimes addressed as "obesity epidemic" (World Health Organization, 2023). Both trends have an impact on the spatial requirements in the cabin.
- An aging society with an increasing portion of seniors in air travel. This has an impact on requested services and support as well as comfort expectations (Hankovska, 2018).

- The commitment to sustainability including accessibility improvements for persons with disabilities in air travel (as in other fields of living) generates requests for extended or new (technical or human) services on board, for example, the use of wheelchairs in the cabin.
- Higher ages of active cabin crew (the median cabin crew age rose from 30 years to 44 years from 1980 to 2007; McMullin et al., 2014). This may have consequences for the design of cabin workplaces (e.g., galleys) and work organization.

These future challenges may require an extension of methods beyond classic ergonomic tools and methods.

References

Airworthiness Standards. (2023). *Transport category airplanes, 14 CFR Part 25 §25.803*. https://www.ecfr.gov/current/title-14/chapter-I/subchapter-C/part-25/subpart-D/subject-group-ECFR88992669bab3b52/section-25.803

Anjani, S., Kühne, M., Naddeo, A., Frohriep, S., Mansfield, N., Song, Y., & Vink, P. (2021). PCQ: Preferred comfort questionnaires for product design. *Work, 68*(S1), S19–S28. https://doi.org/10.3233/WOR-208002

ARINC Report 837. (2011). *Design guidelines for aircraft cabin human machine interfaces*. Aeronautical Radio INC.

European Civil Aviation Conference. (2021). *ECAC policy statement in the field of civil aviation facilitation. ECAC.CEAC DOC No. 30 (PART I) 12th edition/May 2018/Amendment 5*. https://www.ecac-ceac.org/images/activities/facilitation/ECAC-Doc_30_Part_I_Facilitation_12th_Edition_May_2018_Amendment5_20210907.pdf

European Committee for Standardization. (2018). *EN 4730: Aerospace series – anthropometric dimensioning of aircraft seats*. https://www.en-standard.eu/din-en-4730-luft-und-raumfahrt-anthropometrische-dimensionierung-von-flugzeugsitzen-deutsche-und-englische-fassung-en-4730-2018/?nakup=1551347

European Parliament, Council of the European Union. (2006). *Regulation (EC) No 1107/2006 of the European Parliament and of the Council of 5 July 2006 concerning the rights of disabled persons and persons with reduced mobility when traveling by air*. http://data.europa.eu/eli/reg/2006/1107/oj

European Union Aviation Safety Agency. (2021). *Certification specifications and acceptable means of compliance for large aeroplanes, CS 25 §25.803*. https://www.easa.europa.eu/en/document-library/certification-specifications/cs-25-amendment-27

Federal Aviation Administration. (2016). *Transport airplane cabin interiors crashworthiness handbook* (AC 25-17A, Change 1). https://www.faa.gov/documentLibrary/media/Advisory_Circular/AC_25-17A_CHG-1.pdf

Fenety, P. A., Putnam, C., & Walker, J. M. (2000). In-chair movement: Validity, reliability and implications for measuring sitting discomfort. *Applied Ergonomics, 31*(4), 383–393. https://doi.org/10.1016/s0003-6870(00)00003-x

Galea, E. R., Blake, S. J., Lawrence, P., & Gwynne, S. (2002). The airEXODUS evacuation model and its application to aircraft safety. *FAA/JAA Conference in Atlantic City October 2001.*

Gobbin, A., Khosravi, R., & Bardenhagen, A. (2021). Emergency evacuation simulation of commercial aircraft. *SN Applied Sciences, 3*(4), 446. https://doi.org/10.1007/s42452-021-04295-z

Hartung, J. (2006). *Objektivierung des statischen Sitzkomforts auf Fahrzeugsitzen durch die Kontaktkräfte zwischen Mensch und Sitz* [Objectification of static sitting comfort on vehicle seats by contact forces between human and seat] [Doctoral dissertation, Technical University of Munich]. https://d-nb.info/980169062/34

Hankovska, J. (2018). Age of air travellers and its impact on priority of comfort factors. *Transportation Research Procedia, 35*, 64–71. https://www.sciencedirect.com/science/article/pii/S2352146518303533 https://doi.org/10.1016/j.trpro.2018.12.013

International Organization for Standardization. (2006). *Ergonomics of the thermal environment – methods for the assessment of human responses to contact with surfaces – Part 1: Hot surfaces* (ISO Standard No. 13732-1:2006).

International Organization for Standardization. (2008). *Graphical symbols – test methods. Part 2: Method for testing perceptual quality* (ISO Standard No. 9186-2:2008) https://www.iso.org/standard/43484.html

International Civil Aviation Organization. (2013). *Manual on access to air transport by persons with disabilities.* https://de.scribd.com/document/267682223/ICAO-Manual-Doc-9984-1st-Edition-Alltext-en-Published-March-2013

International Organization for Standardization. (2014a). *Graphical symbols – test methods. Part 1: Method for testing comprehensibility* (ISO Standard No. 9186-1:2014). https://www.iso.org/standard/59226.html

International Organization for Standardization. (2014b). *Graphical symbols – test methods – Part 3: Method for testing perceptual quality* (ISO Standard No. 9186-3:2014). https://www.iso.org/standard/59882.html

International Organization for Standardization. (2021). *Ergonomics – manual handling. Part 1: Lifting, lowering and carrying* (ISO Standard No. 11228-1:2021). https://www.iso.org/standard/76820.html

International Organization for Standardization. (2024). *Multi-part document ISO* 9241 – *Ergonomics of human-system interaction.* https://www.iso.org/advanced-search/x/title/status/P/docNumber/9241/docPartNo/docType/0/langCode/en/ics/currentStage/true/searchAbstract/true/stage/stageDateStart/stageDateEnd/committee/sdg

Lanzotti, A., Trotta, M., & Vanacore, A. (2011). Validation of a new index for seat comfort assessment based on objective and subjective measurements. In G. Concheri, R. Menehello, & G. Savio (Eds.), *Proceedings of the IMProVe 2011 International Conference on Innovative Methods in Product Design* (pp. 60–68). Libreria Internazionale Cortina Padova.

Liberty Mutual Insurance. (2012). *Manual materials handling guidelines - tables for evaluating lifting, lowering, pushing, pulling, and carrying tasks.* Liberty Mutual Group.

McMullin, D., Anger, J., Green, R., Stancato, F., Ciaccia, F. Cintra, A. David, A., Morgan, D., Mastaw, M., & Ruggiero, F. (2014). Qualitative and quantitative study of older German and US flight attendants – the BEST AGE study. *Proceedings of the Human Factors and Ergonomics Society Annual Meeting, 58*(1), 140–144. https://doi.org/10.1177/1541931214581030

Mergl, C. (2006). *Entwicklung eines Verfahrens zur Optimierung des Sitzkomforts auf Automobilsitzen* [Development of a procedure for optimization of sitting comfort on car seats] [Doctoral dissertation, Technical University of Munich]. https://d-nb.info/980332915/34

NCD Risk Factor Collaboration (NCD-RisC). (2017a). A century of trends in adult human height. *eLife (5)*, e13410. https://doi.org/10.7554/eLife.13410

NCD Risk Factor Collaboration (NCD-RisC). (2017b). Worldwide trends in body-mass index, underweight, overweight, and obesity from 1975 to 2016: A pooled analysis of 2416 population-based measurement studies in 128·9 million children, adolescents, and adults. *Lancet, 390*, 2627–42. https://doi.org/10.1016/S0140-6736(17)32129-3

Potvin, J. R., Ciriello, V. M., Snook, S. H., Maynard, W. S., & Brogmus, G. E. (2021). The Liberty Mutual manual materials handling (LM-MMH) equations. *Ergonomics, 64*(8), 955–970. https://doi.org/10.1080/00140139.2021.1891297

SAE International. (2020). *SAE ARP 8996. Comprehensibility testing for pictogram-based aircraft signs and placards.* https://www.sae.org/standards/content/arp8996

Shen, W. (1994). *Surface pressure and seated discomfort* [Doctoral dissertation, Loughborough University]. https://hdl.handle.net/2134/11126

Stanton, N., Hedge, A., Brookhuis, K., Sala, E., & Hendrik, H. (2005). *Handbook of human factors and ergonomics methods.* CRC Press. https://doi.org/10.1201/9780203489925

Tan, C. F., Chen, W., Kimman, F., & Rauterberg G. W. M. (2009). Sleeping posture analysis of economy class aircraft seat. In S. I. Ao, L. Gelman, D. W. L. Hukins, A. Hunter, & A. M. Korsunsky (Eds.), *Proceedings of the World Congress on Engineering 2009: Vol. I* (pp. 532–535). Newswood Limited.

Tan, C. F., Chen, W., Rauterberg, G. W. M., & Said, M. R. (2013). The self-reported seat discomfort survey on economy class aircraft passenger in the Netherlands. *Australian Journal of Basic and Applied Sciences, 7*(6), 563–570. http://www.ajbasweb.com/old/ajbas/2013/April/563-570.pdf

U.S. DOT part 382. (2008). *Nondiscrimination on the Basis of Disability in Air Travel, 14 CFR Part 382* (last amended May 7, 2024). https://www.ecfr.gov/current/title-14/chapter-II/subchapter-D/part-382

Vink, P., & Brauer, K. (2011). *Aircraft interior comfort and design.* CRC Press.

Waters, T. R., Putz-Anderson, V., & Garg, A. (1994/2021). *Applications manual for the revised NIOSH lifting equation.* National Institute for Occupational Safety and Health. https://doi.org/10.26616/nioshpub94110revised092021

World Health Organization. (2023). *Controlling the global obesity epidemic.* https://www.who.int/activities/controlling-the-global-obesity-epidemic

Zenk, R. (2008). *Objektivierung des Sitzkomforts und seine automatische Anpassung* [Objectification of sitting comfort and automatic adaptation] [Doctoral dissertation, Technical University of Munich]. http://mediatum.ub.tum.de/doc/656298/document.pdf

Chapter 4

Collect, Understand, Solve, Discuss, Do – The Five Pillars of Assistance Systems

Daniel Dreyer and Alexander Rabl

Abstract

Today's products are evolving with an increase of complexity and capabilities never seen before. This applies for consumer products such as smartphones as well as for professional safety-critical applications such as aircraft cockpits. In contradiction to this ever-growing complexity stands the user expectation of simple, lean and usable user interfaces. One way to cope with these demands is the creation of assistance systems that represent a mediating layer between the user and the core functionalities of a complex product. Unfortunately, the term "assistance system" does not have a common definition or design guidance and has been hijacked by marketing specialists for advertising purposes. It is therefore a deliberate decision to not have a common definition for assistance systems, which puts the system designer in a difficult situation with unclear requirements and boundaries. This chapter presents a framework for assistance system design based on the elicitation of user expectations. The target audience of this method and tool is the system design community, which can benefit from a better understanding of the actual expectations of the users when designing a product. The framework consists of five pillars, each of them dedicated to one essential capability of an assistance system.

Keywords

assistance systems, user expectation, solution space

Introduction

Current assistance systems have widespread use in many facets of daily life. This ranges from assistance on mobile devices for reserving a table in a restaurant (Huffmann, 2019), installing software on a computer (Microsoft, 2023), parking a car (Schmitt, 2019), to even flying an aircraft (Airbus, 2023). These use cases are so diverse that it is difficult to find a commonly accepted and popular definition of the term "assistance system". Nevertheless, almost every industry offers assistance systems or assistance functions in its products. Some have the goal of easing user tasks (Atlassian, 2023) at a low level or offering decision aids in complex processes (Jones et al., 2023). Some of them even seemingly take away responsibility in safety-critical decision-making or monitoring of complex machinery (Dikmen & Burns, 2017). The most infamous examples of assistance systems that do not work as expected or are obviously misused by their owners are the tragic accidents of Tesla drivers faithfully activating the so-called autopilot of their cars (Aydodgu et al., 2018; Dikmen & Burns, 2017). Recent studies revealed that many car owners are not fully aware of the capabilities of their own vehicles (Aydodgu et al., 2018). One reason for that is the non-unified marketing language giving the assistance system attractive names that spread the notion of capability that does not reflect the real technical ability (Eisenstein, 2019). A *Federation Internationale de l'Automobile* (FIA) report (Tsapi et al., 2020) even indicates that 70%–99% of car owners do not fully understand their advanced driver assistance systems (ADAS) and 70% of the driver population overestimates their own understanding of ADAS. This essentially leads to a divergence between user expectation and system performance. The user believes they know the technical capabilities of the system and expect it to behave accordingly. However, a misalignment between user expectations and system capabilities can lead to catastrophic accidents.

Assistance and User Expectations

A common framework for the design of assistance systems on the basis of human information processing in terms of Neisser (1976) and human factors principles (International Organization for Standardization, 2020; Ludwig, 2015; Maedche et al., 2016) does not exist yet, while definitions and taxonomies are manifold. The need for such frameworks is recognised (Maedche et al., 2016), as sophisticated design aids, conceptualisation and evaluation of assistance as well as a definition in greater detail for user assistance in information systems are lacking.

However, assistance definitions are well reflected (e.g. see Wandke, 2005) as assistance *interactively* extends the human natural abilities and helps people do what they cannot do by themselves by providing *functional access given by the machine.* While Wandke (2005) presents one of the most sophisticated taxonomies and definitions in the area of assistance systems, it offers only suggestions for the designers about the different areas to be covered. His suggestions can lead to the sketch of a system but do not go into detail about a functional breakdown.

Further guidance on assistance is given by Ludwig (2015), who identifies six definitive criteria for the technical solutions of assistance systems within information systems: d*iagnosticity*, *interactivity*, *goal orientation*, *error correction*, *explanation*, *and relaxation*. Ludwig's criteria directly relate to action or information-processing stages (Neisser, 1976) and are therefore especially helpful for the analysis and design of assistance. The most recent definition for assistance is given by Kraus et al. (2020) in light of artificial intelligence as a software agent supporting users in their tasks and consisting of multiple properties mirroring a perfect butler.

Considering the above, one can define an assistant system as a combination of functions actively aiding the human to achieve a goal. Therefore, assistance in its essence is best described and analysed within a framework of human information processing to aid a user in achieving a goal from initial perception to action implementation. In this sense, the skills-rule-knowledge acquisition model of Rasmussen (1983) describes user behaviour and can be used to determine what type of assistance should be applied when analysing future system capabilities with regard to users' expectations.

Regarding these theoretical aspects, one needs to review how they are applied in different sectors to aid humans. On the basis of the number of cars, the majority of assistance systems are found in the automotive sector (Bengler et al., 2014). Assistance is defined here either by high-level goals (Bendel, 2021), technical tasks (e.g. adaptive cruise control; Bengler et al., 2014; Narayanan, 2018) or by the technologies and information management concepts utilised (Matthei et al., 2016).

Flemisch and Onken (1998) created a model of cognitive assistance and applied it as *CASSY* (cockpit assistance system), which was even flight tested. Flemisch and Onken report a very high (94%) consensus regarding the situation assessment of pilots and CASSY. The structure and model of cognitive assistance described by Flemisch and Onken is one of the most detailed system plans for an assistance system. However, it lacks abstraction and provides only little advice to a future system designer as it does not explain how to analyse the main system parts.

The most advanced assistance systems are often called "intelligent personal assistants" and are brought on the market by large technology corporations (Tulshan & Dhage, 2019). There are only a few noncorporate assistants, such as Almond (Lam et al., 2019) or Mycroft (Taba, 2019). The user expectations differ regarding the context of use and the service offered. However, little to nothing is known about the inner architecture or conceptual framework – perhaps there is none apart from the idea that what makes it on the market, stays on the market.

User Expectation as Design Driver

The term "socio-technical system" is used to describe the overall system consisting of the user in their organisational, environmental and cultural context and the technical tools to achieve a common goal (Cooper & Foster, 1971). Besides user needs (Lindgaard et al., 2006), user expectations should drive the design of sociotechnical systems. In this context, we define user expectations as the *user's assumed overall capabilities of the socio-technical system in order to achieve certain goals*. Therefore, it should be the designer's duty to close the gap between user expectations and system capabilities.

User expectations are to be considered upfront:

1. from the user's perspective, the subjective appraisal about the system capabilities is based on experience with other systems and therefore not necessarily connected with the actual capabilities of the system itself, and
2. from the designer's point of view as a driver to pave the way during development.

Considering the user's subjective appraisal, practitioners often see these implicitly considered when the user's tasks are taken into account (Stanton et al., 2005, pp. 45–53). However, primarily considering tasks (the "what") as the subject of analysis and design does not explain *how* the system should function from a user's point of view, as this expectation originates from past experiences with other systems. These past experiences go hand in hand with the expectation of how these tasks should be performed with a future system. It is suggested to utilise these expectations for the design of socio-technical systems.

Secondly, if user expectations are taken into account during design, usability can be improved to a satisfactory experience regarding the system's performance and a user's identification with this system. On the contrary, when implicit expectations are not met, confusion, frustration, error and generally inadequate performance or failure overall may be the result. In the worst case, this may lead to the user's rejection of the entire system. In a nutshell, user expectations and task analysis need to be managed in concurrence by the system designer.

The user expectation originates from the user's mental model of the socio-technical system capabilities to achieve a certain goal and their appraisal of their own skills, knowledge and experience (Rasmussen, 1983). In this respect, the question "What can I achieve with the help of this machine?" becomes more comprehensive:

"What Can I (User Skills) Achieve (Goal) With the Help of This Machine (Technical Functions)?"

Only if the subjective user expectation matches reality, users will be able to properly operate the technical system and thus achieve their goals. If users have an invalid assessment of their skills, or the technical capabilities, the expectations are simply wrong and will result in sub-optimal or even potentially dangerous overall system performance. Therefore, it is of vital importance to match the user expectations with the available technical capabilities. As pointed out earlier, the system designer's goal is the matching of user expectations with the actual capabilities of the technical system, thus optimising the total performance of the overall socio-technical system. As long as there is an exact match between user expectation and machine compliance, the technical system is doing exactly what users expect from it. The socio-technical system is delivering optimal results. Higher user expectations than the machine can satisfy will result in disappointed or annoyed users, and consequently lead to poor system performance including errors and slow progress. If the machine is capable of delivering higher performance than the human users expect it to, it will go unnoticed in many cases, but in some rare situation's users may discover new features or be positively surprised. However, unexpected changes to well-trained processes may annoy experienced users, even though the changes are an actual improvement. In most cases, users will not be able to exploit system capabilities, and thus the total system performance will not be utilised. Consequently, expectation matching needs to be in balance in order to elevate the socio-technical system performance to its best achievable results. By contrast, ignoring user expectations may lead to a total lack of system performance, which can be equated to erratic and unpredictable behaviour ultimately resulting in accidents with loss of equipment and potential human fatalities.

Five Pillars of Assistance

This section introduces the framework for the classification and design of assistance systems according to user expectations as, to the best of our knowledge, this gap is not covered by the existing frameworks. The framework (see Figure 4.1) is built on five pillars that define the technical capabilities of the system oriented towards human information processing inspired by Neisser's perception–action cycle (see Neisser, 1976):

1. *Collect* – assist to gather information about all necessary conditions, the *data space*
2. *Understand* – assist to understand and interconnect data and gathered information, the *question space*
3. *Solve* – assist to generate solutions for the task, the *solution space*
4. *Discuss* – assist to give transparency to a generated solution, the *explanation space*
5. *Do* – assist to implement the suggested (and selected) solution

Each pillar represents one step in supporting a user towards the achievement of a socio-technical goal in an expected way. This process may have a strong iterative character with many loops and/or jumps.

The Collect Pillar

The first pillar, Collect (see Figure 4.1, column "collect"), is the technical capability to gather information about the goal, environment, boundary conditions, possible limitations and the overall system state. This includes necessary user input, but it also contains the system's ability to access databases, archives and previously collected data in order to learn from past tasks.

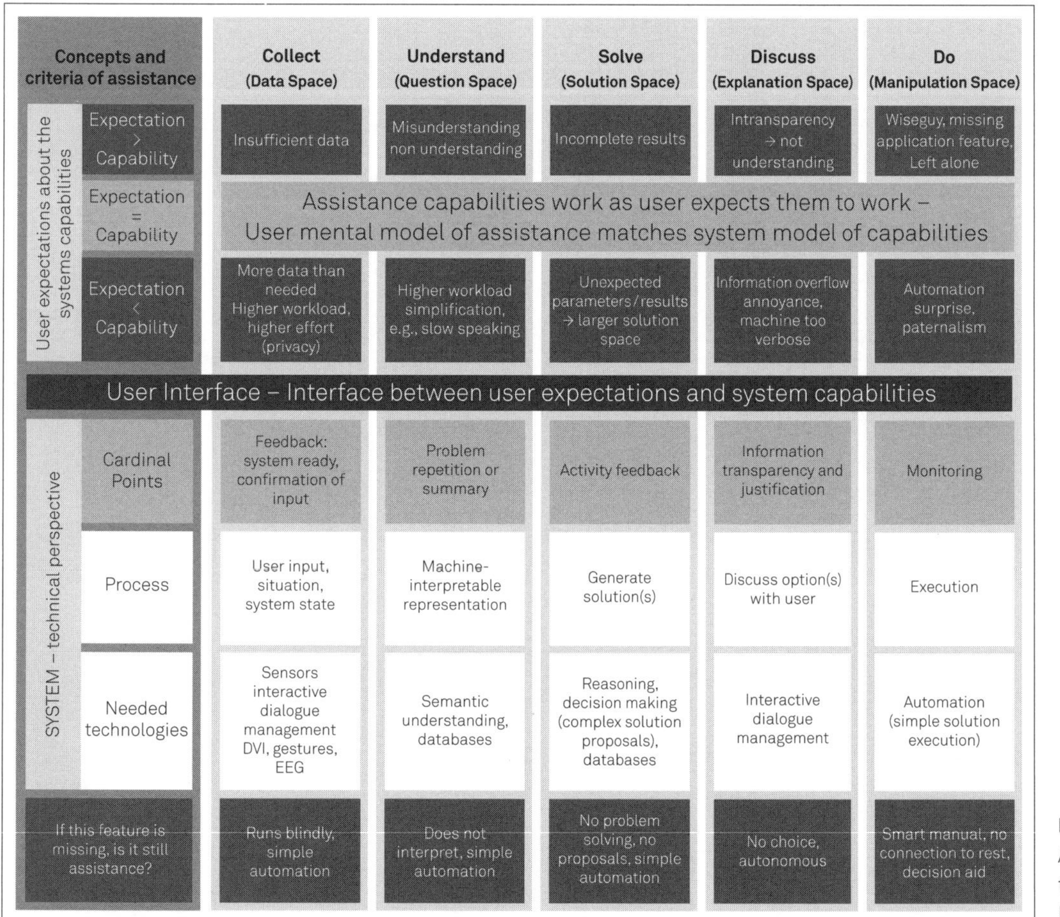

Figure 4.1.
A framework for classification of assistance systems. DVI = direct voice input; EEG = electroencephalogram.

From a user's perspective, it is the main means with which to formulate the intentions of the goals and boundary parameters of the socio-technical system.

User Expectation and Consequences

The user expectations towards the Collect pillar (see Figure 4.1, upper part) have to be analysed regarding information collection. As the socio-technical system has two aspects (human and machine), the expectations of the users will contain the same perspectives: The first perspective is the scope of information they have to, or can, provide in order to feed the machine with the data needed, and the other perspective is their perception of the machine being capable of retrieving additional information on its own.

If the system capabilities match user expectations, users will have the impression that the right number of data, with the right means, at the right time were inserted into the system and that all necessary knowledge to accomplish the task is available. If the expectations are higher than the actual system performance, the users will believe that the system will process more data than it actually does. The users may falsely believe that the system collects supplementary information or has access to other information sources. The users may also be disappointed because they were not able to fully or precisely express the goals. Consequently, as long as the users do not realise the deficiencies, they will have overconfidence in the system's capabilities, but as soon as it becomes clear that the system does not fulfil the Collect capability, the users will not trust the technical system to have fully understood the task or of being able to achieve the demanded results with adequate quality. This lack of trust may lead to behaviour ranging from double-checking of all results to declining the system use at all (disuse). If the expectations are lower, users may be puzzled or annoyed by the abundance of data that are used to perform the given task. Users may wonder why so much information is needed to calculate a simple result. Users may not see the added value of performing this input and they may feel that the number and/or category of collected data is actually not needed for the task at hand. If users have to manually enter data that are perceived to be redundant, they may feel it is a waste of time or even raise privacy concerns.

If the system does not fulfil any of the aforementioned criteria, the machine does not collect input from the user or has no sensors or databases to connect to. It runs *blindly* and follows a hard-coded programme, turning it into a simple reactive automation.

Interaction

Regarding interaction, the technical system (see Figure 4.1, middle part) needs to indicate its operational status and readiness to receive input, and has to confirm the reception of user input. Furthermore, any support for entering the data correctly is quite useful. Future products may even make more extensive use of speech recognition, natural language interpretation, gesture recognition or eye-tracking.

Applying the behaviour model of Rasmussen (1983), the Collect pillar may cover all stages of skill, rule and knowledge acquisition.

- If the user does not know how to solve the task at all, the Collect pillar should already act like a *knowledge-based system*, since the user will not be able to contribute with problem-solving capabilities.
- If the user is not able to solve the task by applying rules known to him or her, or by simple trial-and-error of said rules, the machine will have to compensate for this lack of user knowledge or task flexibility by providing rule-based behaviour. In this context, *task flexibility* refers to a situation that enables trial-and-error approaches. However, safety-critical situations often prohibit such a behaviour.
- If the user is able to solve the task by practising, the technical system can support the user in developing or maintaining the skill.

The Understand Pillar

The second pillar, Understand, is the technical capability to semantically understand and interconnect previously collected data. It translates data into a machine-interpretable representation of the task including scope and assessment of feasibility. The user needs an indication of whether the system understands the task or question and is generally able to support the user in the pursuit of the goal. If not, the user needs instruction about what went wrong and how to refine the data in the Collect pillar.

User Expectation and Consequences

The general user expectation considering the *question space* (see Figure 4.1, upper part) is that the machine has understood the question or task posed. If the expectations are met, the user will have the notion that the technical system knows what to do and how to do it in order to accomplish the given task. The inserted data are correct and complete and the machine has all the necessary capabilities to deliver the expected results.

If the expectations are higher than the actual machine performance, users assume that the machine has understood the task or that sufficient information is available. Depending on the type of system, this misunderstanding may be reported by the technical system immediately, indicating that the available data seem to be incomplete or inconsistent. In the worst case, the machine will not realise such an inconsistency and will start working on false parameters. Comparing it to a human, it will work on false assumptions with poor knowledge, eventually leading to wrong results. The combination of too-high expectations on the user side and poor task-understanding on the machine side can lead to an abundance of trust and dangerous situations.

If the expectations are lower than the actual machine capabilities, users believe that the technical system has difficulties in understanding the task. This will ultimately lead to a higher workload, because the task phrasing will be simplified, abstracted and double-checked until users are convinced that the machine has a proper understanding. This often results in speaking slowly, using simple language and asking questions in order to assess whether the information was properly understood. This unnecessary workload increase may lead to longer task durations and user frustration, but could have a positive side effect of task clarification or error detection on the user side. Simplifying and abstracting complex topics usually helps in understanding not only the problem, but also the machines and tools, albeit at a high cost.

If the machine does not fulfil the aforementioned criteria, it is a simple automation, because it does not interpret or semantically understand the data and context.

Interaction

In order to make sure that the machine has correctly understood the task, users should receive a summary or repetition of the processed parameters (see Figure 4.1, middle part). The perfect user interface would be an acknowledgement of the task by paraphrasing the instructions in *one's own words*, as is usually practised when performing active listening techniques (Levitt, 2002).

Applying the model of Rasmussen (1983), the Understand pillar may cover all stages of skill, rule and knowledge acquisition.

- If users do not know how to sort or interpret the available raw data, the Understand pillar already acts like a knowledge-based system, since they will not be able to contribute with well-formatted or ordered data, which has to be compensated by the machine. The same applies if users do not know what they are actually looking for.
- If the machine expects structured data in order to properly process them, it follows a rule-based approach and can be classified as such.
- If the machine does not need to perform a complex interpretation of the question (e.g. reasoning), it acts on a skill-based level.

The Solve Pillar

The third pillar, Solve, is the technical capability to generate solutions for the task. Depending on the complexity of the task, this capability may range from simple automation to reasoning or even selecting problem-solving strategies in an autonomous manner. The Solve pillar may even create solutions that need to be compared with each other in order to find the best recommendation given the circumstances of use.

Depending on the overall complexity, the result(s) can be presented to the user; especially if multiple solutions are available from which the user may choose.

User Expectation and Consequences

The user's expectations towards the solution-finding mechanism (see Figure 4.1, upper part) manifests itself in the opinion of whether the machine is fit for purpose or not. In the previous pillar the user entered the request, while in this pillar, the user monitors the solution process. If the user's expectations towards the problem-solving process are not fulfilled, the user might judge the tool as insufficient to cope with the task. The Solve pillar has to fulfil at least two criteria. First of all, the user needs to know that the machine is currently processing, especially if reaction times are slower due to high computational workload or network delays. Second, the user might have expectations towards the *solving* process being transparent and want to understand its source data, its weighing and the conclusions drawn or other aspects of the decision-making and problem-solving process.

If the expectations are higher than the actual machine performance, the user assumes that the processing will deliver a solution with higher-quality criteria (user overconfidence) than the system will actually do. In that context, the user may be driven to misuse the system by trusting the results even beyond their intended scope. Depending on the level of transparency, the user may notice these deviations or never realise them at all.

If the expectations are lower than the actual machine performance, the user believes that the algorithm is simpler, slower, takes fewer data sources into account, or will generally deliver doubtful results. In any case, the user will scrutinise the solution and compare it with other tools or knowledge from past experiences. If the assistance system is the only available tool, the user might even decide to disuse it if the results seem too unrealistic.

If the machine does not fulfil the criteria for the Solve pillar, it becomes a simple automation, because it does not actually perform problem solving or generate options or proposals.

Interaction

The system has to show an activity indication as an absolute minimum. Furthermore, it has to inform the user on which data sources and processes are being taken into consideration for the calculation of the solution. Most of those indications will be re-used in the subsequent Discuss pillar, so it will make sense to keep them consistent (see Figure 4.1, middle part).

Applying the behaviour model of Rasmussen (1983), the solve pillar mostly covers the stage of knowledge acquisition, because the user needs to interpret rather complex data relations in order to understand the underlying process.

The Discuss Pillar

The fourth pillar, Discuss, is the technical capability to explain and give transparency and justification to a generated solution. The more complex the solution and the reasoning behind it, the more difficult it will be for the human to understand, judge and select the appropriate suggestion of the technical system. In this context, *discussing* means to comprehensively convey the reasons and background for any solution created in the previous pillar. In order to do so, the technical system has to offer transparency and justification for the solutions presented.

- Explanation is the communication process of the system construing its solution. Explainability as a socio-technical system trait regarding a given solution is "[...] a shared meaning-making process that occurs between an explainer and an explainee. This process is dynamic to the goals and changing beliefs of both parties [...]" (Ehsan et al., 2021).
- Justification goes beyond mere explanation and allows the user to dive deeper into details. The technical system is capable of *defending* its suggestion, when asked by the user. It can *detail* how it came to certain conclusions and *convince* the user about the correctness of its predictions and the solutions offered.
- Transparency is the ability to show the user the related parameters and their influence in the calculation that leads to the suggested solution(s). As such, it is "[...] the user's understanding of why a machine [...] behaved in an [unexpected] way [...]" (Kim & Hinds, 2006).

The fourth pillar is the most debated, because in most cases it is technically very difficult to achieve.

- Some assistance systems are designed in a technology-centred way, which means that the user's understanding or acceptance is deemed unnecessary.
- The system calculates only one solution, which is presented to the user as an unavoidable choice.
- Many systems make use of computational methods that are a black box.
- Assistance systems that are employed in time-critical environments often omit the discussion step in order to save time.

Consequently, the Discuss pillar is not simply about *showing* data, which is commonly referred to as "transparency," but more about *explanation* and *justification* of solutions. Ideally, users can enter into a dialogue with the machine and receive a comprehensible rationale for the selection of certain solutions. *The machine must be able to convince the users of the proposed solutions.*

User Expectation and Consequences

User expectations for the Discuss pillar (see Figure 4.1, upper part) mostly depend on the actual use case. According to Rasmussen (1983), not every assistance system actually requires the ability to justify its decisions to the users, nor is it expedient to waste time and energy on long-lasting discussions in time-critical or rather obvious situations.

In any case, the users will expect the machine to show results and options as well as to justify itself upon user request; they will also expect to have the option of skipping the discussion with the machine. From a user's point of view, the discussion or at least its level of detail should be optional. In addition, the perceived necessity for discussions or justifications may change over time, depending on the user's trust in the system capabilities. Initial reservations may require more details for the bases of a solution or comparisons between several solutions, but in time, with evolving trust, the users may decide to skip this step or to reduce the presented level of detail. However, any disappointment may revert this decision and a higher level of justification may be necessary again.

If the expectations are higher than the system's capabilities for discussing a solution, users may be unconfident concerning the validity or quality of the proposed solution. If users have to choose between different solutions without properly understanding the differences and peculiarities, they might select the wrong solution.

If the expectations are lower, users could be annoyed by the abundance of information the system offers. Especially for time-critical systems or routine tasks, users may want to opt out of any kind of discussion, thus having a lower expectation at this particular point of use. Inexperienced users may be overwhelmed by the system's explanation or justification capabilities, since it requires a certain degree of technical expertise for the user to meaningfully work with it.

If the explanation space is missing and users have no opportunity to critically question the origins of a proposed solution, there is actually no choice for a human to make an informed decision. Thus, even if there are a handful of possible choices, the actual decision may be wrong, because it is not made on a sound basis. The system gets more and more autonomous, leaving users out of the loop.

Interaction

The interaction with the user in the Discuss pillar (see Figure 4.1, middle part) can turn out to be the most demanding user interface of the entire assistance system for the developer, depending on the decision topic at hand and the complexity to convey the explanations or justifications. The first presentation of results lies at the focus of this pillar, which can be achieved through various means, ranging from simple display indications up to multi-modal representations. It is generally advisable to give users the opportunity to scale the depth of the requested information.

In this context, the system developer has to choose between explanation and justification, whichever is more appropriate for the task at hand and the technical capabilities of the system.

Users may select maximum justification, which would basically show all parameters and the decision process that led to the conclusions drawn. Users may also select point justifications, to which the system will then need to provide more explanations. Last, but not least, users may decide to skip the discussion step completely, as explained earlier. The system itself may offer to provide more rationales for its decisions and the user may opt in or out deliberately, as described earlier. In any case, the Discuss-user interface must be designed and assessed with various user types in order to create an efficient and trust-worthy interaction.

Applying the behaviour model of Rasmussen (1983), the Discuss pillar places emphasis on knowledge-based behaviour, because the dialogue focuses on understanding complex relations, decision trees or heuristics. It provides a didactic, interactive dialogue to convey a thorough understanding of the answers. However, it could also be applicable to rule-based behaviour, but would only offer structured answers, which requires previous knowledge on the user side. Finally, it can help the manual execution of labour by highlighting the smallest task steps, which can be then considered a skill-based assistance.

The Do Pillar

The fifth and last pillar, Do *(manipulation space)*, is the technical capability to actually implement the suggested and/or selected solution in the real world, providing more than good advice.

User Expectation and Consequences

The Do pillar (see Figure 4.1, upper part) is often neglected, since the assistance system may not be connected to other systems or devices that actually enable manipulation in a sense that it transforms the solutions to reality due to legal, technical or other reasons. Very often, users are given a hint by the assistance system but will then need to take their own actions in order to achieve the goals proposed by the system.

If the user's expectations are higher than the actual system capabilities, the user will be disappointed and perceive the assistance system as unfinished or as incomplete software, or even as a wisenheimer that can give advice but is not able to support the realisation of the task.

If the user's expectations are lower, the user may be positively surprised and happy to receive this additional level of comfort and support, but it may also turn out that they do not trust the system's capabilities or feel patronised by its power of control.

If the manipulation space is missing, the entire assistance system is not capable of interacting with the environment. It will be able to give advice to users, but will not turn them into action. Actually, such a system may be rather annoying because it will not offer support in improving the situation. In the best case, such a system may act as a smart manual or sophisticated decision aid.

Interaction

The absolute minimum control users must have for the Do pillar (see Figure 4.1, middle part), is an abort function as it can have an effect on the real world. Users must always be able to abort or terminate this process, because assistance systems support the users in a task but still keep the user in authority.

While the system executes the task, users may need means of monitoring the correct execution of the task to check if it is actually meeting the agreed solution characteristics. In the case that the process is behaving unexpectedly, users may wish to interrupt and change the parameters of execution. This could be a complete jump back to the Collect or Discuss, or require an human-machine interface in the Do Pillar.

Applying the behaviour model of Rasmussen (1983), the Do pillar is a knowledge-based system, if it is able to react to unknown or not pre-programmed situations. If it can only react to situations defined during development, it is a rule-based system, and if it relieves users from mechanical or physical tasks, it is a skill-based system.

Summary and Outlook

In this chapter the challenges of contemporary assistance systems and a framework in which to create a better expectation match with such systems is described. In order to improve the current situation, it is necessary to establish an industry-wide definition of the term "assistance system", thus removing the power of interpretation from the marketing departments and creating an objective definition. Secondly, system developers should add the collection of user expectations to their requirement elicitation process, even though such expectations might result in non-functional requirements. The knowledge of these user expectations will focus developments towards the right direction and might push engineers closer to understanding real user needs. Practitioners interested in applying the five pillars of assistance to their daily routines are encouraged to contact the authors, who can provide straightforward checklists with questions that can be used to determine the user expectations for each pillar individually.

References

Airbus. (2023, January 12). *Could the humble dragonfly help pilots during flight?* [Press release]. https://www.airbus.com/en/newsroom/stories/2023-01-could-the-humble-dragonfly-help-pilots-during-flight

Atlassian. (2023). *Im Team schneller vorankommen, an einem Strang ziehen und besser entwickeln* [Move faster as a team, pull together and develop better]. https://www.atlassian.com/de/software/jira

Aydodgu, S., Schick, B., & Wolf, M. (2018). Claim and reality? Lane keeping assistant. In L. Eckstein, S. Pischinger, B. Hammermüller, & R. Wolsfeld (Eds.), *27th Aachen Colloquium Automobile and Engine Technology: The conflict between expectation and customer experience* (pp. 1241–1262). RWTH. https://opus4.kobv.de/opus4-hs-kempten/frontdoor/index/index/docId/141

Bendel, O. (2021). Fahrerassistenzsystem: Definition: Was ist „Fahrerassistenzsystem“? [Driver assistance system: Definition: What is “driver assistance system”?]. *Gabler Wirtschaftslexikon.* Springer Gabler. https://wirtschaftslexikon.gabler.de/definition/fahrerassistenzsystem-53999/version-384609

Bengler, K., Dietmayer, K., Farber, B., Maurer, M., Stiller, C., & Winter, H. (2014). Three decades of driver assistance systems: Review and Future Perspectives. *IEEE Intelligent Transportation Systems Magazine, 6*(4), 6–22. https://doi.org/10.1109/MITS.2014.2336271

Cooper, R., & Foster, M. (1971). Sociotechnical systems. *American Psychologist, 26*(5), 467. https://psycnet.apa.org/record/1972-03897-001 https://doi.org/10.1037/h0031539

Dikmen, M., & Burns, C. (2017). Trust in autonomous vehicles: The case of Tesla Autopilot and Summon. In IEEE (Ed.), *2017 IEEE international conference on systems, man, and cybernetics* (pp. 1093–1098). https://doi.org/10.1109/SMC.2017.8122757

Ehsan, U., Liao, Q. V., Muller, M., Riedl, M. O., & Weisz, J. D. (2021). Expanding explainability: Towards social transparency in AI systems. In Y. A. Quigley, K. Isbister, T. Igarashi, P. Bjørn, & S. Drucker (Chairs), *CHI conference on human factors in computing systems* (pp. 1–19). Association for Computing Machinery. https://doi.org/10.1145/3411764.3445188

Eisenstein, P. A. (2019, July 27). *Safety groups want FTC, state probes of Tesla's autopilot system and its marketing efforts.* CNBC. https://www.cnbc.com/2019/07/26/safety-groups-want-ftc-state-probes-of-teslas-autopilot-system.html

Flemisch, F. O., & Onken, R. (1998). The cognitive assistant system and its contribution to effective man/machine interaction. In North Atlantic Treaty Organization, Research and Technology Organization (Chair), *RTO meeting proceedings 3: The application of information technology (computer science) in mission systems.* https://apps.dtic.mil/sti/tr/pdf/ADA359498.pdf

Huffmann, S. (2019, March 6). *Book a table with the Google Assistant across the country on more devices.* Google. https://blog.google/products/assistant/book-table-google-assistant-across-country-more-devices/

International Organization for Standardization. (2020). *Ergonomics of human-system interaction: Part 210: Human-centred design for interactive systems* (DIN EN ISO 9241-210). https://www.iso.org/standard/77520.html

Jones, V., Chambers, U., & Petit, B. (2023). *WSU decision aid system (DAS).* Washington State University. https://treefruit.wsu.edu/tools-resources/wsu-decision-aid-system-das/

Kim, T., & Hinds, P. (2006). Who should I blame? Effects of autonomy and transparency on attributions in human-robot interaction. In K. Dautenhahn, C. L. Nehaniv, B. Robins, L. Cahamero, & D. Lee (Eds.), *The 15th IEEE international symposium on robot and human interactive communication* (pp. 80–85). IEEE. https://doi.org/10.1109/ROMAN.2006.314398

Kraus, M., Ludwig, B., Minker, W., & Wagner, N. (2020). Assistenzsysteme [Assistance systems]. In G. Görz, U. Schmid, & T. Braun (Eds.), *Handbuch der künstlichen Intelligenz* (pp. 859–906).

De Gruyter. https://epub.uni-regensburg.de/46168/ https://doi.org/10.1515/9783110659948-020

Lam, M. S., Campagna, G., Xu, S., Fischer, M., & Moradshahi, M. (2019). Protecting privacy and open competition with Almond: An open-source virtual assistant. *XRDS: Crossroads, the ACM Magazine for Students, 26*(1), 40–44. https://doi.org/10.1145/3355757

Levitt, D. H. (2002). Active listening and counselor self-efficacy: Emphasis on one microskill in beginning counselor training. *The Clinical Supervisor, 20*(2), 101–115. https://doi.org/10.1300/J001v20n02_09

Lindgaard, G., Dillon, R., Trbovich, P., White, R., Fernandes, G., Lundahl, S., & Pinnamaneni, A. (2006). User needs analysis and requirements engineering: Theory and practice. *Interacting with Computers, 18*(1), 47–70. https://doi.org/10.1016/j.intcom.2005.06.003

Ludwig, B. (2015). *Planbasierte Mensch-Maschine-Interaktion in multimodalen Assistenzsystemen* [Plan-based human-machine interaction in multimodal assistance systems]. Springer Vieweg. https://doi.org/10.1007/978-3-662-44819-9

Maedche, A., Morana, S., Schacht, S., Werth, D., & Krumeich, J. (2016). Advanced user assistance systems. *Business & Information Systems Engineering, 58*(5), 367–370. https://doi.org/10.1007/s12599-016-0444-2

Matthei, R., Reschka, A., Rieken, J., Dierkes, F., Ulbrich, S., Winkle, T., & Maurer, M. (2016). Autonomous driving. In H. Winner, S. Hakuli, F. Lotz, & C. Singer (Eds.), *Handbook of driver assistance systems* (pp. 1519–1556). Springer. https://doi.org/10.1007/978-3-319-12352-3_61

Microsoft. (2023). *Windows 11 installation assistant.* https://www.microsoft.com/software-download/windows11

Narayanan, A. (2018, April 17). *How does ADAS save lives and cars?* ELE Times. https://www.eletimes.com/how-does-adas-save-lives-and-cars

Neisser, U. (1976). *Cognition and reality: Principles and implications of cognitive psychology.* W. H. Freeman.

Rasmussen, J. (1983). Skills, rules, and knowledge; Signals, signs, and symbols, and other distinctions in human performance models. *IEEE Transactions on Systems, Man, and Cybernetics, SMC, 13*(3), 257–266. https://doi.org/10.1109/TSMC.1983.6313160

Schmitt, T. (2019). *BMW Group automated parking.* https://www.press.bmwgroup.com/global/tv-footage/detail/PF0005947/bmw-group-automated-parking?language=en

Stanton, N. A., Salmon, P. M., Walker, G. H., Baber, C., & Jenkins, D. P. (2005). *Human factors methods: A practical guide for engineering and design* (1st ed.). Ashgate Publishing Limited.

Taba, T. (2019). *Personalized AI assistant* [Unpublished bachelor thesis]. Metropolia University of Applied Sciences, Helsinki. https://www.theseus.fi/bitstream/handle/10024/168678/Taba_Tunde.pdf?sequence=2

Tsapi, A., van der Linde, M., Oskina, M., Hogema, J., Tillema, F., & van der Steen, A. (2020). *How to maximize the road safety benefits of ADAS? (Report reference: BH3649-RHD-ZZ-XX-RP-Z-0001* [Update of the 23-10-2020 version]. FIA. https://www.fiaregion1.com/wp-content/uploads/2020/10/FIA-Region-I-_ADAS-study_18122020.pdf

Tulshan, A. S., & Dhage, S. N. (2019). Survey on virtual assistant: Google assistant, Siri, Cortana, Alexa. In S. M. Thampi, O. Marques, S. Krishnan, K.-C. Li, D. Ciuonzo, & M. H. Kolekar (Eds.), *Advances in signal processing and intelligent recognition systems* (pp. 190–201). Springer. https://doi.org/10.1007/978-981-13-5758-9_17

Wandke, H. (2005). Assistance in human-machine interaction: A conceptual framework and a proposal for a taxonomy. *Theoretical Issues in Ergonomics Science, 6*(2), 129–155. https://doi.org/10.1080/1463922042000295669

Chapter 5

Designing System Explainability for Flight Crew

Basic Principles

Denys Bernard and Sonja Biede-Straussberger

Abstract

The regulatory documents in preparation for the certification of artificial intelligence (AI) systems will require the needs for system explainability to be understood and properly fulfilled, by providing relevant explanatory information in operations. After an introduction into the initiatives to drive standardisation, this chapter presents a structured set of concepts to understand and specify explainability, starting from a human user perspective. It then describes how explainability can be handled as a system function throughout the system engineering process.

Keywords

explainability, artificial intelligence, speech act, social theory, diversion, task context

Introduction

In today's society, the push for products based on artificial intelligence (AI) has increased tremendously. For aviation, the European Union Aviation Safety Agency (EASA) published its own roadmap for the deployment of AI (2020, 2023b). In parallel, standardisation working groups (EUROCAE WG-114 and SAE G-34) have been created to elaborate guidelines for the certification of airborne systems implemented by using AI techniques, particularly machine learning (ML). This activity is guided by an EASA concept paper (2021a), with a second issue published in 2023, which points out that AI will give rise to new opportunities and new challenges regarding human–system interaction. Particular attention is granted to explainability.

EASA (2023a, p. 14) defines explainability as a special system capability "to provide the human with understandable, reliable, and relevant information with the appropriate level of details and with appropriate timing on how an AI/ML application" produces its results. It is split into *development explainability* for development and maintenance activities, and *operational explainability*, which enables the operator "to assess the appropriateness" of the outcome of the system.

This chapter focuses on the viewpoint of the operator. In particular, we imagine how our multidisciplinary design methods will initially capture operators' needs for explanations, and then specify explainability functions in an industrial methodology. We first summarise how explanation and explainability are addressed in existing regulatory documents, as a constituent of human-system interaction and cooperation, and position psychological and social theories to describe human explanation behaviour. Then, concepts to ground our common language on explainability are proposed, responding to a similar purpose to the ontology used by Chari et al. (2020) for the biomedical domain: to be more rigorous and efficient in the specification of the explainability subsystem. Finally, the concept of explainability is positioned into aircraft development processes and we conclude by analysing what could be the next steps towards an industrial method for the development of explainability functions in future cooperative/collaborative systems.

System Explainability From the Perspective of the Regulator

This chapter introduces essential documents in the currently existing regulation as well as ongoing initiatives. In particular, the joint working group EUROCAE WG114/SAE G-34 frames the certification of future AI systems following the EASA concept paper (2023a). The essential requirements related to human factors (HF) are covered in a dedicated part of the "Certification Specifications for Large Aircraft: CS 25.1302" (EASA, 2007).

Pre-Regulatory Concepts of AI Explainability

The EASA AI roadmap (EASA, 2020) had first defined three levels of AI applications depending on their autonomy and cooperativeness: Level 1 AI applications only provide information or recommendations, Level 2 cooperate or collaborate with the user, and Level 3 functions are autonomous. These levels have been refined in the EASA concept paper Issue 2 (2023a; Figure 5.1).

Issue 1 of the concept paper (EASA, 2021) guided the elaboration of AS6983 "Process Standard for Development and Certification/Approval of Aeronautical Safety-Related Products Implementing AI", covering Level 1 systems only. It emphasised the role of explainability in development.

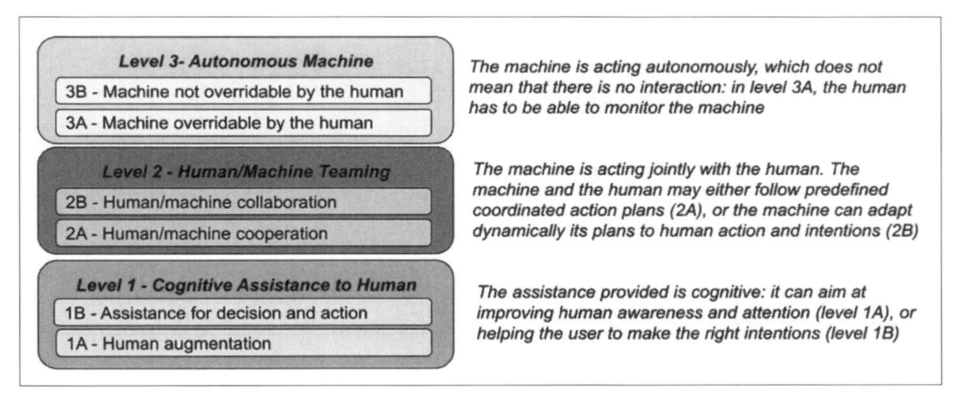

Figure 5.1. Classification of AI applications. Based on EASA roadmap (2020).

Issue 2 (EASA, 2023a) covers Level 2 applications. Because Level 2 applications aim at acting jointly with human users, human–system interaction is a complex and central capability. The levers for an efficient human–system interaction with these new systems are:

- operational explainability;
- human – AI teaming (HAT);
- natural interaction (language, gestures); and
- human error management.

The main goal of *operational explainability* is to get the operators to trust the outcomes of the system and make the best use of the system (whereas the purpose of *explainability in development* is to support system validation and maintenance). Explainability crosses further interaction patterns detailed in the concept paper (EASA, 2023a): HAT assumes mutual awareness, negotiation and adjustment of actions and plans. Regarding natural language, it is acknowledged that the reliability of spoken communication is rooted in our capability to detect errors and recover from them.

EASA (2021, 2023a) requires that the "applicant" of the guidance designs explainability by first understanding the roles held by the human operators and the tasks they perform with the system; then for each output of the system, by identifying the need for an explanation; and finally by specifying the set of necessary explanations to be provided to the human. The applicant has "to ensure the validity of the specified explanation". Dedicated objectives ensure that the explanations provided by the AI systems have required *attributes*: relevance, understandability, level of abstraction, timeliness, and reliability. A *relevant* explanation allows the user "to assess the appropriateness of the decision/action"; hence, it provides reasons to accept or reject them. This definition of "relevance" is restrictive compared with its classic definition in communication theory (Wilson & Sperber, 2012), where it is defined as the ratio between cognitive benefit and cost. In the EASA concept paper, the cognitive cost is mastered through *understandability* and *level of abstraction*, "taking into account the characteristics of the task, the situation, the level of expertise of the end user and the general trust given to the system". Adequate *timeliness* supposes that the time of explanation maximises the utility of explanations and minimises the possible disturbances for the user. It can be obtained through dialogues, where the user specifies conversationally the information required. In Level 2B systems, explanations are needed on different steps of collaborative processes: before the system acts, to be transparent on its intentions and its reasons; while acting, to enable monitoring; and after the action, to report its outcomes. Attribute *reliability* refers both to the reliability of the primary outputs of the AI system and to the validity of the explanations it provides. Finally, as part of its explainability, the AI system must monitor whether it is operating within its specified *operational design domain* (ODD), covering the normal and intended operational situations.

Related Requirements for Aircraft HF Certification

In the latest annex of CS-25 (EASA, 2021b), explainability has not yet been formally introduced in CS 25.1302. However, several rules including means of compliance can be associated with it. The information necessary to accomplish tasks must be provided, in a clear and unambiguous form, at a resolution and precision appropriate to the tasks, accessible and usable in a manner consistent with their urgency, frequency and duration and with acceptable workload. This covers normal and non-normal operational conditions. Information type must be relevant for the task, with suitable speed and precision, consistent across multiple locations, prioritised according to task criticality, and manageable (configurable) with suitable system response times. The crew has to retain enough information about what their action or a changing situation will cause the system to do, avoiding increased time to interpret a function, make a decision, or take appropriate actions that involve undue concentration and fatigue.

Task sharing and distribution requires proper cooperation between the flight crew and the integrated, complex information and control systems. Crews have to be aware of action effects on aircraft and know what the system is doing as a result of their action or a changing situation, and they must be able to enable system activation without confusion. Operationally relevant equipment behaviour has to be predictable and unambiguous, allowing the operator to intervene appropriately. Automated systems require understanding of the current state, of actions taken to achieve or maintain a desired state and of future scheduled states including transitions between system states and modes. Such requirement supports crew coordination and cooperation by ensuring shared awareness of system status and crew inputs, with review and confirmation of the accuracy of commands constructed before being activated. Equipment must enable the management of errors with possible detection and/or recovery, by enabling an understanding of the situation and timely detection of failures and intervention. Effects have to be evident and discouraged by effective means. Finally, integration with other flight deck systems with regard to consistency across information, functions and logics and with the operational environment has to be demonstrated.

Explainability in the Context of Cognitive Processing and Human–Machine Working Together

To understand what an *explainable system* could be for the human user, and what forms of explanation users expect, it seems fruitful to appeal to everyday definitions. In the *Cambridge Dictionary* the explanation is associated with making something clear or easy to understand by describing information about it. For Miller (2019), "explanation" can be considered as a tool or process for an agent (explainer) to explain decisions made to itself or a person (explainee). It can address the *cognitive process* to produce an explanation, the *product* of this process, or the social process to transfer knowledge from explainer to explainee and appears in ontologically very different categories: a logical proposition, a sentence, a speech act, a discourse or a dialogue.

Explanations have to be analysed in the context of how humans perceive, analyse, decide on and act based on information. Also, humans are subject to a number of biases and social expectations, which impacts how people define, generate, select, evaluate and present explanations. The risk of losing trust is often cited as one of the consequences of inadequate explanation, when users cannot explain the observed behaviour or decisions.

In design, explanations have to be addressed as one out of many topics in order to achieve safe, efficient and usable human-machine interaction. This includes most adapted ways of working together (e.g. collaborative vs. cooperative modes). Different types of information are needed to manage the targets and constraints of an operational situation, each of them disposing of specific characteristics (e.g. certainty, complexity). A good explanation only works if it is considered as part of the complete information-processing cycle for the required tasks.

The extensive literature review of Miller (2019), as developed in the section Basic Concepts for Explainability, summarises how people attribute and evaluate social behaviour leading to meaningful explanations, by using concepts such as beliefs, desires, intentions, emotions, and personality traits. Explanations can be required for different reasons: curiosity, learning why certain events occur rather than others, giving meaning by reconciling inconsistencies in the knowledge structure, or managing social interaction by creating shared meaning, changing others' beliefs and impressions (persuasion), emotions, or influencing actions (e.g. assignment of blame). In that context, the notion of action is essential. Actions taken can be explained by goals or intentions, and people consider an action in terms of its precondition, the action itself, and its effects.

People explain actions in two dimensions: intentionality and observability. People have a tendency to explain unobservable events more than observable events, and unintentional behaviour more than intentional behaviour because people are more aware of their own beliefs, desires, feelings and intentions rather than what is observable for others.

For unintentional behaviour, people just assume physical or mechanical causes. To be considered as intentional, a behaviour must be based on desire and a belief that the behaviour can be undertaken, and it requires the ability and awareness of performing the action.

Depending on the circumstances, different modes of explanation reasons are used:

- Links to mental states such as desires, beliefs, and values
- Causal history based on background factors (e.g. motives, emotions, personality)
- Explanations of enabling factors of why the intentional action achieved the outcome it did

Further, humans are biased to explain others' behaviours based on traits rather than situational factors, as situational factors are not easily invisible. People proceed differently if they have to explain the behaviour of individuals or groups, and the need for explanations is influenced by the presence of social norms.

When giving explanations, people infer from observations and prior knowledge. They select some causes based on contrast and probabilities depending on the explanation goal. For the identification of causes of events, people use abductive reasoning and simulations. Where the former relates to inferring causes that explain events by making assumptions about hypotheses and testing these, the latter simulates counterfactual cases to derive explanation.

To select explanations, people use a small subset of identified causes for explanation.

Finally, to evaluate its quality, people go through the process of explanation to determine whether the explanation is satisfactory. The explainee needs to understand the cause; however, the explainee might prefer certain types of explanations over others. As studies have shown, people generalise better from functional than from mechanistic explanations.

The next section, Basic Concepts for Explainability, presents the variety of concepts to be considered that may have an impact on formulation of the content of explanations. People's goals, conditions and next actions have to be taken as a basis to further drive the design of explanations in AI systems.

Basic Concepts for Explainability

Sketching an Ontology of Explainability

Whereas EASA (2023a) refers to explainability as a capability, "explanation" is defined as "information [...] on how an AI/ML application produces its results". But "information" is itself polysemic: Is it a communication act? A piece of knowledge? A sequence of symbols representing logical content, such as text, utterances, images? (the information *bearer*; Ceusters & Smith, 2015). To articulate the social, logical and cognitive dimensions of explanation, we appeal to *communication acts*, derived from speech act theory (Green, 2021; Smith, 1984), as adapted to information systems by Bermúdez et al. (2007). Figure 5.2 summarises the main concepts and relations.

Concept names start with uppercase letters (e.g. *Explanation*). Concepts and relations prefixed by "*dolce:*" are borrowed from the fundamental ontology DOLCE (Borgo et al., 2022). Roles such as *sender* or *receiver* are depicted here as relations for simplicity, but they should be modelled according to a more rigorous strategy (Vieu et al., 2008).

The goal of a communication act is to obtain a certain *intentional state* of its *receiver*. The subclasses are acts of different "forces" (Green, 2021), in particular *constative acts* intend that their *receivers* believe their content. As illustrated in Figure 5.3, we define *explanation acts* as a specialisation of *constative*, whose content are *explanations*.

Concepts in bold will be specialised into lower-level classes when detailing the different kinds of explanation. The receiver of an explanation act is the human *operator*. An explanation relates an *explanans* and an *explanandum*, both of them being *logical content*, that is truth-evaluable constructs (subclass of DOLCE's *proposition*). The explanandum is *about* aspects of the *system outcome* (Hall et al., 2019). The explanans is a set of facts related to the explanandum. Different types of relation between the explanans and the explanandum define different explanation subclasses. For the explanation act to be successful, the following necessary conditions must hold:

1. The explanandum is an aspect of the system outcome.
2. The explanans describes actual facts linked to the explanandum through an appropriate specialisation of explains relation (e.g. in mechanistic explanations, the explanans *causes* the explanandum).
3. The operator knows some *explanation background*.

Relations between explanans and explanandum function similarly to discourse relations, as required in theories of discourse (Asher & Vieu, 2005). They ensure consistency between consecutive discourse segments.

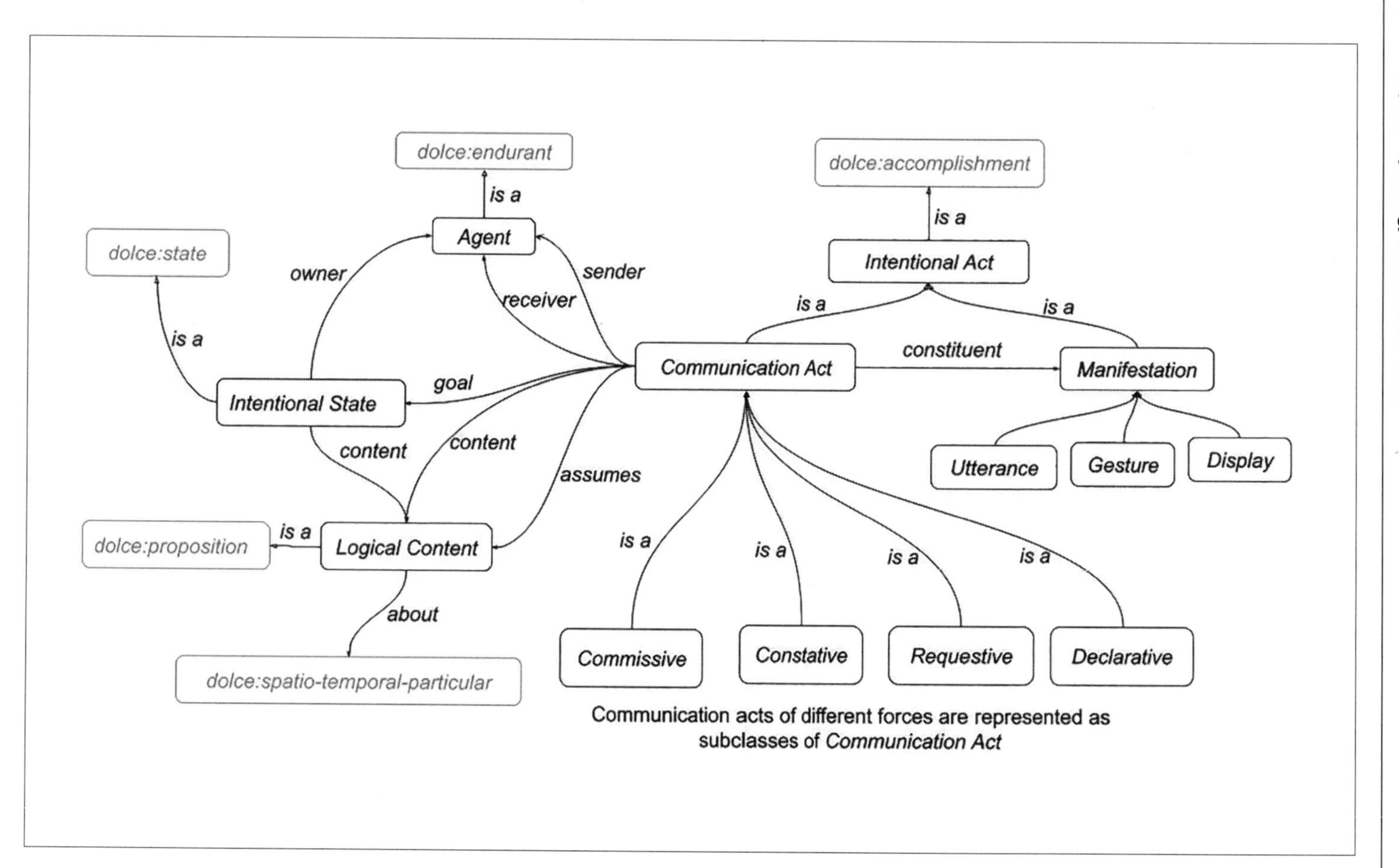

Figure 5.2. Features of the communication act.

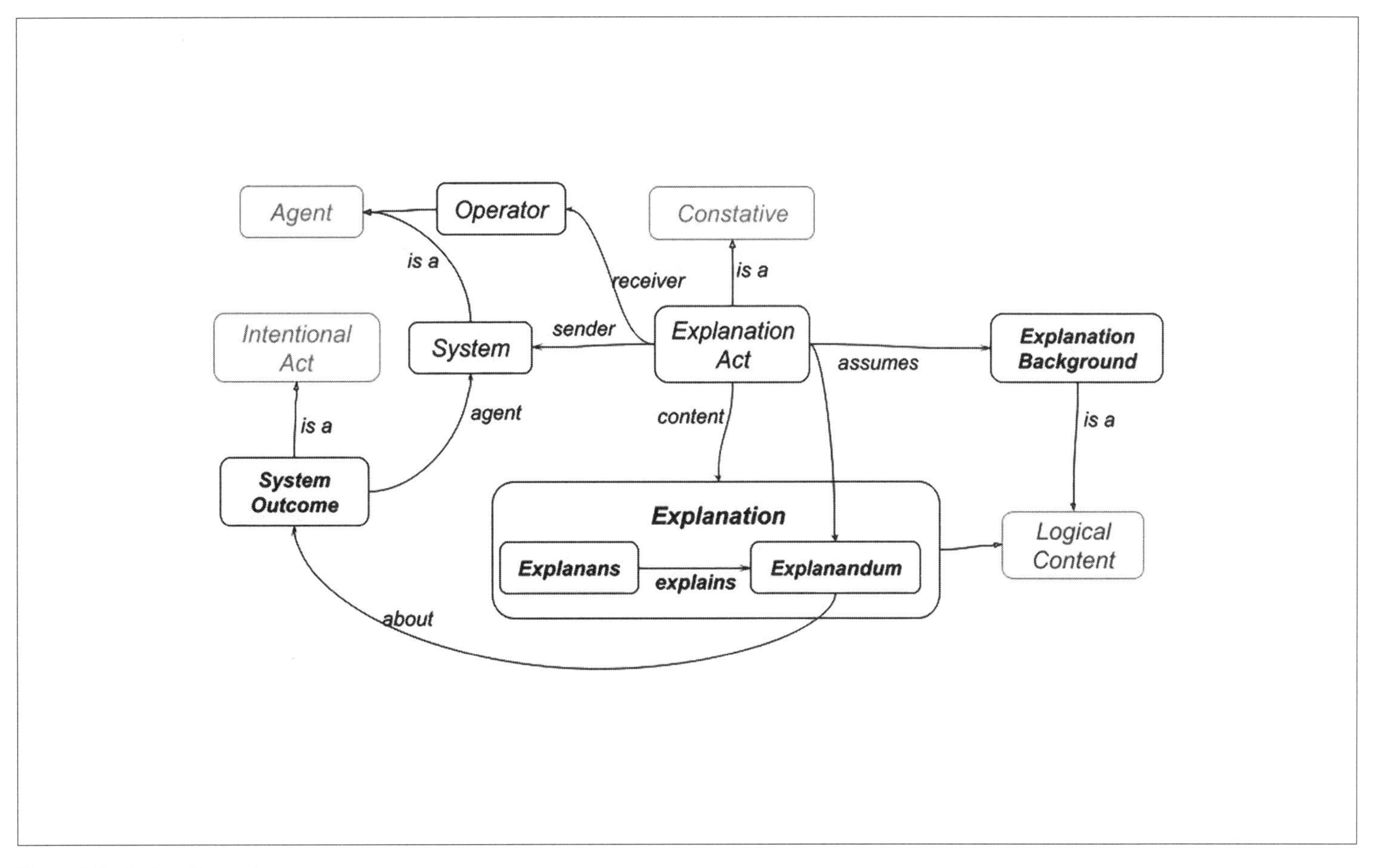

Figure 5.3. Explanation acts.

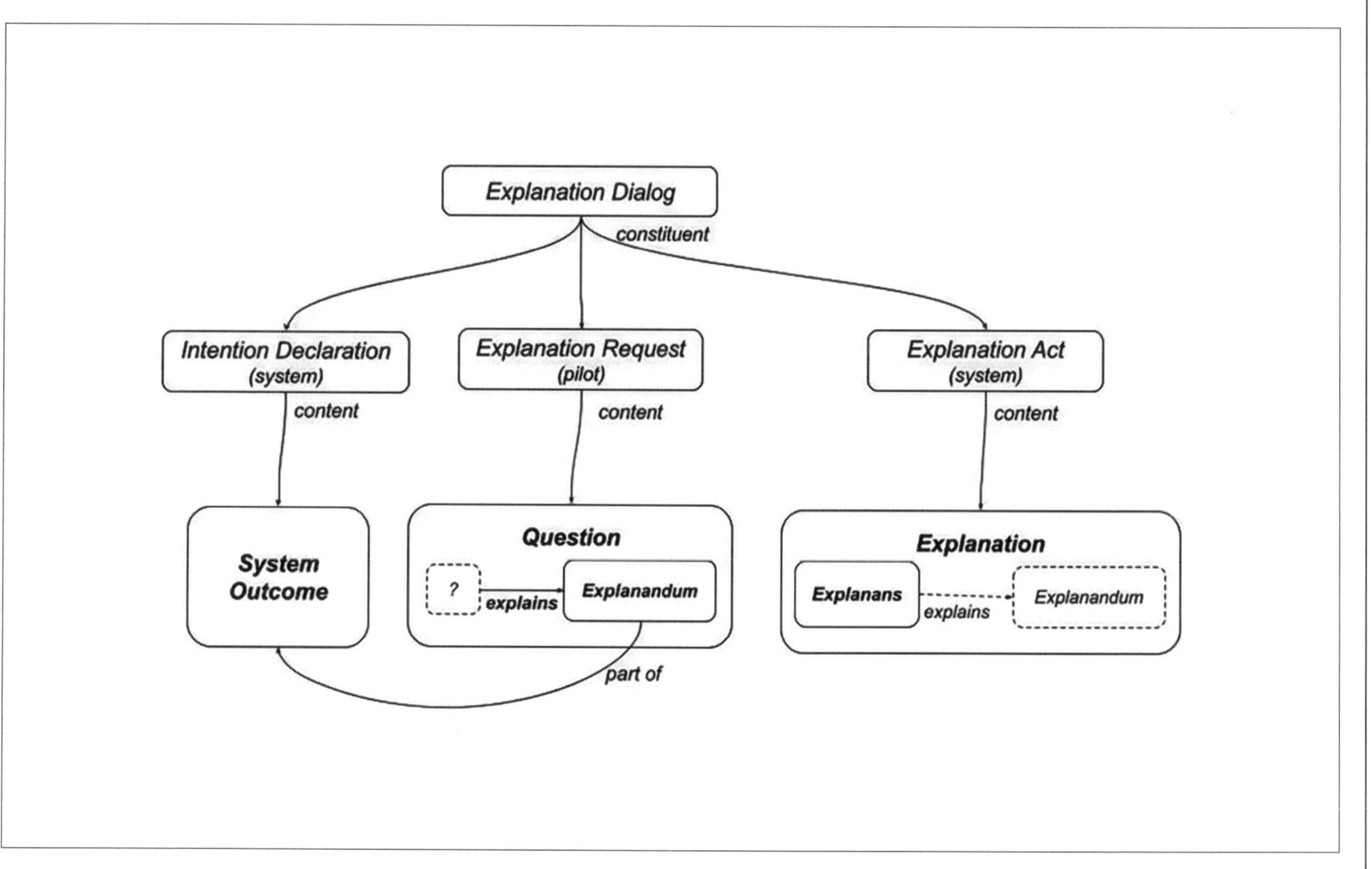

Figure 5.4. Explanation dialogue.

The explanandum is presumed by the explanation act ("presumption condition"; Walton, 2004). The main claim of an explanation act is the relation itself between the explanans and the explanandum. The explanans itself may be known or not known before the explanation act. The type of explanation, including the required types for the explanandum and the explanans, depend on the operator's cognitive need.

The need for explanation is summarised as a question that the user could have asked in an explanatory conversation. This ideation strategy is not restrictive because we focus on the logical content of explanations, and, as noted by Walton (2004), explainability is inherently dialogic. The natural (but not unique) interaction pattern where an explanation can be delivered takes the form of a dialogue initiated by a question, as visualised in Figure 5.4.

The basic pattern for explanation dialogue is the question–answer pair, which operates a "transfer of understanding" (Walton, 2004) to the receiver. The explanation dialogue ends with two communication acts: an explanation request specifies the user's cognitive needs, and the explanation act itself. The content of the explanation request is a question, which specifies the explanandum and the nature of the explains relation in the desired explanation. More formally, the question defines a pattern, in which the explanans is unspecified, for the final explanation.

Example: A Diversion Assistant

Let us assume an aircraft flies from Madrid to Amsterdam and a failure (e.g. fuel leak) forces the pilot to divert to the nearest airport. The purpose of the diversion assistant is to help the pilot to choose the best possible alternate airport. The main outcome of the diversion assistant is a sorted list of alternate airports, for example "The diversion options are Bordeaux, then Toulouse and Montpellier". The pilot might ask: "Why is Bordeaux preferred to Toulouse?" The explanandum is "Bordeaux is preferred to Toulouse", and the explanans could be "Flight time to Bordeaux is shorter than to Toulouse", and the explanation is based on relation "is a reason for". Again, the explanandum (here, "Bordeaux is preferred to Toulouse") is presumed in the explanation act. The relevant content of the explanation is the statement that Bordeaux is closer than Toulouse, and this motivates the recommendation.

The pilot might have had different operational or cognitive needs, and might have expressed them with different questions, leading to other kinds of explanations:

- Is there an ILS (instrument landing system) procedure in use in Bordeaux?
- What is the weather in Bordeaux like?
- How reliable is the recommendation to divert to Bordeaux?
- Does the decision take into account that we have no maintenance team in Bordeaux?
- How safe is the option to divert to Bordeaux?
- Why not Limoges?

We ideate explanations through archetypal explanation dialogues. In this example, the overall interaction sequence would be:

1. An engine shuts down, the pilot decides to divert the flight to one of the closest airports.
2. The diversion assistant proposes a list of diversion options: Bordeaux, Toulouse, Montpellier.
3. Before making a decision, the pilot asks: "Why is Bordeaux preferred to Toulouse?"
4. The system then has to give a reason which is acceptable to the pilot, for example: "The failure can be repaired sooner in Bordeaux than in Toulouse".

By ideating explanation as a dialogue, we do not restrict explanation to language-based conversational interactions. The way to request explanations and to access explanations can be based on a variety of interaction modalities, including graphical interfaces. By ideating the explanation interaction as a question–response pair, we focus on its informative content. Questions are often considered as an efficient description of the actual cognitive needs, and questions may play a role in the process of eliciting the target cognitive needs (Chari et al., 2020; Hoffman et al., 2018).

Taxonomy of Explanation

The expected content of explanations depends (among other factors) on the primary purpose of the system of interest. Following the EASA classification (2023a), AI systems of different levels produce different categories of system outcomes. Consequently, different types of explanandum are required:

- Level 1A systems augment the operational awareness of the user. They typically describe facts or predictions on the task or its context of operations.
- Level 1B systems help the pilot to make decisions. Typically, this kind of system would provide recommendations or suggested actions to be carried by the pilot.
- Level 2 systems contribute to aircraft operations (their contribution is not only cognitive). They have to justify their actions.
- Level 3 systems are autonomous, but they also have to explain their actions, for monitoring purposes.

Figure 5.5 lists the taxonomy of explanation patterns.

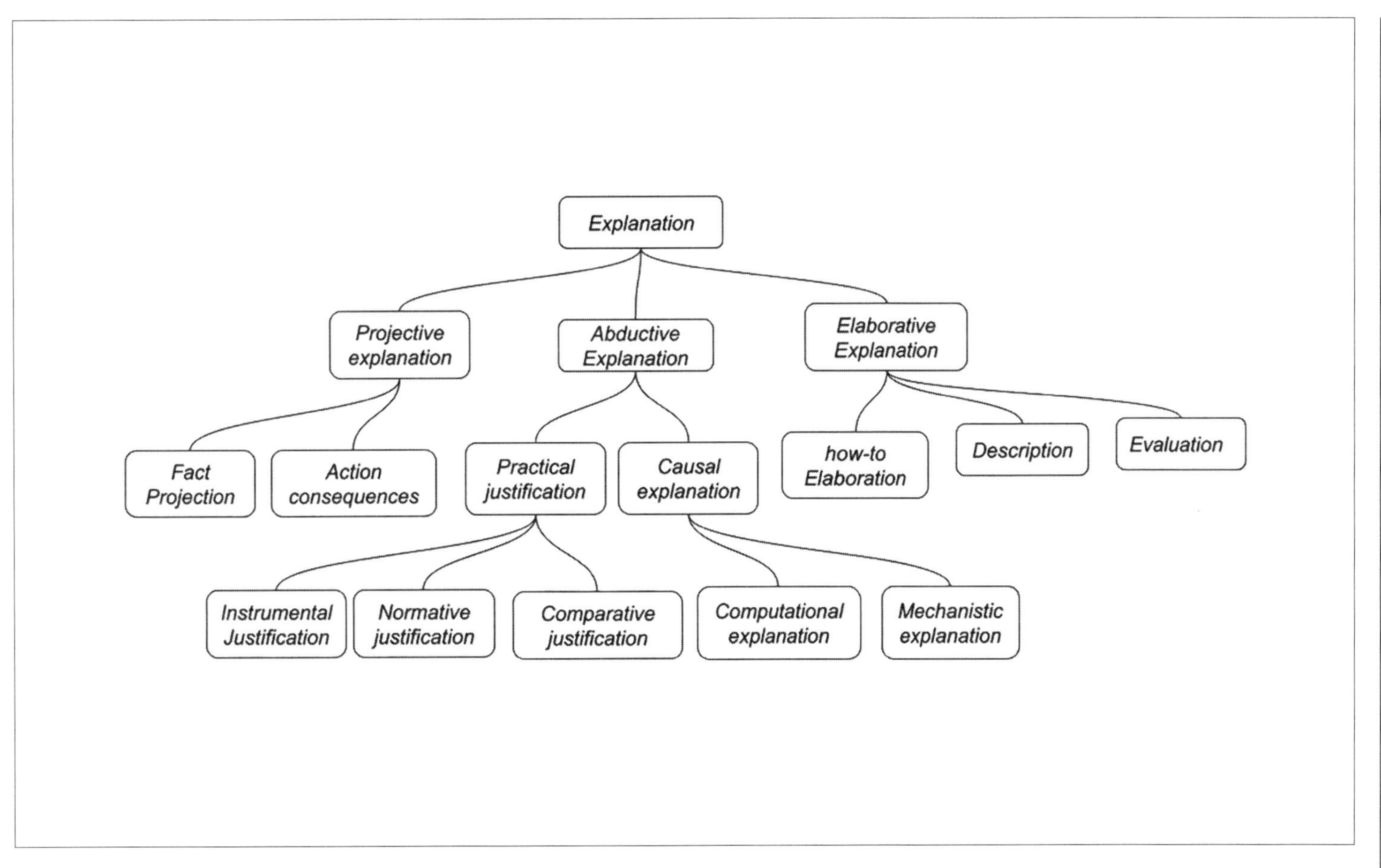

Figure 5.5. The taxonomy of explanation.

Table 5.1. Purpose and content of explanation subclasses

Explanation	Question	AI level	Explanans	Explanandum	Explanation Background
Consequence elaboration	What will A do? What happens after A? What if A?	1B 2 3	Effects of actions (events, states)	Operator actions (1B, 2A) System actions (2B, 3)	Causal knowledge about the action's results Operationally relevant effects
Fact projection	What does F cause? What if F?	1A	Relevant facts or events resulting from *F*	Facts or events, belonging to the informative outcome	Causal rules
Instrumental justification	What is the purpose of A? What for should I do A?	1B 2B 3	Desirable results of an action (events, states)	Operator action (1B) System action (2B, 3)	Causal knowledge about the action's results Desirable outcomes
Normative justification	For what reasons to doA?	1B 2B 3	Reasons to make the action, e.g. norms, procedures, standards.	Operator action (1B) System action (2B, 3)	What are the applicable norms for the decision / intention
Comparative justification	Why A1 rather than A2? Or: Why is A1 preferred?	1B 2B	Features of *A1* and *A2* which justify the recommendation	A1 are A2 operator's actions recommended (1B), or possible system actions (level 2B)	Preference criteria from the perspective of the operator
Computational Explanation	Why does the system show "*F*"? How did the system get to this?	All	Input data used to elaborate the result	The *system outcome* (seen as a physical non-intentional event)	Knowledge of the kind of algorithm/rules implemented

Table 5.1. continued

Explanation	Question	AI level	Explanans	Explanandum	Explanation Background
Mechanistic Explanation	Why did F happen?	1A	Causes of the explanandum	Part or the whole of the system outcome	Causal rules delimit the acceptable explanations in the operational domain
Description	What did the system do?	All	Properties or sub-actions of the system outcome	The outcome of the system, as intentional act	Common action description language
How-to Elaboration	How to do A? How is A done?	1B 2B 3	Action steps to perform A	Operator's possible action (1B) for guidance purpose Systems actions, for monitoring purpose (2B, 3)	The action steps are doable by the operator
Evaluation	How certain is F? How reliable is F?	All levels	System performance indicators (e.g., confidence)	Logical content of the communication acts done by the system (1A, 1B) Operator's actions (2A) Actions planned or done by the system (2B, 3)	Meaning of uncertainty/ performance indicator

Explanation patterns differentiate by: the specific Explanation relation, as described by the question the operator could ask; the type of explanans; the type of explanandum; and the required explanation background. Table 5.1 describes the subclasses of explanation resulting from this analysis and to which level of AI they might be relevant. Note that regarding Level 1B, this does not entail creating specific regulation. Explainability might be a way of complying with existing regulation or improving the performance of the human-machine system.

Explainability in the Design Process

Overview of Design Process Principles

This section highlights how explanation is managed along the engineering phase, with a focus on designing for operational needs, and not the technical capability that a system is able to explain. The need for explanation as information is just one part of the design problem. Already in today's automated systems, machines take actions that require the creation of adequate understanding of what the system is doing. The operator may ignore the detail of automatic routines, but needs to understand why an output occurs, hence, requires an explanation.

AI systems differentiate not only in terms of the complexity of input parameters, but also the complexity of the system to deal with the parameters, leading to a higher variability in potential outputs that still have to make sense for operators. Thus, depending on the associated technology and the drivers and constraints identified, a deeper level of explainability may be required. To achieve the best compromise when designing explanations, a multidisciplinary design team is required:

- HF specialists with a psychological focus use scientific knowledge to address the social and cognitive context of explanation and apply systematic methods to derive user needs by analysing operational tasks and exploring the usefulness of designs against criteria such as task performance, workload, experience, etc. thanks to dedicated studies and end-user tests.
- Operational specialists capture the operational need of an explanation in collaboration with HF specialists, taking into account the diversity of operational contexts.
- System designers provide information handled by the system (input or output) which will be used to explain the primary outcome of the system.
- HMI (human-machine interaction) specialists provide the best form, format, and modality to deliver the explanation.

As highlighted in the section on System Explainability From the Perspective of the Regulator, several passages of the regulation can be connected with explainability. To demonstrate that new products correspond with these requirements, it is necessary to address explainability throughout system engineering, which considers several steps following an adapted V-model of the system development cycle presented in Figure 5.6:

- Identification of needs
- Design and specification
- Validation and verification
- Certification

This process relies on a systematic integration of HF, which also includes how to address explainability, as described next.

Identification of Needs to Be Addressed by Design

Before identifying design concepts, needs to be covered have to be made explicit. Explanation needs may stem from different origins:

- Human needs related to the task or human principles (e.g. cognitive needs, social needs, biological needs, ...)
- Operational needs related to the operational context and events to be covered
- Technical, economic and societal needs

An example of cognitive need would be to reduce ambiguity in information by giving meaning, since humans need stability and balance. Such a cognitive need is on one hand associated with information required for operational tasks, but also information stemming from individual experience or expectations. Among the criteria to be considered is the need for predictability (the more the behaviour is predictable, the less it needs explanation).

After first having identified potential concepts for a design, and which types of contexts, tasks and information shall be addressed by a new product, the need for explanation can be captured:

- Is an explanation needed? – *Can other forms of task allocation avoid uncertainty?*
- Who needs an explanation? – *Only the pilot? The flight dispatcher? The airport?*
- Why? What for? – *Which goal does the pilot need to achieve? Operational tasks? Personal expectations?*
- When? – *At what moment may the pilot request an explanation? Continuously? Event-based?*
- Is there a need that cannot be covered?

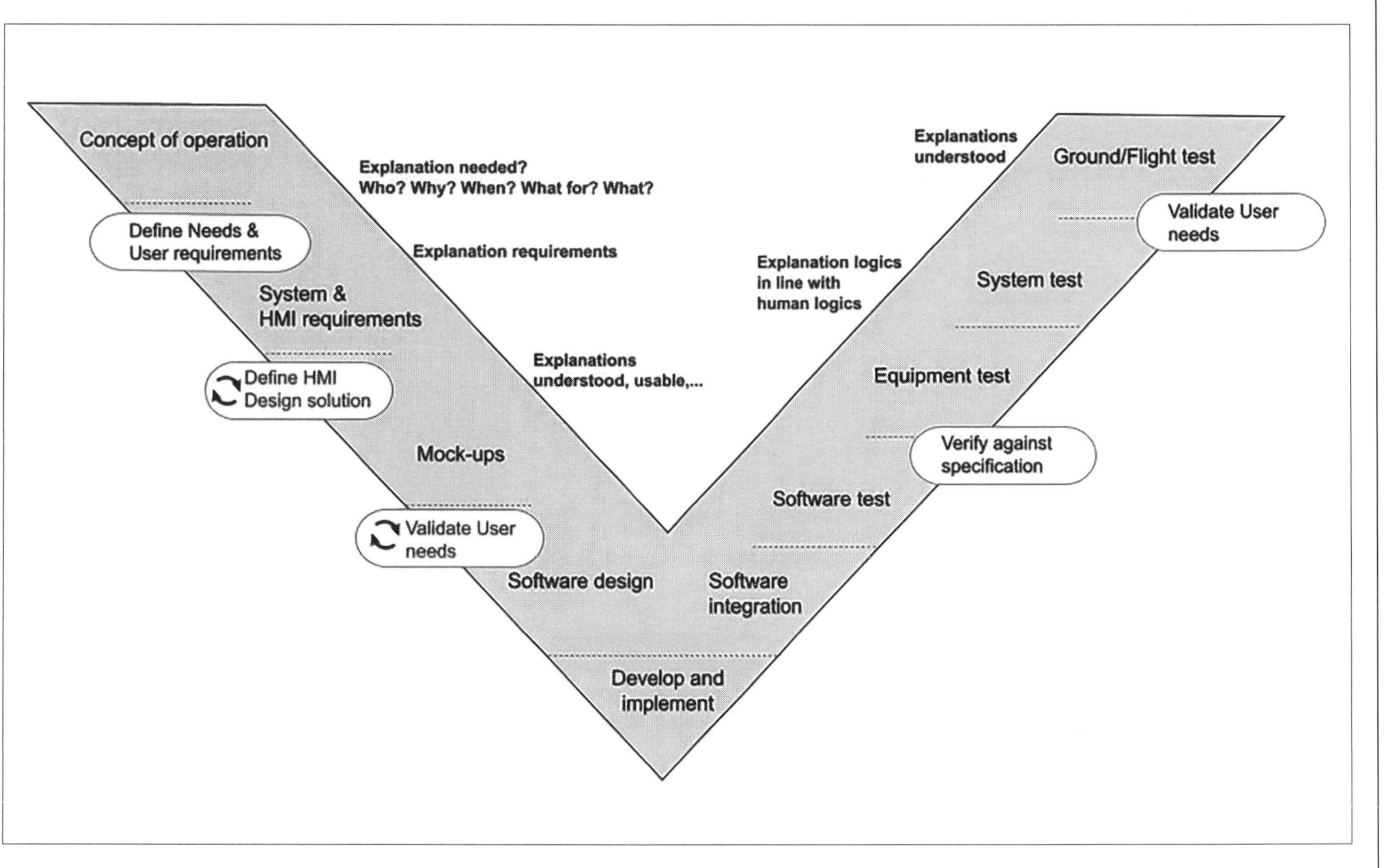

Figure 5.6. Overview of addressing explanation throughout V&V engineering. V&V = validation and verification.

Design and Specification

In this step, design concepts are developed to translate needs into requirements (operational, technical, HMI, safety, performance etc.). They specify the *what* (explanation pattern and content) and *how* (means, data source and reliability, date/time of validity, etc.) of explanations. Requirements must be clear, non-ambiguous, complete, consistent, and verifiable. The goal of requirements is also to reduce risks inherent to explanations having an impact on usability and acceptance, for example an agent being "too talkative", information leading to overload, etc. Hence, the application of HF criteria helps to derive and validate suitable specifications. The goal of this activity is to ensure compliance with Objective EXP-07 for assessing the appropriateness of the decision/action expected (EASA, 2023a). In particular, technical requirements will specify the logical content of explanations based on the conceptual framework presented in the previous chapter: The design team has to identify which part of the system outcomes are worth being explained, what kind of explanation is required, and finally which kind of explanans will make an explanation successful in the targeted operational situation.

Validation, Verification and Demonstration

The objective of this phase is to validate that requirements meet the identified needs and to verify products against requirements, in accordance with applicable regulations. Explanation has to be investigated in the global task and operational context. In the iterative context of system design, first mock-ups are soon proposed to reflect and validate a first level of user needs. Depending on the type of means used, the most suitable approach to investigate explainability may need to be adapted. For example, test protocols may need to focus on specific aspects to cover the understanding of the cognitive background of the user (e.g. Hoffman et al., 2018), while at the same time covering the diversity of the operational context. Also, precise indicators are expected to focus on identifying how rules are managed by a user, for example to be able to accept that A is a valid explanation of B, or which cognitive biases are present to assess the usefulness of a system output.

Summary and Outlook

Achieving explainability for flight crew requires a systematic integration between the system, human, and operational perspectives throughout the engineering process. How to deal with explanation is one of the HF issues to be addressed among others by suitable means in order to respond to the cognitive needs of pilots. Explanation content has to be adapted to different interaction and teaming situations, which highlights the need for a deeper understanding of this notion in order to derive efficient design frameworks and adapted regulation. It has to be further analysed how explanation facilitates collaboration requiring improved tools to capture dynamics. As explainability is not yet part of today's design practices, it requires the development of additional competencies with suitable concepts, methods, and training for design teams, including a systematic integration of linguistic competence. We have proposed a first set of foundational concepts to think about explainability, but in order to obtain complete and rigorous foundations, theoretical work is to be done on ontologies and taxonomies. The mapping between cognitive needs and explanation types will be enriched by practice. Numerous HF research questions remain open, such as the impact of different cultures, the variability of individual experiences to take into account, or possible ethical limits for good explanations.

References

Asher, N., & Vieu, L. (2005). Subordinating and coordinating discourse relations. *Lingua, 115*(4), 591–610. https://doi.org/10.1016/j.lingua.2003.09.017

Bermúdez, J., Goñi, A., Illarramendi, A., Miren, I., & Bagüés, M. I. (2007). Interoperation among agent-based information systems through a communication acts ontology. *Information Systems, 32*(8), 1121–1144. https://doi.org/10.1016/j.is.2007.02.001

Borgo, S., Ferrario, R., Gangemi, A., Guarino, N., Masolo, C., Porello, D., Sanfilippo, E. M., & Vieu, L. (2022). DOLCE: A descriptive ontology for linguistic and cognitive engineering. *Applied Ontology, 17*(1), 45–69. https://doi.org/10.3233/AO-210259

Ceusters, W., & Smith, B. (2015). Aboutness: Towards foundations for the information artifact ontology. In F. M. Couto & J. Hastings (Eds.), *Proceedings of the sixth international conference on biomedical ontology (ICBO), 1515* (pp. 1–5). CEUR-WS.org. https://ceur-ws.org/Vol-1515/regular10.pdf

Chari, S., Seneviratne, O., Gruen, D. M., Foreman, M. A., Das, A. K., & McGuinness, D. L. (2020). Explanation ontology: A model of explanations for user-centered AI. In J. Z. Pan, V. Tamma, C. d'Amato, K. Janowicz, B. Fu, A. Polleres, O. Seneviratne, & L. Kagal (Eds.), *The semantic web - ISWC 2020* (pp. 228–243). Springer International Publishing. https://doi.org/10.1007/978-3-030-62466-8_15

European Union Aviation Safety Agency (EASA). (2007, September 19). *Certification specifications for large airplanes CS-25. Amendment 3*. https://www.easa.europa.eu/sites/default/files/dfu/CS-25_Amdt%203_19.09.07_Consolidated%20version.pdf

European Union Aviation Safety Agency (EASA). (2020, February 7). *Artificial intelligence roadmap 1.0: A human-centric approach to AI in aviation*. https://www.easa.europa.eu/en/downloads/109668/en

European Union Aviation Safety Agency (EASA). (2021a). *EASA concept paper: First usable guidance for Level 1 machine learning applications. A deliverable of the EASA AI Roadmap*, Proposed Issue 01. https://www.easa.europa.eu/en/downloads/126648/en

European Union Aviation Safety Agency. (2021b). *Certification specifications and acceptable means of compliance for large aeroplanes (CS-25, Amendment 27)*. https://www.easa.europa.eu/en/document-library/certification-specifications/cs-25-amendment-27

European Union Aviation Safety Agency (EASA). (2023a). *EASA concept paper: First usable guidance for Level 1&2 machine learning applications. A deliverable of the EASA AI roadmap, Proposed issue 02.* https://www.easa.europa.eu/en/downloads/137631/en

European Union Aviation Safety Agency (EASA). (2023b). *Artificial intelligence roadmap 2.0: A human-centric approach to AI in aviation.* https://www.easa.europa.eu/en/downloads/137919/en

Green, M. (2021). Speech acts. In E. N. Zalta (Ed.), *The Stanford encyclopedia of philosophy.* https://plato.stanford.edu/archives/fall2021/entries/speech-acts/

Hall, M., Harborne, D., Tomsett, R., Galetic, V., Quintana-Amate, S., Nottle, A., & Preece, A. (2019). *A systematic method to understand requirements for explainable AI (XAI) systems.* DAIS ITA. https://dais-legacy.org/doc-4452/

Hoffman, R. R., Mueller, S. T., Klein, G., & Litman, J. (2018). *Metrics for explainable AI: Challenges and prospects.* ArXiv. https://arxiv.org/abs/1812.04608

Miller, T. (2019). Explanation in artificial intelligence: Insights from the social sciences. *Artificial Intelligence, 267*, 1–38. https://doi.org/10.1016/j.artint.2018.07.007

Smith, B. (1984). Ten conditions on a theory of speech acts. *Theoretical Linguistics, 11*(3), 311–330. https://doi.org/10.1515/thli.1984.11.3.311

Vieu, L., Borgo, S., & Masolo, C. (2008). Artifacts and roles: Modeling strategies in a multiplicative ontology. *Frontiers in Artificial Intelligence and Applications, 183*(1), 121–134. https://doi.org/10.3233/978-1-58603-923-3-121

Walton, D. (2004). A new dialectical theory of explanation. *Philosophical Explorations, 7*(1), 71–89. https://doi.org/10.1080/1386979032000186863

Wilson, D., & Sperber, D. (2012). *Meaning and relevance.* Cambridge University Press. https://doi.org/10.1017/CBO9781139028370

Chapter 6

Situation Awareness, Workload and Performance

New Directions

Don Harris, Heikki Mansikka, and Kai Virtanen

Abstract

The assessment of pilot mental workload and situation awareness (SA) is vital for many aerospace applications, for example the validation and verification of designs; evaluation of tactics, techniques and procedures (TTPs); as a component of aircraft certification, or the assessment of task flows on the flight deck. However, the practical utilisation of these measures poses methodological and measurement challenges. Furthermore, pilots do not fly alone: they are a part of a team. In this chapter measurement techniques developed from well-known, commonly used methods of SA and workload measurement (Situation Awareness Global Assessment Technique [SAGAT]; Endsley, 1988; the NASA Task Load Index [TLX] workload scale, Hart & Staveland, 1988; and physiological measures of workload) are described. These developments address some of the measurement issues and shortcomings posed by these commonly used approaches and are extended to describe team performance. The commonly observed dissociation between measures of workload, SA and performance is addressed and the theoretical basis for these sometimes divergent results is described. The chapter concludes with a model describing an integrated approach to the assessment of SA, including team SA, workload, and performance.

Keywords

workload, situation awareness, team situation awareness, output performance, normative performance

Performance, Situation Awareness and Workload

There is an intimate relationship between pilot performance, mental workload and situation awareness (SA); however, this association is not straightforward: Low workload does not necessarily result in high performance, and it is not inevitably associated with high SA (Mansikka et al., 2019a). It is important to understand the relationship between these three concepts when evaluating any new piece of flight deck equipment or procedure. The assessment of performance alone can potentially be misleading (e.g. Mansikka, Virtanen, Harris, & Jalava, 2021a; Mansikka, Virtanen, Harris, & Salomäki, 2021a). Assessment of workload is frequently used in conjunction with measures of SA when comparing between options for new designs or procedures, especially when they produce similar levels of performance. Often the workload question becomes not "Which option produces the best performance" but "What is the cost in terms of information processing to achieve a certain level of performance?" Measures of SA provide a means to describe how successfully pilots can acquire and integrate information in a complex flight environment.

Workload is a measure of the cognitive load experienced by a pilot, in contrast to task load, which is the amount of work that interacting with the system actually requires. It can be conceptualised as the information processing "cost" to perform a given flight task, relating specifically to the finite capacity of cognitive resources (Harris, 2011). However, the level of mental effort invested by the pilot is based upon their subjective assessment of the required performance criteria, *not* the objective task load. Put more simply, from a cognitive standpoint you work as hard as you think that you need to. If pilots are not aware of the actual demands of a task, their workload may be relatively low but ultimately so too will be their performance (Mansikka et al., 2019a). Alternatively, even though the situation does not obviously demand it, a pilot may invest a great deal of cognitive work to attain high SA, hence be under considerably more workload but, in the long run, also attain a superior level of performance.

When performing the same task, in the same aircraft and in identical conditions, two pilots can both produce identical, high levels of performance. However, one pilot may experience lower workload compared to the other (Vidulich & Wickens, 1986) and so have more "spare cognitive capacity" to deal with other issues if required (see Region A in Figure 6.1). If their performance is not limited by other factors, their workload will increase, and performance will eventually degrade when they have no more excess capacity to cope with the increasing task demands (Region B). In Figure 6.1 the vertical bars represent the pilots' overall cognitive capacity, where a darker shading represents more spare capacity and thus lower workload. The solid line represents performance. In Region A, Pilot 1 and Pilot 2 can maintain equal levels of performance as they both have enough cognitive capacity for the task. However, Pilot 2 has less spare capacity than Pilot 1. In Region B, the increase in task load further taxes the pilots' cognitive capacity such that they no longer can maintain their performance. At a similar task load, Pilot 1 still possesses more cognitive capacity and can thus maintain higher performance than Pilot 2. In Region C, both pilots have depleted their cognitive spare capacity, workload (determined by the amount of cognitive spare capacity) is high, and performance is poor, regardless of their efforts. Workload can become dissociated from performance particularly if the task is resource limited (Yeh & Wickens, 1988).

Similarly, the relationship between SA and performance may either be relatively weak (e.g. Endsley, 2019; Fracker, 1991) or complex and unclear (Mansikka et al.,

2019a; Sulistyawati et al., 2009). It has already been stated that awareness of task demands will partially determine workload. This awareness is predicated upon SA. Mansikka et al. (2019a) observed that in simulated air combat scenarios, when awareness of the tactical situation was low, pilots exhibited a combination of low workload and low performance: They were not aware that they should be working harder to attain a higher level of SA. Furthermore, in a highly dynamic, uncertain environment such as air combat, success may occasionally be a product of chance factors and vice versa (Mansikka, Virtanen, Uggeldahl, & Harris, 2021). Pilots' work is often described as a form of the input-process-output (IPO) model, where processes involving both individual and team cognitive activities are converted into outputs. In simple terms, the output is a result of what pilots do and is typically assessed only with output performance measures, irrespective of how they arrived at that output. To understand how the pilots reached their output, measures targeting output performance should be supplemented by adjunct measures such as SA and workload (Mansikka, Virtanen, Harris, & Jalava, 2021a; Mansikka, Virtanen, Harris, & Salomäki, 2021a, 2021b) and other process measures (Mansikka, Virtanen, Harris, & Jalava, 2021). Assessment of SA can help distinguish competent pilots from lucky ones and such adjunct measures can also aid in the evaluation of training interventions and the development of tactics, techniques and procedures (TTPs).

Aviation is also about teamwork. Even pilots of high-performance single-seat military aircraft seldom fly alone. As a minimum they operate as a pair (lead and

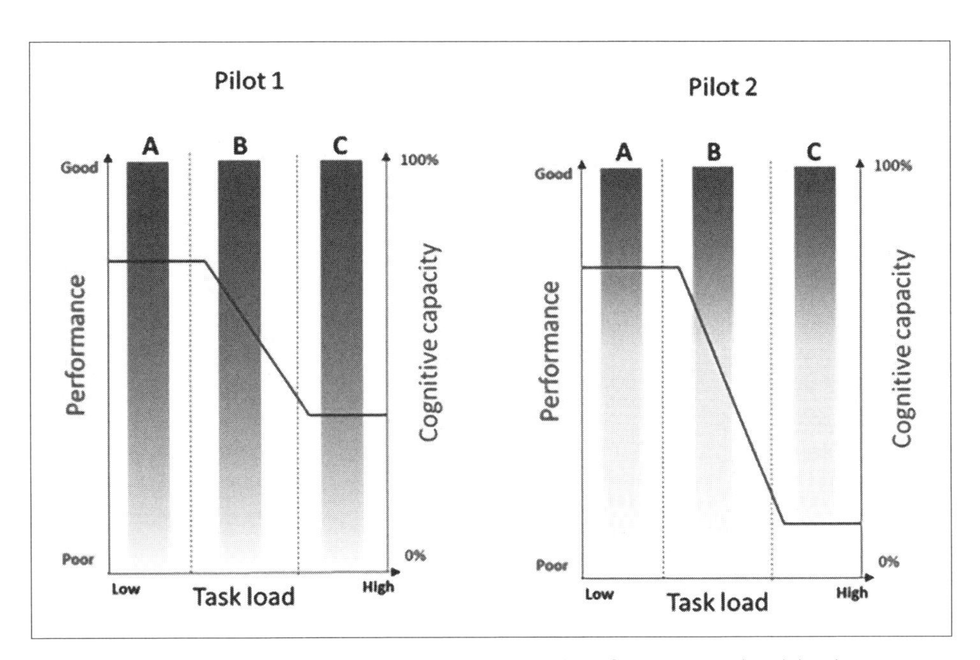

Figure 6.1. Hypothetical relationship between workload, performance, and task load.

wingman) but usually as a four-ship (flight). On the civil flight deck, pilots operate as a crew, not individuals. Consequently, although workload pertains to each individual pilot, both performance and SA need to be evaluated from the perspective of the individual *and* of the team. The performance of the flight is based upon a shared understanding of those involved (team situation awareness [TSA]). Furthermore, evidence now suggests that the relationship between TSA and performance is not a simple linear relationship, but is curvilinear (Mansikka, Virtanen, Uggeldahl, & Harris, 2021; Mansikka et al., 2023).

The measurement of performance alone will only tell part of the story. For a complete picture, performance, workload, SA and/or TSA all need to be assessed, but the measurement of these factors poses practical and theoretical challenges, especially in the highly complex and dynamic environment encountered in aviation.

Measuring Situation Awareness and Team Situation Awareness

What Is SA and TSA?

There are many definitions of SA, but all suggest that it is a dynamically updated mental model containing activated knowledge about a situation. It is a pilot's understanding of "what is going on".

Endsley's three-level model is perhaps the most dominant theory in explaining SA. Endsley (1995a) defines SA as "the perception of the elements in the environment within a volume of time and space [SA level 1], the comprehension of their meaning [SA level 2], and the projection of their status in the near future [SA level 3]" (p. 36). In Endsley's model, each SA level is built upon the level below such that poor SA at a lower level contributes to low SA at higher levels (e.g. Endsley & Garland, 2000). Endsley's approach to the assessment of SA forms the basis of the developments in measurement methodology described in the following section.

Problems in Measurement

The dominant paradigm for the assessment of SA adopts a behavioural approach, usually implemented in a flight simulator. The SAGAT (Situation Awareness Global Assessment Technique) developed by Endsley (1988) uses a series of memory probes developed by subject matter experts (SMEs) who are employed during simulation scenarios. The probes are derived from an SA requirements analysis specific to the task undertaken. At various points, the simulation freezes and screens blank. Pilots are presented with a series of questions relating to Endsley's Level 1, 2 or 3 SA components. Answers are compared to the ground truth, derived from the simulation scenario to provide a measure of SA. The main disadvantage with this approach is that it frequently interrupts the simulation scenario. It has also been criticised as merely a test of memory and not of SA. This technique is also time intensive; it requires dedicated software support and the results produced are scenario specific. Such an approach may also alert pilots to the SA requirements of which they were originally unaware (Stanton et al., 2013).

Table 6.1. Exemplar list of concepts and attributes for beyond visual range air combat mission

Concepts	Attributes
Own flight/ flight members/ other friendly aircraft	Position, flight parameters, offensive capabilities, defensive capabilities, limitations, objectives, tasks, TTPs, weapon effects, electronic warfare effects.
Non-friendly aircraft	Position, types, offensive capabilities, defensive capabilities, targeted/untargeted statuses, declarations, objectives, tactics and manoeuvres, weapon effects, electronic warfare effects.
Friendly and non-friendly forces (other than aircraft)	Positions, types, offensive capabilities, defensive capabilities, limitations, activity, electronic warfare effects.
Environment	Airspace restrictions (air coordination order), terrain, meteorological conditions (visibility, rain, etc.).

Note. From "Team Situation Awareness Accuracy Measurement Technique for Simulated Air Combat: Curvilinear Relationship Between Awareness and Performance", by H. Mansikka, K. Virtanen, U. Uggeldahl, & D. Harris, 2021, *Applied Ergonomics*, *96*, Article 103473, Table 1 on p. 5.

To address some of these measurement issues, Mansikka, Virtanen, Uggeldahl, and Harris (2021) developed an approach based upon Endsley's SAGAT technique and a shortened form of the Critical Decision Method (CDM) structured interview approach (Crandall et al., 2006). The approach also involves undertaking an SA requirements analysis, identifying SA concepts and attributes for a particular task. An *attribute* is the smallest unit of task-related knowledge that a pilot can have awareness of. A *concept* is a functional collective of attributes (Langan-Fox et al., 2004). The CDM-based interviews were undertaken during post-sortie structured interviews to derive SA scores at each level.

Initially, a list of SA concepts and attributes for beyond visual-range air combat was derived from an extensive literature review (see Table 6.1). The content validity of this list was assured by further appraisal using operational test and evaluation pilots. This was followed by a large sample of combat-ready pilots rating each component concerning how important it was to have accurate knowledge about that attribute to develop and maintain SA. To produce the final list of concepts and attributes, experienced weapons instructors and operational test pilots further reviewed the ratings. The final list was organised hierarchically such that there were seven top-level concepts, each consisting of several lower-level attributes (see Mansikka, Virtanen, Uggeldahl, & Harris, 2021).

During a post-sortie debrief, an instructor pilot (IP) reconstructed the mission using facilities such as cockpit video recordings, simulated flight trajectories, sensor tracks, and weapon simulations. At certain times the IP paused the review to analyse a significant decision point. The IP would introduce the decision point and the first attribute associated with it. To establish Level 1 SA, the pilot was asked what they understood about the attribute. Deepening probes were needed if the answer could not be determined unambiguously from the interviewee's responses. Their answers were compared against the ground truth from the simulation to establish the accuracy of their Level 1 SA.

The CDM interview continued to establish the pilot's SA about the attribute's meaning with respect to the overall situation (Level 2). To do this the IP drew heavily on the deepening probes developed (see Table 6.2) as in this case SA could not be determined from observations from the simulation, and the pilot's SA about the attribute's meaning may have been tacit. Based on the pilot's responses to these probes, the IP determined the pilot's Level 2 SA. Finally, Level 3 SA was established by further using the probes to help the pilot to compare and verbalise their expectations and the way in which the situation evolved. At all three levels, accurate SA, that is where the pilot's cognitive model of the situation corresponded to the ground truth was scored "1", whereas inaccurate SA was scored "0". This procedure was repeated until all relevant attributes associated with the DP were dealt with. The review was continued until the next DP was identified. Once the whole mission was reviewed and the pilots' SA of all attributes in the identified DPs was scored, SA scores were aggregated and summed to provide the overall SA index.

This approach assessed individual SA in a well-defined scenario; however, Mansikka et al. (Mansikka, Virtanen, Uggeldahl, & Harris, 2021; Mansikka et al., 2023) extended this approach to address TSA. TSA is more complex than individual SA. Endsley (1995b) argued that team members required the necessary SA for factors relevant for their specific tasks. As a result, good TSA was dependent upon team coordination and communication. Salmon et al. (2008) went further suggesting that TSA comprised the SA of individual members, their shared SA and the combined SA of the team (the "common picture"). The measurement of TSA faces the same challenges as that of individual SA, with many techniques being based around approaches that require pausing the simulation to collect the data (Bolstad &

Table 6.2. CDM deepening probes to elicit SA

Probe type	Probe content
Information	What information were you seeking and from where? What information, if any, did you combine to gain the necessary information? How reliable was the source information? What information, if any, was missing or conflicting? What information, if any, did you misinterpret and how? Did the information change the way you understood the situation, and how?
TTPs and options	How well did the environmental cues match with TTPs? What feasible TTPs did you identify? What TTP did you select and why? / Would you have selected a different TTP than the one that was directed and why? If the TTP was directed to you, did you know what it was? What was your understanding about the flight's TTP adherence and TTP progress? What contingency TTPs, if any, were you prepared to execute and why? What were the cues that you used as triggers for a contingency TTP and why?
Goals, priorities	What were your priorities during this incident and why? What were you trying to achieve and why?
Physical/ time demand	If you experienced time/physical demand, how did it affect you?
Limitations/ alibis	If you experienced perceptual/technical/cognitive limitations, what were they and how did they affect you?
Expectations	Compared to your expectations, how did the status of the attribute or the situation as a whole evolve? How did the mission brief prepare you for this incident?

Note. CDM = critical decision method; SA = situation awareness; TTPs = tactics, techniques, and procedures. From "Team Situation Awareness Accuracy Measurement Technique for Simulated Air Combat: Curvilinear Relationship Between Awareness and Performance", by H. Mansikka, K. Virtanen, U. Uggeldahl, & D. Harris, 2021, *Applied Ergonomics*, *96*, Article 103473, Table 2 on p. 5.

Endsley, 2003; Cooke et al., 1997; Sulistyawati et al., 2009), which is undesirable and also impossible during a live exercise. Furthermore, the emphasis has generally been on determining the accuracy of TSA. However, Mansikka et al. (2023) argue that there are two components to TSA: accuracy, which assesses how closely a team's collective knowledge is aligned with the ground truth, and similarity, which represents the degree of alignment of a team's collective knowledge. If TSA accuracy is high, it will closely resemble the ground truth *and* the SA of each team member will also be very similar. However, if TSA accuracy is low, the pilots may have similarly or dissimilarly inaccurate SA.

Mansikka et al. (Mansikka, Virtanen, Uggeldahl, & Harris, 2021; Mansikka et al., 2023) describe the determination of TSA accuracy and TSA similarity for a flight. Each pilot's SA Level 1 accuracy regarding an attribute is scored by comparing their SA with the ground truth. A higher score reflects a higher accuracy. Next, the similarity of pilots' SA is determined by making pairwise comparisons of SA between all members. Pairwise comparisons are scored such that a higher score is associated with a higher similarity. Both procedures are repeated for every attribute in an incident and for each SA level. Level 1–3 TSA accuracy scores for an incident are determined by calculating the average of individual SA accuracy scores in respective SA levels. Level 1–3 TSA similarity scores are determined in the same fashion by calculating the average of the dyad SA Level 1–3 similarity scores for an incident. Individual SA and TSA scores are calculated for every incident in the mission. Finally, an overall TSA accuracy and TSA similarity indices were determined by averaging the TSA accuracy and TSA similarity scores.

A curvilinear (non-linear) relationship between TSA and flight performance in simulated air combat engagements was demonstrated by Mansikka et al. (Mansikka, Virtanen, Uggeldahl, & Harris, 2021; Mansikka et al. 2023). The rate of gains in offensive and defensive performance both decreased with increases in overall TSA accuracy. The greatest performance benefits accrued with the initial increases in TSA. The relationship between a flight's TSA accuracy and performance was stronger at Level 1 and weaker at Level 3. This is consistent with Endsley's theory (Endsley, 1995a), as SA Level 3 cannot be achieved unless the pilot has Level 2 SA, which itself is predicated upon achieving Level 1. With regard to TSA similarity, successful engagements were characterised by higher degrees of similarity across the flight. When a flight had gained a tactical advantage, they could control the engagement, making it easier for them to maintain a high TSA. By contrast, when a flight had lost the advantage there was a likelihood of them becoming reactive, and regaining lost TSA became difficult meaning that effective decision-making suffered, thereby compromising the flight's performance.

The assessment of TSA provides insights into team performance over and above the measurement of individual SA. Many aeronautical tasks are undertaken as a team and not as an individual. However, to gain a more complete picture of the human factors underpinning performance, the individual cognitive load on each member of the team also needs to be assessed. Doubling the number of team members does not mean that collectively twice as much work can be undertaken as there is a processing overhead involved with communication and coordination, essential tasks to promote TSA. As a result, to understand performance, the concomitant measurement of individual workload is also necessary.

Measuring Mental Workload

There is no universally accepted definition of workload, but it is generally defined as the information processing "cost" of performing a given task, hence it is intimately related to information processing theory and the capacity of cognitive resources in working memory (Harris, 2011).

Moray (1988) suggested that there are three basic approaches to the measurement of workload: behavioural, physiological and subjective. Harris (2011) added a fourth category of "analytical approaches"; however, for practical purposes, the most commonly applied approaches are subjective workload scales and physiological measures. Mansikka et al. (2019b) demonstrated a degree of convergence in workload measurement between physiological and subjective methods.

Workload Measurement Using Subjective Scales

Subjective workload measurement employs less complex uni-dimensional scales, often using a modified Cooper–Harper format (e.g. the Bedford scale; Ellis & Roscoe, 1982), and multi-dimensional measures (e.g. the Subjective Workload Assessment Technique [SWAT]; Reid & Nygren, 1988: NASA Task Load Index [TLX]; Hart & Staveland, 1988). Subjective measures of workload reflect the user's *experience* of workload. The basic assumption is that if a pilot experiences high workload, then they are under high workload regardless of indications from other measures.

The dynamic measurement of workload shares many issues with that of (T)SA. Intrusive workload measurement using complex, multi-dimensional subjective scales that require a simulation scenario to be paused for their completion can negatively impact on primary task performance. If used concurrently with a task, multi-dimensional scales effectively present a secondary task, which has a negative impact on primary task performance. Less complex uni-dimensional scales may be more acceptable to use concurrently with a flight task, but still intrude on primary task performance to a lesser degree. However, as a result of their simplicity to complete, they have poor diagnosticity.

Although multi-dimensional scales are unacceptable to be completed in flight (real or simulated) because of the time and effort to complete them, these types of measures have much enhanced diagnosticity which allows for a more forensic analysis of the determinants of workload. These scales may be completed post-task, but then various issues arise concerning pilots' recollection of the cognitive load that they experienced during the scenario. The ratings provided may represent an assessment of the averaged workload experienced or workload peaks.

NASA-TLX (Hart & Staveland, 1988) is the most commonly used multi-dimensional scale to assess workload. The NASA-TLX requires ratings to be made on three explicit dimensions relating to the sources of workload – *mental demand* (how mentally demanding was the task); *physical demand* (how physically demanding was the task) and *temporal demand* (how hurried or rushed was the pace of the task) – plus three further dimensions concerning the interaction of the pilot with the task: *performance* (how successful were you in accomplishing what you were asked to do); *effort* (how hard did you have to work to accomplish your level of performance) and *frustration* (how insecure, discouraged, irritated, stressed, and annoyed were you)? The contribution of each of these dimensions to workload is then derived by making a series of 15 pairwise comparisons, based upon the premise that different sources of workload contribute different amounts to the overall workload in different circumstances. Participants are required to indicate which of two sources of workload is more important for the task being considered (scored "1" and "0") and the results are summed for each NASA-TLX dimension. The ratings for each sub-scale reflecting the perceived magnitude of workload for a given task are then multiplied by their associated importance weightings derived from the pairwise comparison process and summed to provide an overall workload score. This approach also enhances the diagnosticity of the instrument.

However, despite being used by numerous researchers for many years, some fundamental issues have been identified in the manner by which the weighting factors derived from the pairwise comparison process contribute to the calculation of overall workload and support the diagnosticity of the instrument (Virtanen et al., 2022). As a result of the pairwise comparison process, it is not possible to express two (or more) workload dimensions as being equally important in their contribution to overall workload. If pairwise comparisons are conducted consistently, there exists only one possible order of the relative importance of the dimensions, with a weight of 0.33 (the maximum possible) always being allocated to the most important dimension and a weight of 0.00 to the least important dimension. As a result, the pairwise comparisons, if conducted consistently, essentially ignore one of the dimensions and make the NASA-TLX a five-dimensional rating scale even if the pilot gives the dimension receiving a zero weight a workload rating. The weighting process can also lead to the inconsistent weights, with one dimension being directly considered to be more important than its contrasting dimension in the pairwise comparison process, but being deemed less important when the overall weighting order is derived.

Various enhancements to the NASA-TLX to overcome the shortcomings inherent in the original weighting system have been proposed, for example the Analytic Hierarchy Process (AHP; Saaty, 2000) and Swing method (von Winterfeldt &

Edwards, 1985). Both allow for weights greater than 0.33 for an individual dimension and also avoid the potential for a dimension receiving an unintentional zero weighting (see Virtanen et al., 2022). Both weighting approaches overcome the logical inconsistencies of the original weighting procedure while retaining the diagnostic benefits of the NASA-TLX method and in the case of Swing, it is also easier to administer. Moreover, in a time-dependent decision environment, Swing has been found to provide stable weights over time (Lienert et al., 2016) and has shown test–retest reliability (Bottomley & Doyle, 2001). However, it does produce slightly less variance in the load dimension weights derived, compared to the AHP and the traditional NASA-TLX approach (Virtanen et al., 2022).

The NASA-TLX has also been used widely without using the weighting procedure (so-called raw TLX) with an overall workload score being produced simply by averaging the ratings over the dimensions. This is a valid solution only if it can be assumed that the contribution to overall workload of each of the dimensions is roughly equal. If not, it will result in biased workload estimates. The "raw" TLX is essentially a special case of the "traditional" NASA-TLX where extremely inconsistent pairwise comparisons result in an equal weight for each dimension.

Physiological Workload Measurement

Variations in arousal and the general activation of the autonomic nervous system result in physiological changes, which make them suitable as measures of workload. Physiological measures have the advantage of being able to provide continuous, real-time monitoring of the state of the pilot (Jorna, 1993). Furthermore, they are passive measures that do not intrude upon performance as such, but the associated instrumentation of some measures may still be intrusive. Physiological workload measures include electrodermal activity, electroencephalography, functional near-infrared spectroscopy (fNIR), respiration rate, pupillary diameter and eyeblink (see Mansikka, Simola, et al., 2016). However, many of these measurement approaches are difficult to implement in a simulator as a result of complexities of setting up and calibrating the equipment, and other aspects of the pilot's activities during the flight scenario, for example movement, respiration/talking, and high physiological workload that result in changes in body temperature and sweat production not related to workload. These measures are almost impossible to implement in flight.

The heart is also under the control of the autonomic nervous system. Time domain-based measures of workload are derived from inter-beat intervals (IBIs). The premise is that when a pilot is under higher workload his/her heart beats slightly more quickly as higher brain activity requires a small increase in energy expenditure (these increases in blood flow in the brain can be observed directly using fNIR). Measures based upon IBI can take forms of varying sophistication, from simple measures such as mean IBI, heart rate (HR) and heart rate variation/variability

(HRV), to more sophisticated indices of workload, for instance the square root of the mean-squared differences between successive IBIs, the number of successive IBI pairs that differ by more than 50 ms, or even the integral of the IBI density distribution divided by the maximum of the distribution. IBI data can be collected relatively easily using many commercially available wearable devices, such as smart watches or sensors integrated into chest belts, which may also collect respiration data. While heart rate data collection itself is easy, to be meaningful it must also be linked directly to the pilot's activities in the scenario (e.g. simulator logs), which can be more of a challenge.

Collecting the full electrocardiogram (ECG) waveform allows for further workload measures to be derived from the frequency domain. Frequency-based measures have proven to be sensitive to fluctuations in workload particularly in the mid-frequency band between 0.07 and 0.14 Hz, which is related to the short-term regulation of blood pressure, and in the high-frequency band between 0.15 and 0.50 Hz, associated with respiratory functions. Decreases in power in these bands are associated with increasing workload (Jorna, 1993; Mulder, 1992; Veltman & Gaillard, 1993).

However, collecting any form of ECG data is not straightforward. There are considerable differences in individual cardiac activity and responses to varying task demand. As a result, comparisons are required both within each subject as well as across subjects (Roscoe, 1993). As within-subject comparisons are required, a resting baseline HR/HRV must also be obtained prior to any trial, ideally also followed by a post-trial resting baseline. To make the calculation of HRV meaningful, measurement epochs of at least 3 min are required with a sampling rate in excess of 125 Hz (Lee et al., 2022). The ECG trace needs to be synchronised with events in the flight trial to corroborate the reason for any HR/HRV response observed. The trace will also need careful inspection after data are collected to remove spurious signals/artefacts (e.g. those resulting from movement or electrode motion).

In terms of the study design, differences in workload conditions need to be relatively large to show any significant differences in workload using ECG-derived measures. While being unobtrusive, cardiac-based measures are relatively insensitive. Mansikka et al. (2019b) showed that ECG-derived measures were less sensitive to differences in conditions than were subjective workload scales. Mansikka et al. (Mansikka, Simola, et al., 2016a; Mansikka, Virtanen, et al., 2016) also found the simple measures based upon IBIs to be more sensitive to variations in workload demands than the more sophisticated ECG-derived measures described earlier. Moreover, such measures are not particularly diagnostic and can be confounded by factors unrelated to task demands. Interpretation of cardiac-based measures is often done in retrospect, limiting their utility. Reliability is often poor but can be improved by baselining measures against resting measures (or reference tasks) and by the very careful collection of data. As a result, to have any utility, physiological workload measures need to be supplemented with other performance and/or workload metrics.

Combined (T)SA, Workload and Performance Measurement in Practice

To analyse and understand performance it is essential to have an integrated underlying theoretical model describing the relationship between workload, (T)SA and performance. Mansikka et al. (Mansikka, Virtanen, Harris, & Jalava, 2021; Mansikka, Virtanen, Harris, & Salomäki, 2021a, 2021b) developed a model of team performance and an associated measurement framework to guide the collection of workload, (T)SA and performance data for the analysis of air combat engagements involving a flight of fighter aircraft. The performance aspects of the model encompass both system output performance (OP; "kills" and "survival") and normative performance (NP; based upon adherence and execution of individual components of air combat TTPs). A simplified version of this model is described in Figure 6.2. While ultimately important, OP can be a poor indicator of actual performance as it is a product measure, in contrast to NP, which reflects process; hence both need to be assessed. NP is more diagnostic for training purposes than OP. Mansikka et al. (Mansikka, Virtanen, Mäkinen, & Harris, 2021) describe an approach to the assessment of NP.

The air combat system model and measurement framework presented in Figure 6.2 describes how the selection of the TTPs to be executed is dependent upon the TSA of the flight. The accuracy of TSA is evaluated objectively from a comparison of the flight's mental model of the developing situation versus the objective situation (i.e. the ground truth) as derived from the simulation records. However, there is a cognitive cost to developing TSA, which is reflected in the assessment of workload. Execution of the selected TTP is evaluated against the standards for NP, which also incurs a workload overhead.

Mansikka et al. (Mansikka, Virtanen, Harris, & Salomäki, 2021a, 2021b) described how the measurement framework can be used for the testing, development and evaluation of air combat TTPs. As the approach involves live (L), virtual (V) and constructive (C) simulations, it placed differing demands on the assessment of both SA and workload. At the initial C stage, a simulation model is used to provide estimates of the probabilities of survival (Ps) and kill (Pk) based upon the proposed TTPs but without considering the human component. At the V stage, pilots implement the TTPs against virtual or constructive red (enemy) aircraft in simulators. Pk and Ps are calculated from results which are complemented by measures of pilots' (T)SA and workload. V simulations provide a safe, practical and relatively inexpensive environment for the test and evaluation of the TTPs, and enable the measurement of (T)SA and workload immediately after a simulated engagement, using all the de-brief tools available. V simulations also allow for multiple simulation runs, if required.

After further modification of the TTPs (if required) at the L stage, pilots fly engagements in real aircraft in a real environment. L simulations are expensive and

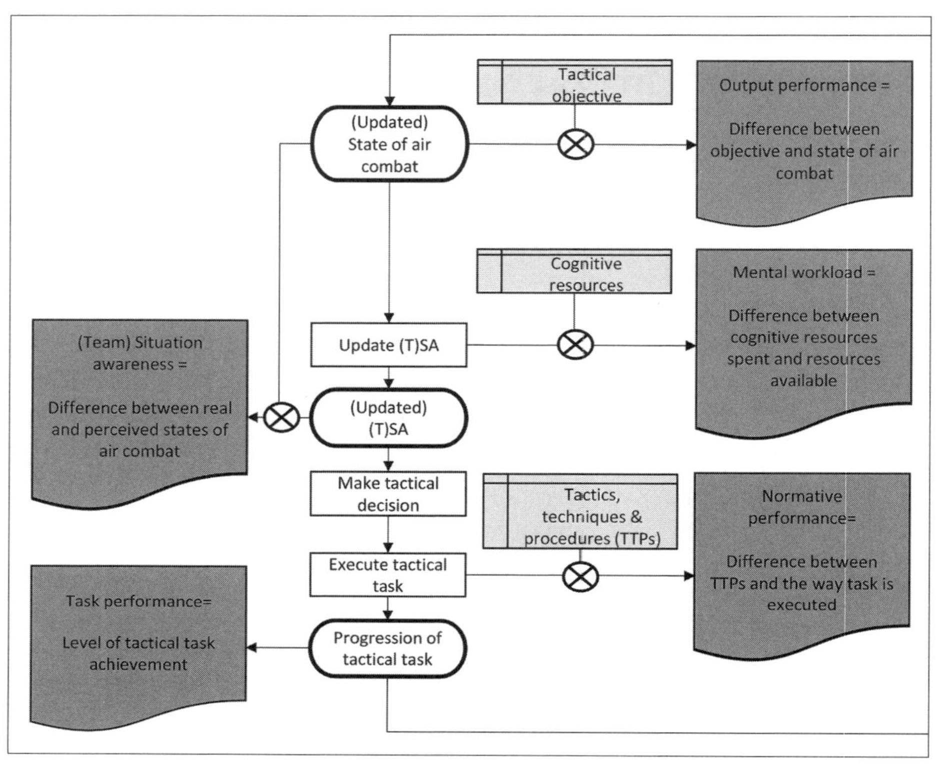

Figure 6.2. Simplified model of the relationship between workload, (T)SA and performance in air combat. (T)SA = team situation awareness; TTPs = tactics, techniques, and procedures. Based on Mansikka et al., 2021a.

resource heavy and are the final stage in the development process. They are important as they present pilots with real-life task complexity and stressors but are essentially used to validate the feasibility of the developed TTP in a near-real-world environment. L simulation provides challenges for the collection of (T)SA and workload data. L simulations cannot be paused and the time between TTP trials and data collection can be considerable. Data collection has to be unobtrusive if undertaken in real time or compromises have to be made if collected post-trial. In L simulations, the emphasis changes away from TTP development (C and V stages) to verification of performance, (T)SA and level of workload imposed on the pilots.

Summary and Outlook

Collecting human performance data alone is of limited utility: It provides very little explanation. Why a certain level of performance has, or has not, been achieved is often unclear. Although the relationship between performance, SA and workload can be complex, collecting all three types of data can provide a much richer description of pilot performance, providing more diagnostic data. Collecting workload and SA data is always a compromise between interfering with an ongoing task and gathering high-quality information that can inform research and development efforts. However, by applying the right data-gathering techniques at the right time and using a variety of complementary measures, a balance between these sometimes conflicting requirements can be found.

References

Bolstad, C., & Endsley, M. (2003). Measuring shared and team situation awareness in the army's future objective force. *Proceedings of the Human Factors and Ergonomics Society Annual Meeting, 47*(3), 369–373. https://doi.org/10.1177/154193120304700325

Bottomley, P., & Doyle, J. (2001). A comparison of three weight elicitation methods: Good, better, and best. *Omega, 29*(6), 553–560. https://doi.org/10.1016/S0305-0483(01)00044-5

Cooke, N., Stout, R., & Salas, E. (1997). Broadening the measurement of situation awareness through cognitive engineering methods. *Proceedings of the Human Factors and Ergonomics Society Annual Meeting, 41*(1), 215–219. https://doi.org/10.1177/107118139704100149

Crandall, B., Klein, G., Klein, G., & Hoffman, R. (2006). *Working minds: A practitioner's guide to cognitive task analysis.* MIT Press. https://doi.org/10.7551/mitpress/7304.001.0001

Ellis, G. A., & Roscoe, A. H. (1982). *The airline pilot's view of flightdeck workload: A preliminary study using a questionnaire. Technical memorandum FS(B) 465.* Royal Aircraft Establishment. https://apps.dtic.mil/sti/pdfs/ADA116314.pdf

Endsley, M. R. (1995a). Toward a theory of situation awareness in dynamic systems. *Human Factors, 37*(1), 32–64. https://doi.org/10.1518/001872095779049543

Endsley, M. R. (1995b). Measurement of situation awareness in dynamic systems. *Human Factors, 37*(1), 65–84. https://doi.org/10.1518/001872095779049499

Endsley, M. R. (1988). Situation Awareness Global Assessment Technique (SAGAT). In *Proceedings of the IEEE 1988 National Aerospace and Electronics Conference* (pp. 789–795). IEEE. https://doi.org/10.1109/NAECON.1988.195097

Endsley, M. R. (2019). A systematic review and meta-analysis of direct objective measures of situation awareness: A comparison of SAGAT and SPAM. *Human Factors, 63*(1), 124–150. https://doi.org/10.1177/0018720819875376

Endsley, M. R., & Garland, D. J. (2000). *Situation awareness: Analysis and measurement.* Lawrence Erlbaum Associates. https://doi.org/10.1201/b12461

Fracker, M. L. (1991). *Measures of situation awareness: An experimental evaluation. Report No. AL-TR-1991*-0127. Wright-Patterson Air Force Base. Human Engineering Division, Armstrong Laboratory. https://apps.dtic.mil/sti/pdfs/ADA262732.pdf

Harris, D. (2011). *Human performance on the flight deck.* Ashgate.

Hart, S. G., & Staveland, L. E. (1988). Development of NASA-TLX (task load index): Results of empirical and theoretical research. In P. A. Hancock & N. Meshkati (Eds.), *Human mental workload* (pp. 139–183). Elsevier Science Publishers. https://doi.org/10.1016/S0166-4115(08)62386-9

Jorna, P. G. A. M. (1993). Heart rate and workload variations in actual and simulated flight. *Ergonomics, 36*(9), 1043–1054. https://doi.org/10.1080/00140139308967976

Langan-Fox, J., Anglim, J., & Wilson, J. (2004). Mental models, team mental models, and performance: Process, development, and future directions. *Human Factors and Ergonomics in Manufacturing & Service Industries, 14*(4), 331–352. https://doi.org/10.1002/hfm.20004

Lee, K. F. A., Chan, E., Car, J., Gan, W.-S., & Christopoulos, G. (2022). Lowering the sampling rate: Heart rate response during cognitive fatigue. *Biosensors, 12*(5) 315. https://doi.org/10.3390/bios12050315

Lienert, J., Duygan, M., & Zheng, J. (2016). Preference stability over time with multiple elicitation methods to support wastewater infrastructure decision-making. *European Journal of Operational Research, 253*(3), 746–760. https://doi.org/10.1016/j.ejor.2016.03.010

Mansikka, H., Simola, P., Virtanen, K., Harris, D., & Oksama, L. (2016). Fighter pilots' heart rate, heart rate variation and performance during instrument approaches. *Ergonomics, 59*(10), 1344–1352. https://doi.org/10.1080/00140139.2015.1136699

Mansikka, H., Virtanen, K., & Harris, D. (2019a). The dissociation between mental workload, performance and task awareness in fast jet pilots. *IEEE Transactions on Human-Machine Systems, 49*(1), 1–9. https://doi.org/10.1109/THMS.2018.2874186

Mansikka, H., Virtanen, K., & Harris, D. (2019b). Comparison of NASA-TLX scale, modified Cooper–Harper Scale and mean inter-beat interval as measures of pilot mental workload during simulated flight tasks. *Ergonomics, 62*(2), 246–254. https://doi.org/10.1080/00140139.2018.1471159

Mansikka, H., Virtanen, K., & Harris, D. (2023). Accuracy and similarity of team situation awareness in simulated air combat. *Aerospace Medicine and Human Performance, 94*(6), 1–8. https://doi.org/10.3357/AMHP.6196.2023.

Mansikka, H., Virtanen, K., Harris, D., & Jalava, M. J. (2021). Measurement of team performance in air combat – have we been underperforming? *Theoretical Issues in Ergonomic Science, 22*(3), 338–359. https://doi.org/10.1080/1463922X.2020.1779382

Mansikka, H., Virtanen, K., Harris, D., & Salomäki, J. (2021a). Live-virtual-constructive simulation for testing and evaluation of air combat tactics, techniques and procedures, Part 1: Assessment framework. *Journal of Defense Modeling and Simulation, 18*(4), 285–293. https://doi.org/10.1177/1548512919886375

Mansikka, H., Virtanen, K., Harris, D., & Salomäki, J. (2021b). Live-virtual-constructive simulation for testing and evaluation of air combat tactics, techniques and procedures, Part 2: Demonstration of framework. *Journal of Defense Modeling and Simulation, 18*(4), 295–308. https://doi.org/10.1177/1548512919886378

Mansikka, H., Virtanen, K., Harris, D., & Simola, P. (2016). Fighter pilots' heart rate, heart rate variation and performance during an instrument flight rules proficiency test. *Applied Ergonomics, 56*(September), 213–219. https://doi.org/10.1016/j.apergo.2016.04.006

Mansikka, H., Virtanen, K., Mäkinen, L., & Harris, D. (2021). Normative performance measurement in simulated air combat. *Aerospace Medicine and Human Performance, 92*(11), 908–912. https://doi.org/10.3357/AMHP.5914.2021

Mansikka, H., Virtanen, K., Uggeldahl, U., & Harris, D. (2021). Team situation awareness accuracy measurement technique for simulated air combat – curvilinear relationship between awareness and performance. *Applied Ergonomics, 96(October),* 103473. https://doi.org/10.1016/j.apergo.2021.103473

Moray, N. (1988). Mental workload since 1979. *International Reviews of Ergonomics, 2*, 123–150. https://doi.org/10.1007/978-1-4757-0884-4

Mulder, L. J. M. (1992). Measurement and analysis methods of heart rate and respiration for use in applied environments. *Biological Psychology, 34*(2–3), 205–236. https://doi.org/10.1016/0301-0511(92)90016-N

Reid, G. B., & Nygren, T. E. (1988). The subjective workload assessment technique: A scaling procedure for measuring mental workload. In P. A. Hancock & N. Meshkati (Eds.), *Human mental workload* (pp. 185–214). Elsevier Science Publishers. https://doi.org/10.1016/S0166-4115(08)62387-0

Roscoe, A. H. (1993). Heart rate as psychophysiological measure for in-flight workload assessment. *Ergonomics, 36*(9), 1055–1062. https://doi.org/10.1080/00140139308967977

Saaty, T. (2000). *Fundamentals of decision making and priority theory with the analytic hierarchy process.* RWS Publications. https://doi.org/10.1007/978-94-015-9799-9_2

Salmon, P., Stanton, N., Walker, G., Baber, C., Jenkins, D., McMaster, R., & Young, M. (2008). What really is going on? Review of situation awareness models for individuals and teams. *Theoretical Issues in Ergonomic Science, 9*(4), 297–323. https://doi.org/10.1080/14639220701561775

Stanton, N., Salmon, P. M., & Rafferty, L. A. (2013). *Human factors methods: A practical guide for engineering and design.* Ashgate.

Sulistyawati, K., Wickens, C., & Chui, Y. (2009). Exploring the concept of team situation awareness in a simulated air combat environment. *Journal of Cognitive Engineering and Decision Making, 3*(4), 309–330. https://doi.org/10.1518/155534309X12599553478791

Veltman, J. A., & Gaillard, A. W. K. (1993). Indices of mental workload in a complex task environment. *Neuropsychobiology, 28*(1), 72–75. https://doi.org/10.1159/000119003

Vidulich, M. A., & Wickens, C. D. (1986). Causes and dissociation between subjective workload measures and performance. *Applied Ergonomics, 17*(4), 291–296. https://doi.org/10.1016/0003-6870(86)90132-8

Virtanen, K., Mansikka, H., Kontio, H., & Harris, D. (2022). Weight watchers: NASA-TLX weights revisited. *Theoretical Issues in Ergonomic Science, 23*(6), 725–748. https://doi.org/10.1080/1463922X.2021.2000667

von Winterfeldt, D., & Edwards, W. (1985). *Decision analysis and behavioural research.* Cambridge University Press.

Yeh, Y.-Y., & Wickens, C. D. (1988). The dissociation of subjective measures of mental workload and performance. *Human Factors, 30*(1), 111–120. https://doi.org/10.1177/001872088803000110

Chapter 7

Understanding Pilot Attention and Awareness With Eye-Tracking

Mickaël Causse, Julia Behrend, and Randall J. Mumaw

Abstract

A large number of air accident analyses highlighted the fact that crews sometimes fail to take into account important visual information displayed in the cockpit. In response, aviation authorities in various countries have issued recommendations for improving cockpit monitoring. The recent elevation of the pilot monitoring function was also meant to stress the importance of the role of monitoring. The most promising tool to better understand and characterize cockpit monitoring is eye-tracking. However, this technology has yet to be adopted in cockpits because of the relative complexity of obtaining reliable data outside the laboratory, as well as the difficulty in interpreting the results. In this chapter, we review the fundamentals of attention and awareness in the cockpit, the eye-tracking technology itself, and what such devices can deliver in terms of meaningful information, especially for airlines, pilots, and flight instructors. We also list the limitations and propose several avenues for future research and application.

Keywords

cockpit monitoring, visual scanning strategies, eye-tracking, aviation, awareness, attention

Why We Should Care About Pilots' Awareness/ Attention

Operating an airplane requires the flight crew to maintain awareness of a large amount of information concerning the current airplane state, weather and traffic, flight planning and updated clearances, flight path management, coordination with ground operations, etc. Almost all of this information is presented visually on a small set of physical displays. In addition, the flight crew must determine – in a highly dynamic environment – which tasks require attention. It is not uncommon for an unexpected event (e.g., non-normals, calls from air traffic control [ATC] or the cabin) to require immediate attention, causing disruption of on-going tasks.

The flight crew, thus, must both maintain a broad and up-to-date awareness of operationally relevant information and place attention on the "most appropriate task" at the current time. During quieter portions of the flight, this is easily done, but during busier flight segments performance can degrade: Information is not updated and there are errors in task prioritization. In this chapter, we discuss theories about how attention is allocated to support awareness, and we describe effective methods for measuring how visual attention is being allocated. An important element of this discussion is the use of and limitations on eye-tracking measures.

Sitting in the pilot's seat, it is easy to be overwhelmed by the number of indications, controls, and displays presented across the flight deck interface: in the pilot's central vision, on side displays, on the overhead panels, and even on panels behind the pilot's position (as well as the view out the window; see Figure 7.1). Indeed, Geratewohl (1987) estimated that 80% of flight information is processed visually. Pilots must extract information from multiple locations and integrate it into a coherent characterization of the airplane's position and state. Maintaining this awareness is sustained by specific gaze strategies, choreographies performed by the eyes to pick up relevant information at the appropriate moment.

Figure 7.1. An illustration of a modern aircraft cockpit (Boeing 777): A plethora of instruments and displays must be monitored by the crew. Image courtesy of Olivier Lefrançois. Reprinted with permission.

The record of accidents and serious incidents in commercial aviation has pointed to a breakdown in flight crew awareness of basic flight path parameters or autoflight state. Standard examples are Turkish Airways 1951 and Asiana 214 (loss of awareness of decreasing airspeed and increasing pitch on approach); Atlas Air 3591 and Armavia 967 (loss of awareness of excessive nose-down pitch); Flash Airlines 604 and Kenya Airways 507 (loss of awareness of large bank angle changes); and Bhoja Air 213 and the Thomsonfly incident at Bournemouth (loss of awareness of autoflight state). In these events, loss of awareness can lead to a surprise, inappropriate control inputs, and a loss of control in-flight (LOC-I). In some accidents, awareness was lost because attention was over-committed to some element of airplane state that was unrelated to flight path management.

This is not just a recent discovery; the National Transportation Safety Board (NTSB, 1994) identified poor monitoring as a factor in 84% of major accidents in the United States from 1978 to 1990. A Commercial Aviation Safety Team (CAST) panel also identified loss of awareness of airplane state as a major contributor to a set of 18 accidents and incidents (Mumaw et al., 2019). Recently, a report issued by the British Loss of Control Action Group – a collaboration between civil aviation authorities and industry (supported by British Airways, easyJet, flybe, Jet2, Thomas Cook Airlines, Thomson Airways, and Virgin Atlantic Airways) – details 10 crashes with a total of 651 fatalities in which poor monitoring was a contributing factor (Loss of Control Action Group, 2013). In fact, since the period covered by the NTSB study (1978–1990), Sumwalt et al. (2015) estimated that 17 accidents involving monitoring problems have occurred. These findings suggest that less-than-adequate awareness can have tragic consequences, and, thus, understanding how skilled pilots use their attentional resources to maintain a broad operational awareness has become an important research objective.

It is worth taking a more detailed look at the Turkish Airways flight 1951 accident on approach to Amsterdam's Schiphol Airport (Dutch Safety Board, 2010) to see how this type of accident evolves. A faulty radio altimeter caused the 737's autothrottle to set thrust to idle on approach, meaning that the autothrottle would not manage airspeed; the flight crew was aware of the fault but unaware of its effect. The flight crew, struggling to meet stabilized approach criteria, initiated the landing checklist quite late. Then, while running that checklist, they were distracted by unexpected indications (tied to the fault) when they armed the speedbrakes. According to the cockpit voice recorder, both pilots were focused on the speedbrake indications just as airspeed dropped below the airspeed target. The low airspeed alert for this airplane was a visual-only alert, but again, both pilots were focused elsewhere and did not see the airspeed tape's visual alerting. The approach-to-stall alert (stick shaker) finally made them aware of the airspeed problem. Their subsequent actions failed to increase thrust in time to recover the aircraft before it stalled and crashed. In addition to illustrating inadequate monitoring, this may also be an example of how overtrust in automation (Antonovich, 2008) can contribute to reduced flight path monitoring. Boeing has since issued recommendations to remind pilots of the importance of monitoring airspeed and altitude during approach.

A number of empirical studies have corroborated that flight crews can fail to maintain full awareness. For example, Sarter et al. (2007), using eye-tracking, revealed that pilots frequently failed to check manual autopilot mode selections or failed to notice autopilot mode changes, demonstrating that inadequate monitoring is an important contributor to breakdowns in pilot–automation interaction. In response to this collection of findings, the world's aviation safety institutions have issued multiple recommendations to improve cockpit monitoring. The importance of monitoring is also reflected in the shift in labeling flight crew roles: the pilot flying (PF) and the pilot monitoring (PM). In 2003, the Federal Aviation Administration (FAA) issued recommendations to remove the term "pilot non-flying" and replace it with the term "pilot monitoring" (see AC 120-71A, Standard Operating Procedures for Flight Deck Crewmembers, Federal Aviation Administration, 2003). Subsequently, in 2013, the FAA stipulated that the development of monitoring skills become a priority and required airlines to incorporate an explicit training program to improve cockpit monitoring skills. In 2013, the *Bureau d'Enquetes et d'Analyses pour la securite de l'aviation civile* (BEA; Bureau of Enquiry and Analysis for Civil Aviation Safety) conducted a study with eye-tracking and recommended to manufacturers to study pilots' visual scanning in order to validate procedures (BEA, 2013). The report mentioned that the evaluation of the visual scanning of the pilots is fundamental in the development of flying procedures, in particular with regard to the go-around (recommendation FRAN-2013-030).

Models of Pilots' Attention

Before presenting studies of pilots' eye fixations and scanning, we introduce the range of models that have been developed to describe the drivers of pilots' attention. Existing models can be characterized using two performance dimensions:

1. Level of conscious control – Some models specify that the pilots consciously "direct attention," while others see visual attention as a more "automated" process, in the sense of what is often called "skill automaticity" (Oppici et al., 2018).
2. Knowledge-driven versus stimulus-driven – Knowledge-driven (see Wickens, 2021) models start with the pilot's need to develop or complete a coherent understanding of current airplane position and state, which determines which information should be attended. Stimulus-driven models focus more on characteristics of the stimulus (e.g., flight deck interface) to determine how attention will be allocated. While these dimensions are useful in discussing models of attention, existing models do not typically place themselves at a dimension endpoint; theories offer a mixed set of concepts to describe the drivers and mechanisms of allocating attention.

An early effort on modeling attention is represented by the mathematical models developed in the 1960s (e.g., Carbonell, 1966) to predict the performance of a human operator performing a control task, such as would be encountered in process control (e.g., a nuclear power plant control room); that is, a human monitoring a small set of indicators for deviations. The modeling attempted to predict eye behavior based largely on the characteristics of the display elements (e.g., Senders, 1983). The rate of change of an indicator (called "bandwidth") proved to be a good predictor of how frequently a human fixated it; that is, more change led to more fixations. Some models also considered the "cost" of looking away from an indication, where cost refers to the momentary loss of awareness of an operationally relevant indication.

In a more sophisticated update of the early mathematical models of monitoring, Wickens (e.g., Wickens et al., 2003) expanded the factors considered by the model and then applied his model to a more realistic aviation context. Wickens proposed four factors that drive monitoring behavior: the salience (S) of the indication, the pilot's expectancy (E) of the indication, which is an expression of bandwidth, the effort (E) needed to shift attention to an indication, and the task relevance or value (V) of the indication (which can capture cost). Wickens uses the acronym SEEV to represent these four factors. The SEEV model, like the older models, makes fairly accurate predictions of eye fixations (at an aggregated level) over time.

A contrast to these modeling efforts are descriptions of monitoring that are more strongly knowledge driven (e.g., Ebbatson, 2009). These models start with an evolving mental representation of current airplane position and state: A mental model or situation model that supports a pilot in projecting into the future and anticipating airplane performance and system behavior. Monitoring, in this framework, is the primary means to confirm or update the representation. This approach has also been referred to as "sense-making" (Billman et al., 2020; Mumaw et al., 2020), which emphasizes the role of monitoring to ensure coherence in the pilot's mental representation. Shiferaw et al. (2019) consider that gaze behavior reflects the interaction between external information and the individual's internal state. The internal state includes information in working memory, which is influenced by the mental model. They propose that this interaction is underpinned by a closed-loop process driven by a bottom-up input on gaze orientation (external influences), and an open-loop process driven by top-down prediction. Gaze control would be the product of spatial predictions based on bottom-up and top-down interaction.

The overall goal in the modeling work is to identify what drives skilled monitoring – how interface design and pilot training can best support a broad awareness and responsive attention management. Empirical studies of pilot performance, described below, show that skilled performance reflects both a better understanding of which information is relevant and increased efficiency in finding and extracting that information from the interface. Further, much of this expertise is captured in automated scanning strategies (Yu et al., 2016) that are unavailable to conscious introspection.

Aviation Studies Using Eye-Tracking to Examine Attention and Awareness

The scientific literature reveals a long history of attempts to better capture eye movements, and thus the attentional processes behind piloting activity. As early as the 1950s, eye-tracking studies (Fitts et al., 1950/1980; Milton et al. 1949) were conducted using primitive recording systems such as a motion picture camera that directly photographed a pilot's head and eyes as reflected in a small mirror located on the instrument panel. These pioneering works showed that pilots monitored instruments with varying durations and frequencies, which helped to establish the strength of the relationships between different instruments. These simple data aided flight deck designers in making visual scans more efficient, for example, by placing frequently observed instruments in the center of the field of view or by placing them in close proximity (e.g., attitude and velocity indicators). Eye-tracking measures can benefit aviation safety more broadly (Peißl et al., 2018). Eye-tracking provides information that could not be obtained by simple observation of behavior (e.g., actions on flight controls), and eye-tracking systems have become more convenient to implement and use.

The Links Between Pilots' Monitoring Strategies and Flying Performance

Cockpit monitoring can be described as the methodical visual scanning of instruments necessary to control aircraft trajectory or to make appropriate changes in aircraft attitude. Effective cockpit monitoring is a critical input to manual flight control, which requires continuous adjustments, mainly through fine-motor inputs on the flight controls (Haslbeck & Hoermann, 2016). Sensing and interpreting flight data and the subsequent flight control adjustments to achieve the desired flight path (Benson, 2002) have been described as "closed-loop" control.

Monitoring allows pilots to update their awareness of the aircraft's current state (Lounis et al., 2021). Scanning patterns that include a large proportion of primary fight parameters (e.g., attitude indicator, speed, altimeter, heading indicator) have been found to be critical for high-precision maneuvers. Lefrançois et al. (2021) showed that pilots performing suboptimal visual scanning showed less accurate flying and, in some cases, unstabilized approaches. Indeed, these "unstabilized" pilots were either under- or over-focused on various instruments, and their number of visual scanning patterns was lower than those of more accurate pilots when

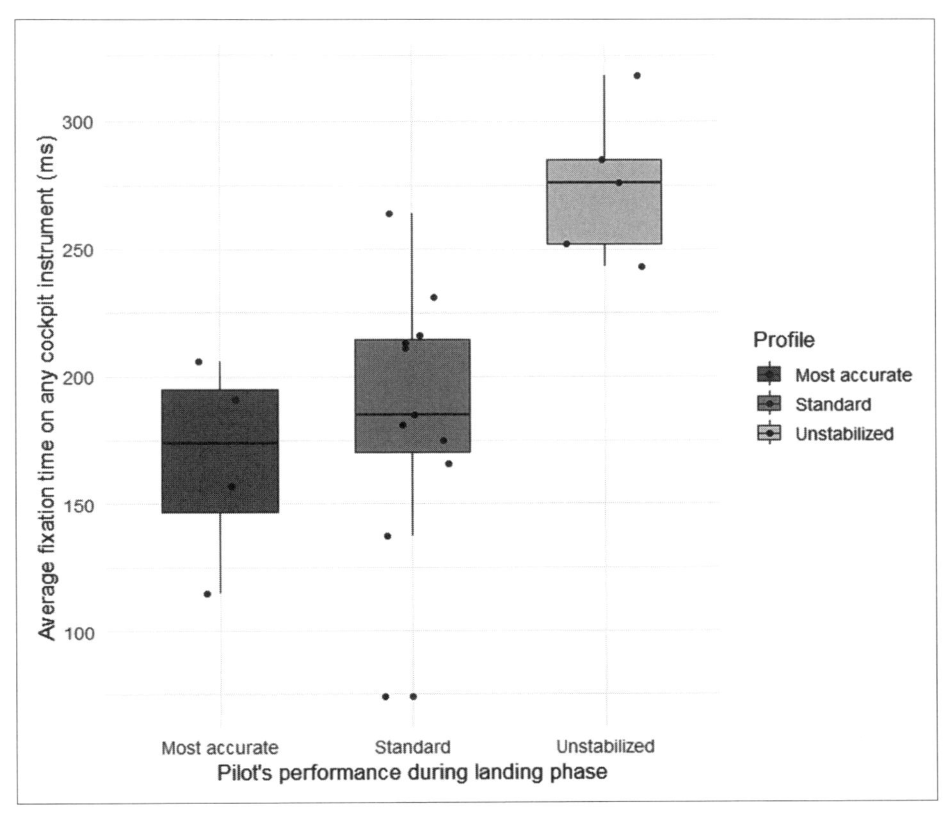

Figure 7.2. Average fixation time on cockpit instruments according to flying performance during landing. Each dot represents a pilot. Adapted from Lefrançois et al., 2021, p. 15. (Text on x- and y-axis adapted, legend added.)

performing a nominal approach. The most accurate pilots showed faster information extraction (lower average fixation times on instruments) than the standard and unstabilized pilots. The unstabilized pilots who missed their approach had the longest fixation times on instruments (see Figure 7.2).

Pilot role (PF or PM) naturally impacts monitoring activity (Behrend & Dehais, 2020; Reynal et al., 2016). A proper coordination of monitoring between PF and PM appears to be an important aspect of flight crew performance. Dehais et al. (2017) have shown that cross-checking between the PF and PM is not always effective in detecting trajectory deviations, which, according to the authors, may explain why two thirds of the crews in their study made errors, including critical trajectory deviations during go-arounds. According to Jarvis (2017), aviation accidents with failures in awareness are largely the result of parallel monitoring of PF and PM rather than complementary monitoring.

Overtrust in automation may also contribute to inadequate attention to critical flight parameters. A study by Mercier et al. (2022) showed that pilots pay too much attention to the flight director when it is engaged at the expense of more direct monitoring of raw flight parameters (engines, trajectory). Despite improved flying performance, such a change could make pilots more vulnerable to loss of situational awareness, especially when faced with suddenly unreliable or inconsistent flight guidance functions (Endsley & Kiris, 1995).

Using Eye-Tracking to Track Expertise and "Augment" Training

Several studies showed that pilots' monitoring performance (e.g., fixation duration and frequency) evolves with level of expertise (Brams et al., 2018; Haslbeck & Zhang, 2017; Peißl et al., 2018; Robinski & Stein, 2013; Yang et al., 2013; Ziv, 2016). Kasarskis et al. (2001) noticed that expert pilots make more fixations, have shorter fixation dwell times, and have more structured scanning patterns than novices. In addition, Lorenz et al. (2006) showed that expert pilots looked outside the cockpit more, compared to novices, during taxiing. Similarly, a study involving fighter pilots found that the best-performing pilots made shorter fixations on the heads-down tactical display and alternated more frequently between the tactical display and the outside world (Svensson et al., 1997). These findings suggest that eye-tracking could measure practice effects and aid in tracking the adoption of efficient monitoring strategies during training.

Reduced fixation durations as an indicator of skill have appeared in several studies and are associated with broader scanning behavior. During a navigation simulation in a helicopter simulator, Sullivan et al. (2011) showed that pilots' median fixation times were reduced by 28 ms for every 1,000 hr of flight. In addition, the number of transitions between areas of interest (external world and navigation map) significantly increased. In their review of the literature on eye movements in medicine and chess, Reingold and Sheridan (2011) termed this greater perceptual efficiency of experts "superior perceptual encoding of domain-related patterns" (p. 524). Similarly, Gegenfurtner et al. (2011) showed that experts fixate more on relevant regions while spending less time on each fixation. This visual strategy of "information reduction" (Haider & Frensch, 1996) – optimizing visual information processing by separating relevant and irrelevant tasks – has also been found in studies of experienced and novice general aviation pilots (Bellenkes et al., 1997). In another study, Schriver et al. (2017) compared the distribution of attention between expert and novice pilots during a problem-solving task in a flight simulator using an eye-tracker. The faster decisions of the experts were accompanied by more relevant fixations, for example, toward system faults when they occurred.

One potentially valuable approach to training monitoring is gaze following, which asks the novice to track an expert's scan pattern. Humans have an innate ability to learn, including causal learning, from gaze following (Jarodzka et al., 2012; Van Hecke & Mundy, 2007). This technique allows the learner to follow the expert's gaze to discover relevant information and guide his/her attention to the current goal. Visual scene recordings with gaze position overlays could be used during example-based learning programs. Visual patterns of trainee pilots can be compared with those of the experts, making it possible to identify scanning errors and provide individualized feedback (Lefrançois et al., 2021). The videos should be broadcast quickly after the performance to be effective. It has long been known that training is most effective when feedback on performance is delivered quickly after the performance (Bourne, 1957).

Pilots often benefit from feedback after an unusual event to avoid making the same mistake in the future. Generally, fixations are automatic and performed without conscious input (Robinski & Stein, 2013), and it is highly unlikely that eye movements will be remembered. Hence, eye movement videos could be replayed and viewed after the flight session to review monitoring weakness. For example, the replay could show a failure to fixate a piece of critical information, or, conversely, an over-focalization on a particular flight instrument. In this sense, F16 instructors have used eye-tracking in real time to make novices aware of their own visual scanning and mistakes (Wetzel et al., 1998).

The use of eye-tracking for training purposes can be done in real time. Dubois et al. (2015) trained 15 ab initio military pilots to focus on the outside view and much less on flight instruments. They created an alert that was triggered by a "head-down" position of more than 2 s. The researchers compared three groups: a control condition, with no alerting, a condition with alerting (beep), and a condition in which the head-down condition led to instrument masking. The results showed that beeping and masking both significantly reduced the 2-s violation rule.

An experiment in a B737-700 full flight simulator used four operational scenarios to show that pilots look less often at the primary flight display (PFD) when the flight director is on (Zaal et al., 2021). Pilots who successfully completed monitoring challenges were more focused on the most-relevant information. In addition, better pilots adapted their monitoring strategies to the task at hand, focusing on either the PFD or the navigation display (ND), depending on the challenge. The findings of this study confirm the importance of adjusting gaze allocation according to the context, raising the question of whether and to what degree prespecified scanning behaviors can be identified and trained.

Aviation Design

A lesser-known use of eye-tracking in aviation is to evaluate the ergonomics of the cockpit interface (Hebbar et al., 2021; Wang et al., 2019). By analyzing eye movement patterns, researchers can identify which areas of the cockpit are receiving the most visual attention and assess whether critical information is being processed effectively. This information helps to determine the optimal placement of instruments, redesign cockpit layouts, and ensure that pilots can quickly shift their gaze between important elements. Ultimately, this minimizes cognitive load and improves situational awareness and understanding of the airplane state. Despite the potential advantages of this use of eye-tracking and its relatively wide and well-established applications in human–computer interfaces (HCI) and usability research (Jacob & Karn, 2003; Sharma & Dubey, 2014), there are few studies of the cockpit. One notable study was conducted by Hanson (2004). He examined how knowing the focus of attention can be used to assess an enhanced ground proximity warning system (EGPWS) with the aim of reducing pilot errors. He combined eye-tracking measures of information acquisition with subjective reports of information misinterpretation and inability to project. The eye-tracking data showed that information misinterpretation was reduced by the EGPWS prototype, with an intensification of information gathering when the EGPWS was used in hazardous conditions. Subjective data supported this eye-tracking finding and showed that the EGPWS improved the interpretation of lateral distance to terrain.

Potential Problems With the "Point of Fixation"

Eye-tracking studies are largely based on the eye–mind hypothesis (Just & Carpenter, 1980), which states that the point of fixation is closely related to what is processed. The eye–mind hypothesis clearly has value for describing reading, as it is quite logical that the eye remains focused on the words being read. However, in the cockpit, the organization of visual information is complex and dynamic, and the pilot's mind may not always be aligned with what is being fixated. That is, thoughts may be on past, present, or future tasks or events, making it more difficult to establish a direct link between the point of fixation and current attention and awareness. In fact, it is quite possible for a pilot to be fixating on an instrument when they are already planning actions or thinking about the tasks they need to perform for their approach and landing, for example.

Several researchers have demonstrated that, even when fixations are very near an indication, the observer can fail to perceive an unexpected stimulus. This phenomenon has been termed "inattentional blindness" (Mack & Rock, 1998). The famous "gorilla" study is an excellent example of how objects in our central focus can be missed when visual attention is narrowly focused. In this study, Simons & Chabris (1999) asked participants to watch a video of a basketball game. Participants were instructed to count how many times players wearing white shirts passed the basketball. During the video, a person wearing a gorilla suit crossed the scene, and across all conditions, 46% failed to notice the gorilla. Interestingly, this study was replicated with eye-tracking to ensure that participants' eyes were indeed on the gorilla (Gelderblom & Menge, 2018). Although 43% of the participants fixated the gorilla, only 22% reported noticing it. The phenomenon of inattentional blindness has also been observed in the field of aviation, with up to 70% of participants failing to notice runway incursions (Kennedy et al., 2017). However, this study did not use eye-tracking to ensure that participants were actually looking in the direction of the aircraft crossing the runway.

Another issue with the tracking fixations is that attention can be drawn by objects or events in the periphery. Peripheral vision can detect objects in up to 99.9% of the visual field (Rosenholtz, 2016), whereas central or foveal vision, which provides the highest level of visual acuity, is approximately 2° of eccentricity (Strasburger et al., 2011). A study by Ramón Alamán et al. (2020) attempted to determine the extent to which the pilot can detect events around the fixation point. In their experiment, participants had to look at the center of a PFD (at the center of the attitude indicator) and indicate whether changes in the surrounding parameters, such as speed or altitude, had occurred. The study confirmed that pilots' perceptual span extends to peripheral vision; that is, they are able to perceive parameter changes outside the foveal vision. Further, the percentage of parameter changes successfully perceived depends on the distance of the change from the fixation point and the conspicuousness of the change. Some research has attempted to capture attention outside the fovea. This is the case, for example, in the study by Wang et al. (2023), in which they compute a machine learning-based gaze measure called the "visual attention index," which introduces attention to peripheral vision. In addition, Brams et al. (2020) developed a technique that allowed them to take into account both foveal and parafoveal vision. Such attempts may help refine eye-tracking measures in aviation in the future.

What Can Be Done Currently With Eye-Tracking in Aviation (Especially Airliners)?

Eye-tracking has made considerable progress in recent years, and its use is becoming increasingly easier in environments such as a jet transport cockpit, including actual aircraft (e.g., Dehais et al., 2008). Recent eye-trackers rely less on cumbersome calibrations (e.g., Pupil Labs' Pupil Invisible), and data processing and analysis are becoming accessible to a wide range of users. It is also possible to obtain real-time data (e.g., Lounis et al., 2020; Ratwani & Trafton, 2011) with software development kits provided by most of the current manufacturers. However, there are still limitations to eye-tracking usage, in particular in operational conditions. Among them, light variations can produce data loss, and individuals wearing glasses or lenses can be difficult to record. Also, some devices have limitations in the measurement of eye movements, especially when the angle of the eyes becomes very important or when the individual turns their head. For a setting such as an aircraft cockpit, precision, reliability, and accuracy are not yet ideal. Regarding the type of eye-tracking, while wearable devices are extremely practical for research because they do not require physical integration in the cockpit, they will probably be difficult to accept by pilots because they should be worn permanently. Thus, remote eye-trackers in the form of cameras integrated into a full-flight simulator are probably much more realistic in daily use for airliners. Knabl-Schmitz et al. (2023) highlighted the benefits and challenges via three studies in a full-flight simulator equipped with remote eye-trackers. In addition to the correct equipment integration and adaptations to instructor training (Rudi et al., 2020), ethical considerations and data management (use, storage, and protection) remain a challenge for safety and training departments.

The first integrations should probably not be too ambitious regarding accuracy and usage. Which part of the screen is gazed? Does the pilot look outside? A paper by Peysakhovich et al. (2018) formalizes four stages of integrating eye-tracking to improve aviation safety. The stages are ranked according to feasibility levels, with the fourth stage being the most complex. According to the authors, Stage I concerns the use of eye-tracking for pilot training and flight performance analysis on the ground. Stage II evokes on-board eye recordings as additional data for the "black box" recorders. This may prove useful in the case of an abnormal event or accident. Stage III proposes gaze-contingent flight deck adaptation, which could include triggering an alert when visual monitoring becomes suboptimal. Finally, Stage IV would involve gaze-based aircraft adaptation, including control by the aircraft in the event of abnormal pilot behavior, but this would be a later stage due to the various technical issues to resolve.

Summary

We have reviewed the long and rich history of efforts to use eye-tracking measures to understand pilot attention and awareness; more specifically, to understand the markers of skilled performance. Over time, this work has shifted to the analysis of performance in more ecologically valid operational settings, and improvements in the eye-tracking technology have made less intrusive measurements possible. While these improvements have been shown to be beneficial in understanding performance, there is still a need to employ eye-tracking conservatively and to validate it with other measures of attention and awareness. The greatest strengths of eye-tracking are that it is typically strongly linked with visual attention and that it does not require non-operational actions, such as verbal reports, from the pilot.

References

Antonovich, B. (2008). The flight crew and automation. *Journal of Aviation/Aerospace Education & Research, 17*(3), 3. https://doi.org/10.58940/2329-258X.1453

Behrend, J., & Dehais, F. (2020). How role assignment impacts decision-making in high-risk environments: Evidence from eye-tracking in aviation. *Safety Science, 127*, Article 104738. https://doi.org/10.1016/j.ssci.2020.104738

Bellenkes, A. H., Wickens, C. D., & Kramer, A. F. (1997). Visual scanning and pilot expertise: The role of attentional flexibility and mental model development. *Aviation, Space, and Environmental Medicine, 68*(7), 569–579.

Benson, A. J. (2002, April 15–17). *Spatial disorientation – a perspective* [Paper presentation]. RTO HFM Symposium on Spatial Disorientation in Military Vehicles: Causes, Consequences and Cures, La Coruña, Spain. https://skybrary.aero/sites/default/files/bookshelf/3510.pdf

Billman, D., Mumaw, R., & Feary, M. (2020). A model of monitoring as sensemaking: Application to flight path management and pilot training. *Proceedings of the Human Factors and Ergonomics Society Annual Meeting, 64*(1), 244–248. https://doi.org/10.1177/1071181320641058

Bourne, L. E., Jr. (1957). Effects of delay of information feedback and task complexity on the identification of concepts. *Journal of Experimental Psychology, 54*(3), 201. https://doi.org/10.1037/h0044684

Brams, S., Hooge, I. T., Ziv, G., Dauwe, S., Evens, K., De Wolf, T., Levin, O., Wagemans, J., & Helsen, W. F. (2018). Does effective gaze behavior lead to enhanced performance in a complex error-detection cockpit task? *PloS one, 13*(11), Article e0207439. https://doi.org/10.1371/journal.pone.0207439

Brams, S., Rejtman, R. F., Levin, O., Ziv, G., Hooge, I. T., & Helsen, W. F. (2020). Eye-tracking in aviation: A new method for detecting learned visual scan patterns of cockpit instrument in simulated flight. In V. Peysakhovich, D. Rudi, M. Causse, P. Kiefer, F. Dehais, & M. Raubal (Eds.), *Eye-tracking in aviation. Proceedings of the 1st international workshop* (pp. 69–77). ISAE-SUPAERO, Université de Toulouse, Institute of Cartography and Geoinformation (IKG). https://doi.org/10.3929/ethz-b-000407654

Bureau d'Enquetes et d'Analyses pour la securite de l'aviation civil. (2013, August). *Study on aeroplane state awareness during go-around*. https://bea.aero/etudes/asaga/asaga.study.pdf

Carbonell, J. R. (1966). A queueing model of many-instrument visual sampling. *IEEE Transactions on Human Factors in Electronics, HFE-7*(4), 157–164. https://doi.org/10.1109/THFE.1966.232984

Dehais, F., Behrend, J., Peysakhovich, V., Causse, M., & Wickens, C. D. (2017). Pilot flying and pilot monitoring's aircraft state awareness during go-around execution in aviation: A behavioral and eye tracking study. *The International Journal of Aerospace Psychology, 27*(1–2), 15–28. https://doi.org/10.1080/10508414.2017.1366269

Dehais, F., Causse, M., & Pastor, J. (2008). Embedded eye tracker in a real aircraft: New perspectives on pilot/aircraft interaction monitoring. In *Third international conference on research in air transportation* (pp. 303–309). https://www.icrat.org/previous-conferences/3rd-international-conference/icrat2008_proceedingscomplete.pdf

Dubois, E., Blättler, C., Camachon, C., & Hurter, C. (2015). Eye movements data processing for ab initio military pilot training. In R. Neves-Silva, L. C. Jain, & R. J. Howlett (Eds.), *Intelligent decision technologies* (pp. 125–135). Springer International Publishing. https://doi.org/10.1007/978-3-319-19857-6_12

Dutch Safety Board. (2010, May). *Crashed during approach, Boeing* 737-800, *near Amsterdam Schiphol Airport, 25 February 2009* (Project number M2009LV0225_01). https://catsr.vse.gmu.edu/SYST460/TA1951_AccidentReport.pdf

Ebbatson, M. (2009). *The loss of manual flying skills in pilots of highly automated airliners* [Doctoral thesis, Cranfield University]. https://dspace.lib.cranfield.ac.uk/bitstream/1826/3484/1/Ebbatson_Thesis%20_2009.pdf

Endsley, M. R., & Kiris, E. O. (1995). The out-of-the-loop performance problem and level of control in automation. *Human Factors, 37*(2), 381–394. https://doi.org/10.1518/001872095779064555

Federal Aviation Administration. (2003). *Standard operating procedures for flight deck crewmembers (Advisory circular, AC 120-71A)*. https://www.faa.gov/regulations_policies/advisory_circulars/index.cfm/go/document.information/documentid/23216

Fitts, P. M., Jones, R. E., & Milton, J. L. (1980). Eye movements of aircraft pilots during instrument-landing approaches. *Aeronautical Engineering Review, 9*(2), 1–6. (Original work published 1950)

Gegenfurtner, A., Lehtinen, E., & Säljö, R. (2011). Expertise differences in the comprehension of visualizations: A meta-analysis of eye-tracking research in professional domains. *Educational Psychology Review, 23*, 523–552. https://doi.org/10.1007/s10648-011-9174-7

Gelderblom, H., & Menge, L. (2018). The invisible gorilla revisited: Using eye tracking to investigate inattentional blindness in interface design. In Association for Computing Machinery. *Proceedings of the 2018 international conference on advanced visual interfaces*. Association for Computing Machinery. https://doi.org/10.1145/3206505.3206550

Geratewohl, S. (1987). *Guidance for aviation psychology*. Verlag für Wehrwiss.

Haider, H., & Frensch, P. A. (1996). The role of information reduction in skill acquisition. *Cognitive Psychology, 30*(3), 304–337. https://doi.org/10.1006/cogp.1996.0009

Hanson, E. K. S. (2004). Focus of attention and pilot error. In *Proceedings of the 2004 symposium on eye tracking research & applications*. Association for Computing Machinery. https://doi.org/10.1145/968363.968377

Haslbeck, A., & Hoermann, H.-J. (2016). Flying the needles: Flight deck automation erodes fine-motor flying skills among airline pilots. *Human Factors, 58*(4), 533–545. https://doi.org/10.1177/0018720816640394

Haslbeck, A., & Zhang, B. (2017). I spy with my little eye: Analysis of airline pilots' gaze patterns in a manual instrument flight scenario. *Applied Ergonomics, 63*, 62–71. https://doi.org/10.1016/j.apergo.2017.03.015

Hebbar, A., Pashilkar, A., & Biswas, P. (2021). Using eye tracker to evaluate cockpit design – a flight simulation study. *ArXiv*. https://arxiv.org/ftp/arxiv/papers/2106/2106.07408.pdf

Jacob, R. J. K., & Karn, K. S. (2003). Commentary on Section 4 – eye tracking in human-computer interaction and usability research: Ready to deliver the promises. In J. Hyönä, R. Radach, & H. Deubel (Eds.), *The mind's eye* (pp. 573–605). Elsevier. https://doi.org/10.1016/B978-044451020-4/50031-1

Jarodzka, H., Balslev, T., Holmqvist, K., Nyström, M., Scheiter, K., Gerjets, P., & Eika, B. (2012). Conveying clinical reasoning based on visual observation via eye-movement modelling examples. *Instructional Science, 40*, 813–827. https://doi.org/10.1007/s11251-012-9218-5

Jarvis S. R. (2017). Concurrent pilot instrument monitoring in the automated multi-crew airline cockpit. *Aerospace Medicine and Human Performance, 88*(12), 1100–1106. https://doi.org/10.3357/AMHP.4882.2017

Just, M. A., & Carpenter, P. A. (1980). A theory of reading: From eye fixations to comprehension. *Psychological Review, 87*(4), 329–354. https://doi.org/10.1037/0033-295X.87.4.329

Kasarskis, P., Stehwien, J., Hickox, J., & Wickens, C. D. (2001). Comparison of expert and novice scan behaviors during VFR flight. In *Proceedings of the 11th international symposium on aviation psychology*. https://www.semanticscholar.org/paper/COMPARISON-OF-EXPERT-AND-

NOVICE-SCAN-BEHAVIORS-VFR-Kasarskis-Stehwien/d8f89a7499e10f8a502bc2f13e9091d1059d263f

Kennedy, K. D., Stephens, C. L., Williams, R. A., & Schutte, P. C. (2017). Repeated induction of inattentional blindness in a simulated aviation environment. *Proceedings of the Human Factors and Ergonomics Society Annual Meeting, 61*(1), 1959–1963. https://doi.org/10.1177/1541931213601969

Knabl-Schmitz, P., Cameron, M., Wilson, K., Mulhall, M., Da Cruz, J., Robinson, A., & Dahlstrom, N. (2023). Eye-tracking: From concept to operational training tool. *Aviation Psychology and Applied Human Factors, 13*(1), 47–57. https://doi.org/10.1027/2192-0923/a000240

Lefrançois, O., Matton, N., & Causse, M. (2021). Improving airline pilots' visual scanning and manual flight performance through training on skilled eye gaze strategies. *Safety, 7*(4), Article 70. https://doi.org/10.3390/safety7040070

Lorenz, B., Biella, M., Teegen, U., Stelling, D., Wenzel, J., Jakobi, J., Ludwig, T., & Korn, B. (2006). Performance, situation awareness, and visual scanning of pilots receiving onboard taxi navigation support during simulated airport surface operation. *Human Factors and Aerospace Safety, 6*(2), 135–154.

Loss of Control Action Group. (2013). *Monitoring matters, guidance on the development of pilot monitoring skills: CAA Paper 2013/02 (*2nd ed.). https://publicapps.caa.co.uk/docs/33/9323-CAA-Monitoring%20Matters%202nd%20Edition%20April%202013.pdf

Lounis, C., Peysakhovich, V., & Causse, M. (2020). Flight eye tracking assistant (FETA): Proof of concept. In N. Stanton (Ed.), *Advances in human factors of transportation* (Vol. 964, pp. 739–751). Springer International Publishing. https://doi.org/10.1007/978-3-030-20503-4_66

Lounis, C., Peysakhovich, V., & Causse, M. (2021). Visual scanning strategies in the cockpit are modulated by pilots' expertise: A flight simulator study. *PLoS one, 16*(2), Article e0247061. https://doi.org/10.1371/journal.pone.0247061

Mack, A., & Rock, I. (1998). Inattentional blindness: Perception without attention. In R. D. Wright (Ed.), *Visual attention* (pp. 55–76). Oxford University Press. https://doi.org/10.1093/oso/9780195126938.003.0003

Mercier, M., Lefrançois, O., Matton, N., & Causse, M. (2022, June 1–2). *Effects of the automation level on gaze behavior: A full flight simulator campaign with professional airline pilots* [Conference presentation]. The International Conference on Cognitive Aircraft Systems (ICCAS), Toulouse, France. https://events.isae-supaero.fr/event/14/contributions/398/attachments/63/101/Effects_of_the_automation_level_on_gaze_behavior_-_a_full_flight_simulator_campaign_with_professional_airline_pilots.pdf https://doi.org/10.5220/0012127500003622

Milton, J., Jones, R., & Fitts, P. (1949). *Eye movements of aircraft pilots: II. Frequency, duration, and sequence of fixation when flying the USAF instrument low approach system* (ILAS) (USAF Technical Report No. 5839). United States Air Force, Wright-Patterson Air Force Base.

Mumaw, R. J., Billman, D., & Feary, M. S. (2019). *Factors that influenced airplane state awareness accidents and incidents. CAST SE-210 Output 2: Report 2 of 6* (Report NASA/TM—20205010985). https://ntrs.nasa.gov/api/citations/20205010985/downloads/TM20205010985.pdf

Mumaw, R. J., Billman, D., & Feary, M. S. (2020). *Analysis of pilot monitoring skills and a review of training effectiveness* (Report NASA/TM-20210000047). https://ntrs.nasa.gov/api/citations/20210000047/downloads/NASA_TM_20210000047.pdf

National Transportation Safety Board. (1994). *A review of flightcrew-involved, major accidents of US air carriers, 1978 through 1990* (Safety Study Report NTSB/SS-94/01). https://www.ntsb.gov/safety/safety-studies/Documents/SS9401.pdf

Oppici, L., Panchuk, D., Serpiello, F. R., & Farrow, D. (2018). The influence of a modified ball on transfer of passing skill in soccer. *Psychology of Sport and Exercise, 39*, 63–71. https://doi.org/10.1016/j.psychsport.2018.07.015

Peißl, S., Wickens, C. D., & Baruah, R. (2018). Eye-tracking measures in aviation: A selective literature review. *The International Journal of Aerospace Psychology, 28*(3–4), 98–112. https://doi.org/10.1080/24721840.2018.1514978

Peysakhovich, V., Lefrançois, O., Dehais, F., & Causse, M. (2018). The neuroergonomics of aircraft cockpits: The four stages of eye-tracking integration to enhance flight safety. *Safety, 4*(1), Article 8. https://doi.org/10.3390/safety4010008

Ramón Alamán, J., Causse, M., & Peysakhovich, V. (2020). Attentional span of aircraft pilots: Did you look at the speed? In V. Peysakhovich, D. Rudi, M. Causse, P. Kiefer, F. Dehais, & M. Raubal (Eds.), *Eye-tracking in aviation. Proceedings of the 1st international workshop* (pp. 108–113). ISAE-SUPAERO, Université de Toulouse, Institute of Cartography and Geoinformation (IKG). https://doi.org/10.3929/ethz-b-000407659

Ratwani, R. M., & Trafton, J. G. (2011). A real-time eye tracking system for predicting and preventing postcompletion errors. *Human-Computer Interaction, 26*(3), 205–245. https://doi.org/10.1080/07370024.2011.601692

Reingold, E. M., & Sheridan, H. (2011). Eye movements and visual expertise in chess and medicine. In S. P. Liversedge, I. Gilchrist, & S. Everling (Eds.), *The Oxford handbook of eye movements* (pp. 523–550). Oxford University Press eBooks. https://doi.org/10.1093/oxfordhb/9780199539789.013.0029

Reynal, M., Colineaux, Y., Vernay, A., & Dehais, F. (2016). Pilot flying vs. pilot monitoring during the approach phase: An eye-tracking study. In *International conference on human-computer interaction in aerospace*. Association for Computing Machinery. https://hal.science/hal-01682792 https://doi.org/10.1145/2950112.2964583

Robinski, M., & Stein, M. (2013). Tracking visual scanning techniques in training simulation for helicopter landing. *Journal of Eye Movement Research, 6*(2), Article 3. https://doi.org/10.16910/jemr.6.2.3

Rosenholtz, R. (2016). Capabilities and limitations of peripheral vision. *Annual Review of Vision Science, 2*, 437–457. https://doi.org/10.1146/annurev-vision-082114-035733

Rudi, D., Kiefer, P., & Raubal, M. (2020). The instructor assistant system (iASSYST) – utilizing eye tracking for commercial aviation training purposes. *Ergonomics, 63*(1), 61–79. https://doi.org/10.1080/00140139.2019.1685132

Sarter, N. B., Mumaw, R. J., & Wickens, C. D. (2007). Pilots' monitoring strategies and performance on automated flight decks: An empirical study combining behavioral and eye-tracking data. *Human Factors, 49*(3), 347–357. https://doi.org/10.1518/001872007X196685

Schriver, A. T., Morrow, D. G., Wickens, C. D., & Talleur, D. A. (2017). Expertise differences in attentional strategies related to pilot decision making. In D. Harris (Ed.), *Decision making in aviation* (pp. 371–386). Routledge. https://doi.org/10.4324/9781315095080-25

Senders, J. W. (1983). *Visual sampling processes* [Doctoral thesis, Tilburg University]. https://pure.uvt.nl/ws/portalfiles/portal/1204640/3955241.pdf

Sharma, C., & Dubey, S. K. (2014). Analysis of eye tracking techniques in usability and HCI perspective. In *2014 international conference on computing for sustainable global development* (pp. 607–612). IEE. https://doi.org/10.1109/IndiaCom.2014.6828034

Shiferaw, B., Downey, L., & Crewther, D. (2019). A review of gaze entropy as a measure of visual scanning efficiency. *Neuroscience & Biobehavioral Reviews, 96*, 353–366. https://doi.org/10.1016/j.neubiorev.2018.12.007

Simons, D. J., & Chabris, C. F. (1999). Gorillas in our midst: Sustained inattentional blindness for dynamic events. *Perception, 28*(9), 1059–1074. https://doi.org/10.1068/p281059

Strasburger, H., Rentschler, I., & Jüttner, M. (2011). Peripheral vision and pattern recognition: A review. *Journal of Vision, 11*(5), Article 13. https://doi.org/10.1167/11.5.13

Sullivan, J., Yang, J. H., Day, M., & Kennedy, Q. (2011). Training simulation for helicopter navigation by characterizing visual scan patterns. *Aviation, Space, and Environmental Medicine, 82*(9), 871–878. https://doi.org/10.3357/asem.2947.2011

Sumwalt, R., Cross, D., & Lessard, D. (2015). Examining how breakdowns in pilot monitoring of the aircraft flight path. *International Journal of Aviation, Aeronautics, and Aerospace, 2*(3), Article 8. https://doi.org/10.15394/ijaaa.2015.1063

Svensson, E., Angelborg-Thanderez, M., Sjöberg, L., & Olsson, S. (1997). Information complexity-mental workload and performance in combat aircraft. *Ergonomics, 40*(3), 362–380. https://doi.org/10.1080/001401397188206

Van Hecke, A. V., & Mundy, P. (2007). Neural systems and the development of gaze following and related joint attention skills. In R. Flom, K. Lee, & D. Muir (Eds.), *Gaze-following: Its development and significance* (pp. 17–51). Lawrence Erlbaum Associates Publishers.

Wang, F. S., Lohmeyer, Q., Duchowski, A., & Meboldt, M. (2023). Gaze is more than just a point: Rethinking visual attention analysis using peripheral vision-based gaze mapping. *Proceedings of the 2023 Symposium on Eye Tracking Research and Applications, 1-7*, Article 66. https://doi.org/10.1145/3588015.3589840

Wang, Y., Liu, Q., Lou, W., Xiong, D., Bai, Y., Du, J., & Guo, X. (2019). Ergonomics evaluation of large screen display in cockpit based on eye-tracking technology. In S. Long & B. S. Dhillon (Eds.), *Man-machine-environment system engineering* (pp. 347–356). Springer Singapore. https://doi.org/10.1007/978-981-13-2481-9_40

Wetzel, P. A., Anderson, G. M., & Barelka, B. A. (1998). Instructor use of eye position based feedback for pilot training. *Proceedings of the Human Factors and Ergonomics Society Annual Meeting, 42*(20), 1388–1392. https://doi.org/10.1177/154193129804202005

Wickens, C. D., Goh, J., Helleberg, J., Horrey, W. J., & Talleur, D. A. (2003). Attentional models of multitask pilot performance using advanced display technology. *Human Factors, 45*(3), 360–380. https://doi.org/10.1518/hfes.45.3.360.27250

Wickens, C. (2021). Attention: Theory, principles, models and applications. *International Journal of Human–Computer Interaction, 37*(5), 403–417. https://doi.org/10.1080/10447318.2021.1874741

Yang, J. H., Kennedy, Q., Sullivan, J., & Fricker, R. D. (2013). Pilot performance: Assessing how scan patterns & navigational assessments vary by flight expertise. *Aviation, Space, and Environmental Medicine, 84*(2), 116–124. https://doi.org/10.3357/asem.3372.2013

Yu, C.-S., Wang, E. M., Li, W.-C., Braithwaite, G., & Greaves, M. (2016). Pilots' visual scan patterns and attention distribution during the pursuit of a dynamic target. *Aerospace Medicine and Human Performance, 87*(1), 40–47. https://doi.org/10.3357/AMHP.4209.2016

Zaal, P., Lombaerts, T., Mumaw, R., Billman, D., Torron, I., Jamal, S., Shyr, M. C., & Feary, M. (2021). Eye-tracking analysis from a flight-director-use and pilot-monitoring study. *In AIAA Aviation 2021 Forum*. American Institute of Aeronautics and Astronautics. https://doi.org/10.2514/6.2021-2995

Ziv, G. (2016). Gaze behavior and visual attention: A review of eye tracking studies in aviation. *The International Journal of Aviation Psychology, 26*(3–4), 75–104. https://doi.org/10.1080/10508414.2017.1313096

Chapter 8

Safety Performance Indicators – Enabling a Data-Driven Approach to Fatigue Risk Management Systems

Matthew J. W. Thomas

Abstract

Fatigue risk management systems (FRMS) provide an important framework for identifying and mitigating fatigue-related risk in aviation operations worldwide. FRMSs are often described in terms of being "data-driven" frameworks for managing fatigue-related risk. As an enabling mechanism to assist in data collection and facilitate monitoring, safety performance indicators (SPIs) are used within an FRMS. An SPI sets out acceptable performance criteria to indicate adequate mitigation of fatigue-related risk, and typically draws on metrics relating to factors such as: rostering and scheduling; sleep/wake patterns; subjective fatigue and neuro-behavioural performance; operational performance; occurrences of fatigue; and the contribution of fatigue to operational incidents. This chapter explores, in practical terms, the sources of data that can be used to develop a set of new SPIs built into an FRMS, and provides an overview demonstrating how to construct a set of SPIs to enable the effective identification and monitoring of fatigue-related risk. This chapter also discusses new approaches in the way in which SPIs can be used to inform change management and demonstrate acceptable levels of safety in new types of operations.

Keywords

fatigue, safety, risk management, safety performance indicators

Introduction to Fatigue Risk Management Systems

Fatigue is a ubiquitous human factors issue that presents acute safety-related risk across all facets of aviation operations. Implicated in accidents and incidents in all transport modes worldwide, fatigue poses an insidious problem due to its wide range of causal factors, including both work and non-work-related origins, rendering fatigue a complex human factors issue to manage. Fatigue is defined in terms of *performance impairment*, specifically decreased mental and physical capability that can impact an operator's alertness and performance of safety critical tasks. Fatigue can result from extended wakefulness, insufficient sleep, circadian phase, or workload and task-related factors (Akerstedt, 2000; International Air Transport Association [IATA] et al., 2015). The impacts of fatigue on performance are wide-ranging, from basic neuro-behavioural impairment, such as slowed reaction time, with flow-on negative impacts on higher-order cognitive processes such as reasoning and decision-making, as well as psychosocial impacts on team functioning through changes in communication and mood (Thomas et al., 2023).

For many decades, the risks associated with fatigue were mitigated primarily through hours of work rules, designed to prevent the occurrence of fatigue by restricting hours of work and mandating minimum periods free of duty (Gander et al., 2011). For instance, for commercial flight crew, regulatory bodies worldwide have established rules relating to flight and duty time limitations that set out maximum lengths for duty periods, minimum lengths for rest periods, and other restrictions relating to cumulative work and rest. However, given the highly individual sources of fatigue, in combination with the limitations of simple rule-sets, the traditional prescriptive (rule-based) approach has been shown to have significant short-comings with respect to effectively mitigating fatigue-related risk (Sprajcer et al., 2023). Within the context of prescriptive rule sets, it is entirely feasible that individuals may experience fatigue due to factors such as restricted sleep or extended wake, even when working with schedules that are permissible under the prescriptive rule sets. Similarly, patterns of work that would not be permissible under these rule sets might be entirely safe, if operators are able to obtain sufficient rest (Dawson & McCulloch, 2005).

In response to these limitations of a purely prescriptive rule-based approach to managing fatigue, the international aviation community has developed a more sophisticated risk-based approach to identifying fatigue-related hazards and mitigating fatigue-related risk. Drawing on our learnings from safety management systems (SMS), a similar approach to managing fatigue has been developed in the form of fatigue risk management systems (FRMS). FRMS takes the structures of the SMS, and the processes of risk management embedded in international standards such as ISO 31000 (International Organisation for Standardization [ISO], 2018,) and applies them to the risks associated specifically with fatigue. The International Civil Aviation Organisation (ICAO, 2020 p. xvi) defines an FRMS as: "A data-driven means of continuously monitoring and managing fatigue-related safety risks, based upon scientific principles and knowledge as well as operational experience that aims to ensure relevant personnel are performing at adequate levels of alertness."

In accordance with ICAO (2020), an FRMS comprises several key elements, including a clear *policy framework*, organisational *governance structures* that establish core roles and responsibilities, processes for ongoing *safety assurance* and

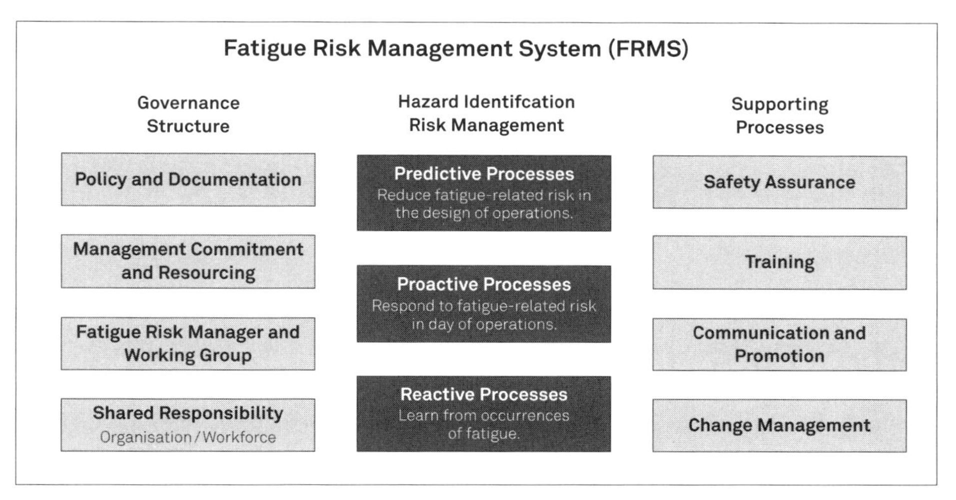

Figure 8.1. The core elements of an FRMS.

change management as well as *training* and *communication* processes. However, absolutely central to the FRMS are processes for *hazard identification* and *risk management*. Each of the core components of an FRMS are outlined in Figure 8.1.

Within an FRMS, the core hazard identification and risk management processes are often described in three layers, described in terms of (1) predictive, (2) proactive, and (3) reactive hazard identification and risk management processes (IATA et al., 2015; Sprajcer et al., 2022).

- Predictive Processes
  First, *predictive* hazard identification and risk management processes set about to design systems of work to minimise the inherent risks associated with fatigue. In the aviation context, these primarily involve adequate workforce resourcing and the design of working time arrangements to ensure the likelihood of fatigue is as low as reasonably practicable (IATA et al., 2015).
- Proactive Processes
  Next, *proactive* hazard identification and risk management processes set out to manage the risks of fatigue as they might manifest during day-of-operations. In the aviation context these processes revolve around fitness for duty and ensure operators are sufficiently alert to maintain the safety of operations (IATA et al., 2015).
- Reactive Processes
  Finally, *reactive* hazard identification and risk management processes set out to respond to instances where fatigue may have occurred and potentially resulted in reduced margins of safety. In the aviation context these processes include fatigue occurrence reporting, safety event reporting, and incident and accident investigation (IATA et al., 2015).

Compared to the traditional *prescriptive* rule-based approach to managing fatigue, an FRMS is a *performance-based* approach to managing fatigue-related risk. As such, continuous measurement and monitoring of fatigue-related hazards and of the effectiveness of FRMA employed by an organisation are essential. To this end, the development of sources of data and metrics to evaluate performance within an FRMS is critical to its success. Within the existing structures of safety management, these metrics are called "safety performance indicators" (SPIs).

Safety Performance Indicators

SPIs are an integral part of a SMS, and as such they have in turn become core elements of an FRMS due to the similar frameworks seen in both SMS and FRMS. SPIs utilise data from a range of sources to facilitate monitoring of fatigue-related hazards and the effectiveness of fatigue-related risk mitigation within the FRMS.

In general terms, an SPI is defined as "a data-based safety parameter used for monitoring and assessing safety performance" (ICAO, 2018, p. vii). Within the context of an FRMS, an SPI sets out acceptable performance criteria to indicate adequate mitigation of fatigue-related risk, which are usually aligned with the three levels of hazard identification and risk management.

A variety of SPIs can be defined within each of the levels of hazard identification and risk mitigation within an FRMS, in turn representing inherent levels of fatigue in the operations, exposure to fatigue during day of operations, and any consequences of fatigue in terms of operational safety. In accordance with IATA (IATA et al., 2015, p. 77) within the FRMS, SPIs typically draw on metrics relating to factors such as:

- FRMS predictive processes
  - previous experience
  - evidence-based rostering and scheduling
  - bio-mathematical modelling of planned work
- FRMS proactive processes
  - fitness-for-duty tools
  - surveys relating to sleep/wake patterns
  - subjective fatigue and neuro-behavioural performance
- FRMS reactive processes
  - operational performance
  - occurrences of fatigue
  - the contribution of fatigue to operational incidents

SPIs specify acceptable performance within the FRMS and enable continuous monitoring and evaluation of the effectiveness of risk mitigation.

Anatomy of a Safety Performance Indicator Within an FRMS

An effective SPI within an FRMS relies on (1) a reliable and valid source of data, (2) a target for demonstrating acceptable performance, (3) an alert threshold that indicates elevated risk, and (4) a process for ongoing monitoring and review. These elements are illustrated, with an example SPI relating to subjective levels of fatigue experienced during normal operations by flight crew (a proactive SPI), in Figure 8.2.

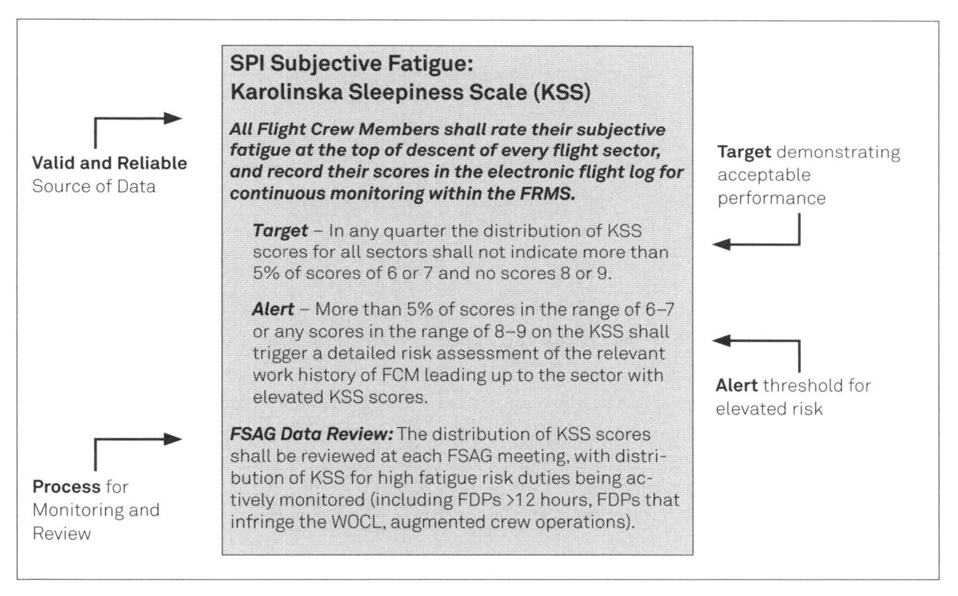

Figure 8.2. Anatomy of a safety performance indicator (SPI). FDPs = flight duty periods; FSAG = Fatigue Safety Action Group; WOCL = window of circadian low.

Reliable and Valid Sources of Data

First, to be effective, every SPI depends on having a reliable and valid source of data to ensure that decisions relating to fatigue-related risk can be made with confidence by an organisation. Validity refers to whether the data reflects what is actually trying to be measured, and reliability refers to factors of measurement such as consistency and accuracy. As an example, if a bio-mathematical model is to be used to evaluate crew rosters, it must have been validated for use in the operational context in which it is being deployed, such as trans-meridian operations. Similarly, if a measure of subjective fatigue is to be used, the measure must have been scientifically evaluated to demonstrate it is fit for purpose.

Of the SPIs that collect data from day-of-operations monitoring, it has been suggested that metrics relating to actual sleep obtained by individuals are arguably the most valuable, given the direct relationship between sleep and waking function (Gander et al., 2014). For instance, it has been demonstrated that when flight crew have obtained less than 5 hr sleep in the 24 hr prior to a flight, on average they commit nearly twice as many errors as crews that are well rested (Dawson et al., 2021; Thomas & Ferguson, 2010). However, collecting actual sleep data from flight crew using objective measures (such as wrist actigraphy) is to some degree an onerous and obtrusive form of data collection, and not without financial burden in terms of the equipment used, data analysis, and reporting. To this end, self-report measures of sleep and subjective measures of alertness are often used as an effective substitute (Signal et al., 2005).

Targets and Alerts – Thresholds and Action

Typically, an SPI establishes an acceptable level of performance, referred to as a *target*, and a trigger point for action, often referred to as an *alert*, that indicate the need for additional forms of risk management. Both targets and alerts need to specify threshold metrics at which point fatigue-related risk is deemed to be acceptable, or the point at which they are elevated to a degree that indicates the need for further investigation and potential changes made to the forms of risk mitigation that have already been put in place. As an example, an SPI relating to subjective fatigue experienced by flight crew must set a threshold value (for instance a score of 6 or above on the Karolinska Sleepiness Scale [KSS]) and a frequency of occurrence (for instance 5% of flight duty periods [FDPs]) for the SPI. Often, if performance on the SPI is below this threshold value, fatigue-related risk is deemed acceptable, but if the SPI is raised above this threshold, it serves as an alert for further investigation and action.

Process for Monitoring and Review

Every SPI should also be associated with a clear process for ongoing monitoring and review, such that indications of elevated risk can be promptly identified and subjected to review. Within an FRMS, SPIs are often subjected to continuous monitoring in the form of a "dashboard" for the FRMS manager to monitor on an ongoing basis, and/or subjected to periodic regular review by the FRMS working group. Indeed, the agenda of the FRMS working group meetings are often structured around a review of each SPI, in order to ensure the FRMS remains a "data-driven" process and not influenced unduly by industrial or other priorities.

Safety Performance Indicators in Operation

More Than Ticking Boxes

SPIs ensure that an FRMS is more than a set of policies and procedures, and they actually form the "engine room" of the data-driven FRMS. Although elements of FRMS have been criticised for an overly burdensome and bureaucratic imposition on organisations in the aviation industry (Bourgeois-Bougrine, 2020), a well-designed set of SPIs provides the evidence base for informed decision-making and also ensures that the critical elements of the FRMS are operating in practice.

Continuous Monitoring

Well-crafted sets of SPIs enable continuous monitoring of FRMS performance. To achieve this, an SPI requires a source of data that facilitates continuous monitoring and regular review. In today's world of data analytics and business information systems, organisations frequently develop a data dashboard for the continuous monitoring of SPIs within an FRMS, which offers the FRMS manager and working group enhanced visibility with respect to the overall operation and effectiveness of the FRMS.

Triangulation

It has been suggested that no single SPI should be relied upon to make a determination of fatigue-related risk. Within an FRMS, multiple SPIs should be utilised and compared. In this context, when one SPI reaches a threshold, performance across other SPIs should be reviewed in the evaluation of fatigue (Gander et al., 2014).

Driving Improvement and Enhanced Risk Mitigation

Perhaps most importantly, the set of SPIs within an FRMS enable the identification of areas where the FRMS is not operating as designed or where levels of fatigue-related risk are not being mitigated to the desired degree. To this end, SPIs form an important safety assurance function and help pinpoint any areas where enhancements need to be made to the processes within the FRMS to better manage fatigue-related risk.

Safety Change and Equivalence Testing

Within the FRMS, SPIs also provide a critical source of data to support the process of managing change and of demonstrating acceptable levels of safety are maintained when significant changes to operations are made. Often, measuring the effects of changes to scheduling or introducing new types of operations is a critical component of the change management process. Within the context of an FRMS, aircraft with longer-range capabilities, new routes, or changes to flight crew rostering all introduce changes that may be associated with increases in fatigue-related risk. The data collected from SPIs can be effectively utilised to measure any impacts of these changes and to demonstrate an equivalent level of safety is being achieved.

Traditionally, the approach to statistical analysis of *difference testing* has been the predominant approach used. Difference testing sets out to establish whether the means or distributions of a particular measure are different between two or more conditions (Wu et al., 2018). For instance within the context of an FRMS, difference testing would establish whether flight crew are obtaining on average more or less sleep prior to a duty before and after a change in rostering. However, this approach is influenced to a large degree by sample size, with larger samples able to detect statistically significant differences at small levels of actual difference. Accordingly, using this approach might determine that crew obtain on average significantly less sleep on a new roster, without that difference being practically significant in terms of elevated fatigue-related risk.

More recently, within the context of an FRMS, the analytic approach of *equivalence* and *non-inferiority* testing has been adapted for use as a tool in FRMS to determine whether a change is able to achieve the same levels of safety performance (Lamp et al., 2019). Equivalence and non-inferiority testing adopt a different approach to statistical analysis, and set out to determine whether a change is associated with equivalent or non-inferior performance on a specific metric. This approach has been commonly used in the pharmaceutical industry to determine whether a new therapeutic agent is associated with the same or better level of treatment outcome than an existing regimen.

Using this approach, equivalence on a given metric is established if the mean and 95% confidence intervals fall within $\pm$ a threshold value (δ). The value of δ is determined to represent a difference from baseline that is *practically* significant. In the context of fatigue-risk management, δ values for metrics such as sleep or wake are determined with reference to scientific evidence that might suggest a change in performance (such as reaction time or error rates) that is likely to significantly increase fatigue-related risk. This approach is illustrated in Figure 8.3.

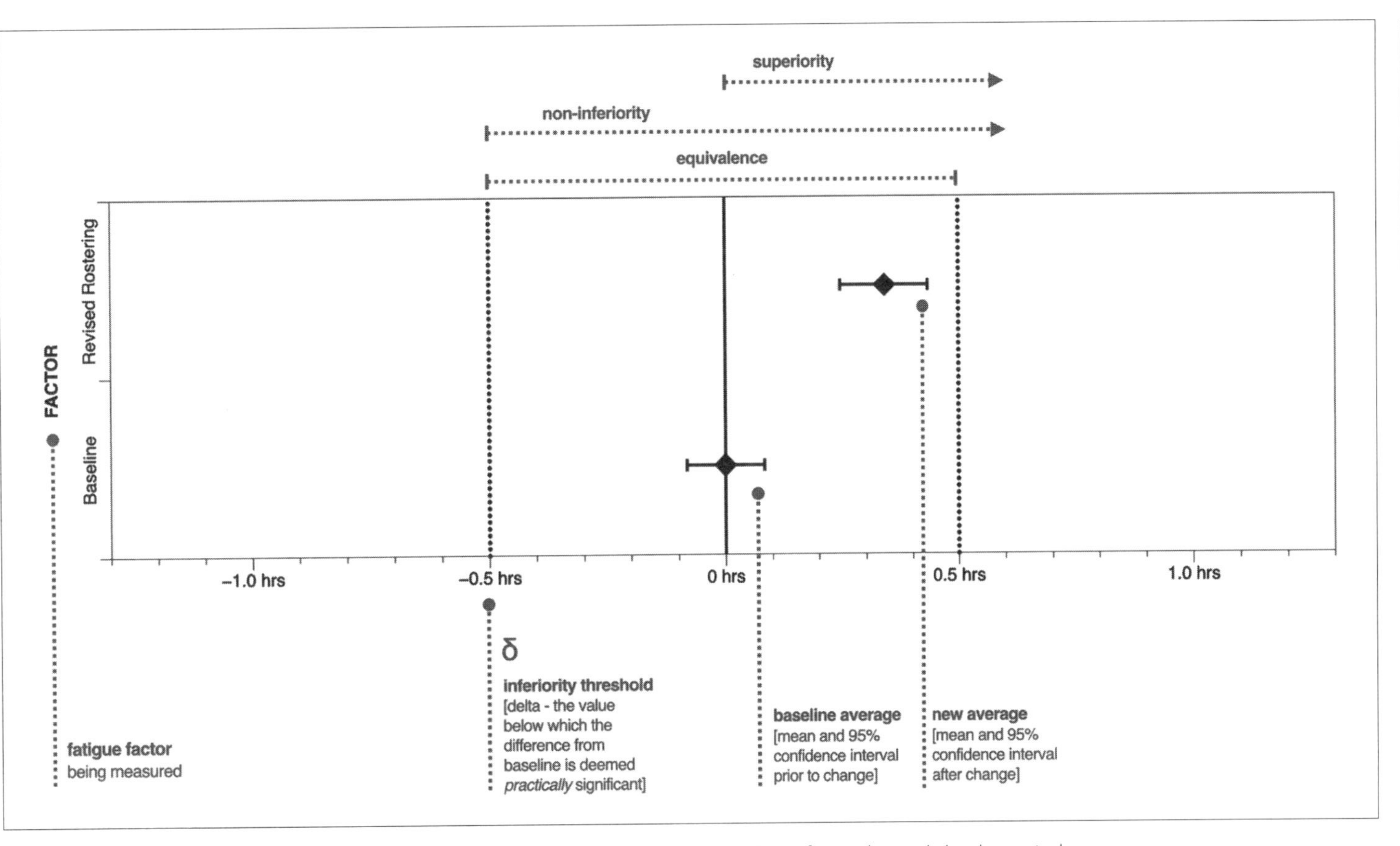

Figure 8.3. Annotated non-inferiority chart, showing how measures differ from baseline after a change is implemented.

Summary and Outlook

As FRMSs evolve, there will be a need to explore new and innovative sources of data for SPIs, and new ways to automate the monitoring of performance against these SPIs. New technologies provide significant opportunity to develop enhanced SPIs, especially with novel wearable technologies and automated data collection with respect to sleep and performance of individual crew members. However, these forms of data collection give rise to new issues relating to privacy and ethical considerations that must be taken into account (Dawson et al., 2021). Similarly, real-time operator monitoring through the use of fatigue-detection technologies has become commonplace in other industries, such as road transport. These technologies have potential to enhance SPIs within the aviation industry as well as in other domains (Dawson et al., 2014). A modern FRMS is defined in terms of being a data-driven approach to managing risks associated with fatigue in aviation operations. To enable this, SPIs are used within an FRMS to assist in data collection and facilitate monitoring of performance. Within the context of an FRMS, an SPI sets out acceptable performance criteria to indicate adequate mitigation of fatigue-related risk and the effectiveness of the FRMS overall.

References

Akerstedt, T. (2000). Consensus statement: Fatigue and accidents in transport operations. *Journal of Sleep Research, 9*(4), 395–395. https://doi.org/10.1046/j.1365-2869.2000.00228.x

Bourgeois-Bougrine, S. (2020). The illusion of aircrews' fatigue risk control. *Transportation Research Interdisciplinary Perspectives, 4*, Article 100104. https://doi.org/10.1016/j.trip.2020.100104

Dawson, D., & McCulloch, K. (2005). Managing fatigue: It's about sleep. *Sleep Medicine Reviews, 9*, 365–380. https://doi.org/10.1016/j.smrv.2005.03.002

Dawson, D., Searle, A. K., & Paterson, J. L. (2014). Look before you (s)leep: Evaluating the use of fatigue detection technologies within a fatigue risk management system for the road transport industry. *Sleep Medicine Reviews, 18*(2), 141–152. https://doi.org/10.1016/j.smrv.2013.03.003

Dawson, D., Sprajcer, M., & Thomas, M. (2021). How much sleep do you need? A comprehensive review of fatigue related impairment and the capacity to work or drive safely. *Accident Analysis & Prevention, 151*, Article 105955. https://doi.org/10.1016/j.aap.2020.105955

Gander, P. H., Hartley, L., Powell, D., Cabon, P., Hitchcock, E., Mills, A., & Popkin, S. (2011). Fatigue risk management: Organizational factors at the regulatory and industry/company level. *Accident Analysis & Prevention, 43*, 573–590. https://doi.org/10.1016/j.aap.2009.11.007

Gander, P. H., Mangie, J., van den Berg, M. J., Smith, A. A. T., Mulrine, H. M., & Signal, T. L. (2014). Crew fatigue safety performance indicators for fatigue risk management systems. *Aviation, Space, and Environmental Medicine, 85*(2), 139–147. https://doi.org/10.3357/asem.3748.2014

International Air Transport Association, International Civil Aviation Organisation, & International Federation of Air Line Pilots' Associations. (2015). *Fatigue management guide for airline operators* (2nd ed.). https://www.iata.org/en/publications/fatigue-management-guide/

International Civil Aviation Organisation ICAO. (2018). *Safety management manual (SMM)* (4th ed.). https://www.icao.int/safety/safetymanagement/pages/guidancematerial.aspx

International Civil Aviation Organisation ICAO. (2020). *Manual for the oversight of fatigue management approaches* (2nd ed.). https://www.icao.int/safety/fatiguemanagement/FRMS%20 Tools/Doc%209966.FRMS.2016%20Edition.en.pdf

International Organisation for Standardization (ISO). (2018). *Risk management: ISO 31000*. https://www.iso.org/files/live/sites/isoorg/files/store/en/PUB100426.pdf

Lamp, A., Chen, J. M., McCullough, D., & Belenky, G. (2019). Equal to or better than: The application of statistical non-inferiority to fatigue risk management. *Accident Analysis & Prevention, 126*, 184–190. https://doi.org/10.1016/j.aap.2018.01.020

Signal, T. L., Gale, J., & Gander, P. H. (2005). Sleep measurement in flight crew: Comparing actigraphic and subjective estimates to polysomnography. *Aviation, Space, and Environmental Medicine, 76*(11), 1058–1063.

Sprajcer, M., Thomas, M. J., & Dawson, D. (2023). Approaches to fatigue management: Where we are and where we're going. In C. M. Rudin-Brown & A. J. Filtness (Eds.), *The handbook of fatigue management in transportation* (pp. 259–271). CRC Press. https://doi.org/10.1201/9781003213154-23

Sprajcer, M., Thomas, M. J. W., Sargent, C., Crowther, M. E., Boivin, D. B., Wong, I. S., Smiley, A., & Dawson, D. (2022). How effective are fatigue risk management systems (FRMS)? A review. *Accident Analysis & Prevention, 165*, Article 106398. https://doi.org/10.1016/j.aap.2021.106398

Thomas, M. J., Sprajcer, M., & Dawson, D. (2023). The effects of fatigue on performance in transportation operations. In C. M. Rudin-Brown & A. J. Filtness (Eds.), *The handbook of fatigue management in transportation* (pp. 53–64). CRC Press. https://doi.org/10.1201/9781003213154-6

Thomas, M. J. W., & Ferguson, S. A. (2010). Prior sleep, prior wake, and crew performance during normal flight operations. *Aviation, Space and Environmental Medicine, 81*, 665–670. https://doi.org/https//doi.org/10.3357/asem.2711.2010

Wu, L. J., Gander, P. H., van den Berg, M., & Signal, T. L. (2018). Equivalence testing as a tool for fatigue risk management in aviation. *Aerospace Medicine and Human Performance, 89*(4), 383–388. https://doi.org/10.3357/AMHP.4790.2018

Chapter 9

Hazards of Human Space Exploration

Research Methods

Cheryl Agyei, Anna Fogtman, Adrianos Golemis, Tobias Weber, and Sylwia Kaduk

Abstract

Human space exploration is associated with five main hazards: radiation, isolation and confinement, distance from Earth, altered gravitational fields, and enclosed and hostile environments. This chapter describes research methods used to study some of these hazards both on Earth and in space.

Keywords

microgravity, spaceflight, radiation, isolation, psychology, human health

Introduction

Humans always strove to explore despite the risks. Today, with the Earth well mapped, exploratory instincts lead humans to space, which comes with many new risks, such as microgravity, and enhances risks experienced also during terrestrial expeditions, such as isolation. However, through careful risk management, people can live and work in space for prolonged periods with limited health consequences (Vernikos et al., 2016). The National Aeronautics and Space Administration (NASA) has identified five primary hazards of human space exploration: radiation, isolation and confinement, distance from Earth, hostile and closed environments, and altered gravity (Lewis, 2022). These hazards are interdependent and can have detrimental effects on the health and well-being of astronauts (Childress et al., 2023).

To investigate these hazards and their consequences, researchers have developed methods to simulate and study them on Earth and in space, mainly on the International Space Station (ISS). On Earth, scientists use space analogues such as isolated research stations in extreme environments, bed rest or dry immersion studies, and underwater habitats to simulate the effects of long-duration spaceflight. These analogues allow for the study of some constraints of the space environment separately from others, and of larger groups of people and conditions that are more easily controlled than the environment of space stations (Posselt et al., 2021). Concurrently, many observations are conducted on the ISS, giving access to a true space environment.

This chapter provides an overview of the research methods used in widely understood human factors studies related to human space exploration. Specifically, it describes the methods used to study the hazards of radiation, isolation and confinement, distance from Earth, and altered gravity, both on Earth and the ISS. It also discusses the strengths and limitations of these methods and their implications for future space exploration. The hazard of a hostile environment is not described in this chapter, as it is a substantial category already incorporating the majority of risks mentioned in other subsections such as Radiation and Isolation and Confinement, for example isolation, reliance on support systems or hypogravity. It contains unique factors, such as microbial contamination or excessive noise, but the methods to study them are beyond the scope of this chapter.

Hazard 1: Radiation

Ionising radiation (IR) is one of the key risks of spaceflight and can be a single limiting factor for human space exploration (Fogtman et al., 2023). This is due to the gaps in our understanding of how complex space radiation can affect human health and performance in the short and long term. The major obstacles in addressing these knowledge gaps derive from the scarcity of (a) human data from spaceflight and (b) challenges in the replicability of the space radiation environment on Earth. From an operational viewpoint, understanding the health effects of spaceflight must be accompanied by the ability to predict the health risks of a particular mission as well as throughout the entire lifetime of an astronaut (Figure 9.1).

Such a risk model necessitates solid scientific evidence coming from large cohorts of exposed populations. In the case of space radiation, risk models are based on epidemiological evidence from the Life Span Study (LSS; Hafner et al., 2021), which studies the life-long health effects of about 120,000 people exposed to IR during atomic bombings (A-bomb) in Hiroshima and Nagasaki. However, there are serious limitations to the clinical conclusions one can draw from these data on the health effects of exposure to space radiation due to the differences in ionising spectra between space and A-bomb radiation, which lead to different clinical outcomes in different timeframes. Another challenge particularly connected to the LSS is the high risk of bias of this study, as there is limited availability of crucial metadata on other risk and confounding factors, such as other toxic exposures (e.g., smoking). Therefore, limitations of Earth-bound epidemiology are complemented by meticulous observations and analysis of current space crews' health and performance.

This is done through sophisticated space dosimetry assessing crew-personal doses in real time, with the resolution of measuring different components of IR in space, including high energetic particles (HZE) and neutrons. Physically measured radiation doses are then confronted with biological effects measured in the human body in response to the spaceflight environment, as compared to pre-flight baselines. These include measurements of the oxidative stress and inflammation markers, as well as deoxyribonucleic acid (DNA) damage and repair pathways. The gold standard method is to test these parameters in blood; however, with the view of using these tests as biomarkers of IR exposure for operational purposes, finding non-invasive ways of testing, such as in urine or saliva, is consequently of utmost demand. As IR leads to DNA damage through indirect (via radiolysis) or direct actions, testing for chromosomal aberrations in the blood of exposed humans has been set as a standard. Yet the use of this assessment has not been applied, since it showed little benefit for low earth orbit (LEO) operations.

There is also an urgent need for studying pathways that lead to other-than-cancer effects of space radiation in astronauts, with a particular focus on effects that

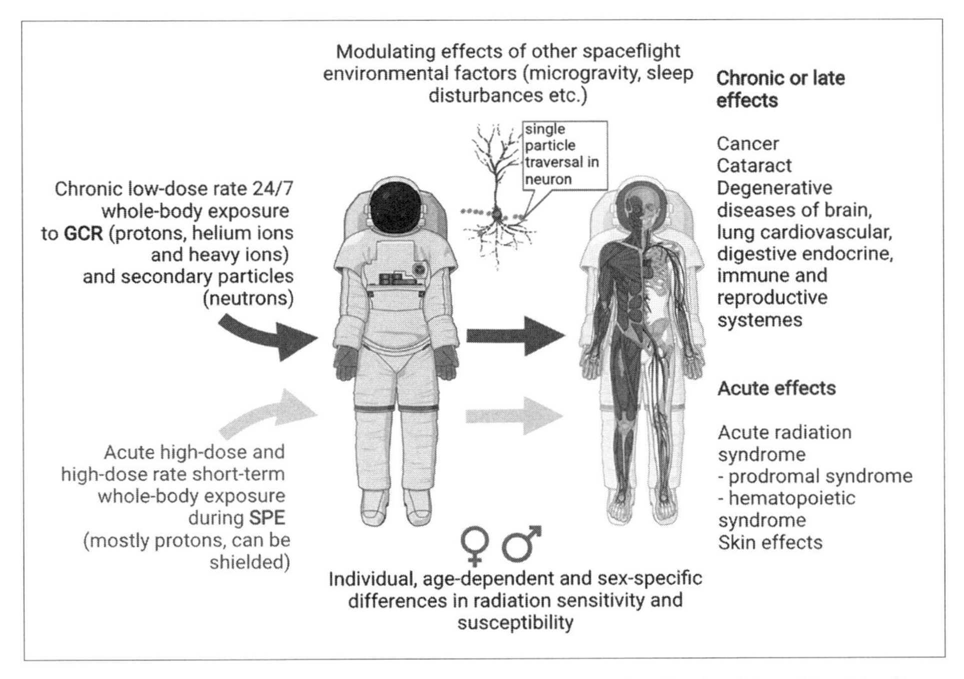

Figure 9.1. Radiation exposure during space missions beyond low Earth orbit and health effects of space radiation. From "Towards Sustainable Human Space Exploration – Priorities for Radiation Research to Quantify and Mitigate Radiation Risks" by A. Fogtman, S. Baatout, B. Baselet, T. Berger, C. E. Hellweg, P. Jiggens, C. La Tessa, L. Narici, P. Nieminen, L. Sabatier, G. Santin, U. Schneider, U. Straube, K. Tabury, W. Tinganelli, L. Walsh, and M. Durante, 2023, *npj Microgravity*, *9*, Article 8, Figure 2 on p. 3. © The Author(s). CC BY license (http://creativecommons.org/licenses/by/4.0/)

may be operationally relevant and stemming from individual predispositions, as well as mechanistic ground for those effects. Here, large-throughput technologies such as OMICS open a plethora of possibilities for such studies, assessing not only expression profiles of individual responses to IR, but also their epigenomic environments and metabolomic and proteomic phenotypes. Studying those cellular phenomena in astronauts before, during and after their space missions, as well as throughout their entire careers, is the core of the operational health risk assessment for human space exploration. However, as roughly 600 people have reached outer space to date, even with the assumption that since the beginning of spaceflight a set of standard measures was used on every individual to assess the health effects of IR (which is not the case), these scientific data would not be sufficient to draw clinical conclusions for the entire population of current and future space crews. Therefore, the lack of human cohorts exposed to IR in space is complemented by molecular epidemiology and animal research in earth analogues of space radiation. These facilities reproduce the complex composition of radiation in space and currently constitute the best proxy on Earth for the space radiation environment, also because of the accessibility of such facilities from both technical and financial sides. An example of such a facility is the GSI (Helmholtz Centre for Heavy Ion Research) accelerator in Darmstadt, Germany (Durante, 2014), which provides access to heavy ion beams with energies that represent the heavy ion component of the space radiation spectra. The GCR simulator concept exploits the possibility of fast switching between different energies of a single heavy ion species (^{56}Fe) interacting with complex material beam modulators that reproduce the spectrum of secondary irradiations behind the spacecraft's shielding (Schuy et al., 2020). In such conditions cell cultures or small animals (rodents) are irradiated, and the space radiation effects are studied at molecular, tissue and organ levels. The European Space Agency collaborates with the GSI facility as well as with other irradiation facilities in Europe within the Investigating the Biological and Physical Effects of Radiation (IBPER) programme's Open Space Innovation Platform (OSIP) channel. The purpose of the program is to grant beam time for space research dedicated to investigating the biological and physical effects of space radiation, with a focus on individual responses and differences between females and males. Such research will not only benefit the safety of human space exploration but can also provide insight into the lifelong health effects of patients undergoing particle therapies as well as of other groups exposed occupationally to ionising radiation, such as medical professionals and nuclear workers.

Hazard 2: Isolation and Confinement

While psychological studies have been conducted on different space platforms, most of the research on isolation and confinement comes from ground-based

Figure 9.2. Concordia Polar Research Station. © ESA/IPEV/ENEA/Adrianos Golemis. Reprinted with permission.

analogues varying in the level of fidelity and available space mission scenarios reflected. For example, the Concordia Station (see Figure 9.2; Healey, 2020) or the desert space analogues organised by the Austrian Space Forum (Gruber et al., 2020) provide a relatively high-fidelity environment with harsh conditions, remoteness, and long-duration stays. On the other hand, projects like the Human Exploration Research Analog (HERA; Neigut, 2015) or the Mars500/SIRIUS studies facility in Moscow offer opportunities to study crew confinement, albeit without the remote environment. The use of analogue environments is primarily focused on simulating the psychological and social conditions that individuals may experience in space, rather than on replicating the exact physical setting.

Researchers have employed various methodologies to study the psychosocial impact of isolation and confinement:

- Questionnaires are frequently used as self-report measures to understand the experiences of individuals living in such environments (McMenamin et al., 2020). Such questionnaires assess the crew's mood and other psychological variables, and attempt to track their trends during the different phases of isolation. They include the Evaluation of Personal Parameters questionnaire, the Profile of Mood States (POMS) and the Positive and Negative Affect Schedule (PANAS). Mood and other psychological variables may be evaluated in relation to environmental constraints or factors reported by questionnaires such as the Lake Louise Acute Mountain Sickness (AMS) questionnaire or the Karolinska Sleepiness Scale (KSS).

- Prospective and longitudinal designs (Musson & Helmreich, 2005) are commonly used to examine adaptation and the responses of individuals over time. For instance, studies conducted in Concordia (Nicolas et al., 2016) and Mars500 (Ushakov et al., 2014) have employed psychological questionnaires.
- Cognitive testing has also garnered considerable interest in understanding the impact of isolation and confinement. Cognition test batteries, such as the Spaceflight Cognitive Assessment Tool for Windows (WinSCAT), can be utilised in isolation and confinement to assess cognitive functions, as is being done on the ISS (Strangman et al., 2014).
- In recent years, there has been a growing interest in complementing psychological data with information from biomarkers. Biomarkers serve as signals that help understand experiences related to stress, resilience, and team dynamics. Examples of biomarkers are blood and saliva samples for cortisol level (Strewe et al., 2019), pupillometry or heart rate variability (HRV) monitored using wearable devices for sleep and sleepiness (Groemer et al., 2010), stress or workload.
- Furthermore, the linguistic analysis of communication has proven useful in understanding the dynamics of interactions (Alcibiade et al., 2020).
- Observational analyses have been conducted to further understand isolation and confinement. Researchers code data captured from video and audio recordings, enabling a comprehensive examination of behaviours and interactions (Tafforin, 2015; Tafforin et al., 2021).
- Qualitative methods, primarily interviews, have been employed in numerous studies to capture rich detail about individuals' experiences in isolation and confinement. Typically, such interviews are conducted face to face, between the researcher, usually a psychologist, and each member of the crew, separately. They are usually held at the end of the mission (isolation period), but may be compared to an interview conducted before the start of isolation. The latter may focus on the motivation and expectations of the crewmember, while the interview at the end of the mission discusses the actual experience, differences from original expectations, reflections on the self and the rest of the crew, as well as lessons learnt. Less frequently the researcher may hold one interview or more *during* the isolation period as well. These methods may be used to gain insights into the psychological and emotional journey of each crewmember. Examples from polar expedition research are provided by Leon (2005) and by Van Puyvelde et al. (2022).

One of the challenges faced in studying isolation and confinement is the limited sample size. However, advancements in data analysis techniques have enabled researchers to generate large quantities of nested data, such as time series data for individuals. This allows for the application of complex models to examine the collected data, for example with the application of mixed models on the nested data (Zuur et al., 2009). Data collected in ground-based analogues are complemented by studies conducted on ISS. Many of them employ research methods to investigate the issues surrounding sleep loss, circadian disruption, workload, and performance management as well as behavioural changes. For example, one longitudinal study (Jones et al., 2022) has investigated how sleep deficiency during the mission is linked with elevated stress and deficiencies in neuro-behavioural performance, by employing the use of the computerised Reaction Self-Test (RST), with data collection occurring pre-, during and post-flight. A Psychomotor Vigilance Test–Brief (PVT-B) was used to assess behavioural alertness via reaction time, a sensitive neuro-behavioural metric of sleep loss; finally, a visual analogue scale (VAS) was used by astronauts to rate their psychological state.

A further investigation utilised psychophysiological monitoring techniques such as actigraphy along with photometry to assess how circadian misalignment affects sleep and medication use, with data collected before and during missions. Participants used wrist-worn actiwatches during data collection periods, providing data on sleep/wake activity and exposures to light intensity. Astronauts also completed a sleep log, including a VAS, providing subjective ratings of sleep quality (Flynn-Evans et al., 2016).

Additionally, to analyse the psychological effects of isolation-induced stress in spaceflight, several studies have utilised behavioural assessments. These evaluations assess the crew members' mood, interpersonal dynamics and cognitive function, being delivered in the form of standardised questionnaires, mood scales and measures of performance.

Hazard 3: Distance From Earth

Distance from Earth is a multidimensional hazard. Consequences of distance from Earth include communication delays, reduced ability to respond in emergencies, reliance on life support systems, and homesickness. Communication delays can result in miscommunication and delayed responses to emergencies (Kanas, 2012). Even though each one of these consequences imposes a different set of difficulties and requires different countermeasures, similar research methods are used to investigate them. They include simulation studies, studies of similar working environments, surveys, and questionnaires (McMenamin et al., 2020), interviews (Tafforin, 2015; Tafforin et al., 2021), case studies, focus groups, analysis of communication logs, and technology development. Often a few of these methods are combined.

- Simulation studies attempt to recreate some parts of the space environment on the ground, for instance communication delays in a controlled environment to study their impact on human performance. For example, analogue Mars mis-

sions organised by the Austrian Space Forum used from 10- to 20-min delays between analogue astronauts and ground support (Groemer et al., 2014; Groemer et al., 2016). Such studies rarely rely on retrospective memory and can go beyond self-reported effects, because measurements are taken during the simulated mission. However, they do not fully recreate the psychological state of actual astronauts. Another challenge related to delayed communication is potential data fragmentation, possibly leading to further communication disturbance that cannot be easily simulated during the analogue mission. Simulation studies can also imitate situations of life support system failure to study human reactions.
- Studies conducted within similar work environments use some analogies between space travel and other professions. For example, submarine crews also experience high reliance on life support systems and their reaction patterns to a system failure might be similar (Shepanek, 2005).
- Technology development studies commonly use intervention study protocols. Human factors metrics are compared between a group using new technology and a control group. Examples of technological countermeasures might be the development of communication protocols intended for space exploration (Mosier & Fischer, 2021), the development of dedicated software that could support delayed communication (Fischer et al., 2023) or the development of telemedicine which could improve ground support during an emergency (Cermack, 2006).

As humans venture further into space, the ISS stands as a pivotal platform for investigating the challenges that come with distance from Earth. Building on the work of ground-based analogue studies, various ISS research methods are employed to investigate how cognitive ability, team dynamics and crew performance are affected.

On the ISS, a common way of monitoring cognitive ability is via standardised neurocognitive tests (Casario et al., 2022) including attention, memory, decision-making and problem-solving skills. Researchers can track changes in cognitive function over the course of a mission and make an evaluation of the influence that the distance from Earth has on psychological functions. Such tests have provided practical objective data, while also allowing for longitudinal analysis. However, these tests may not entirely capture the full complex cognitive demands.

Another method is the monitoring of *physiological reactions*. Verheyden et al. (2009) investigated the adaptation of heart rate and blood pressure to short- and long-duration spaceflight using an electrocardiogram together with beat-to-beat finger arterial blood pressure. To overcome the challenge of accurate data collection, astronauts were trained to perform the in-flight measurements by themselves and were guided through the experiment using dedicated software, allowing for standardisation of the testing procedures (Verheyden et al., 2009).

With these methods to address the issues associated with living at a distance from one's home planet, researchers and crew must also contend with additional factors potentially exacerbating these issues, such as communication delays, which can result in miscommunication and delayed responses to emergencies (Kanas, 2012). One study assessed how communication delays *impact individual and team performance and behaviour*. Both crewmembers and mission support personnel were asked to perform tasks with and without a communication delay during an ISS mission. Following each task, both groups completed post-task questionnaires, which included questions to assess individual and team performance, behaviour, and mood through self-reported ratings and mood scales. Semi-structured interviews were also conducted post-flight with participating astronauts (Kintz & Palinkas, 2016).

Crucian, Babiak-Vazquez et al. (2016) also explored the incidence of *clinical symptoms* during long-duration spaceflight, by examining medical records from 6-month missions to the ISS. They assessed incidence data as well as clinical symptoms and medical events to better understand how to support astronauts clinically during missions away from Earth, and also analysed the frequency of medical conditions which are reported (Crucian, Babiak-Vazquez et al., 2016; Crucian, Johnston et al., 2016). Such studies revealed that skin rashes followed by upper respiratory symptoms were the most frequently reported clinical symptoms in flight. The same has been done with medication use data from clinical records. The analysis revealed that medication to aid sleep had the highest rate of usage during missions, followed by medication for pain relief (Wotring, 2015).

Hazard 4: Altered Gravity Fields

Human anatomy and physiology have adapted to Earth's gravitational environment over millions of years of evolution (Osborn,1894). When leaving Earth's gravitational field, the human body is exposed to microgravity in orbit or to altered gravity environments such as hypogravity on the surface of other celestial bodies (e.g. 0.16G on the Lunar surface). Exposure to micro- (Hawkey, 2003) and hypogravity (Richter et al., 2017) leads to various short- and long-term physiological adaptations affecting almost all physiological systems. These changes occur in LEO missions since gravitational forces acting in the G_z axis of the human body are almost entirely diminished (microgravity), whereas in the case of missions on the lunar surface, gravitational forces are reduced as compared to Earth (hypogravity). Almost all human cells are sensitive to mechanical forces acting on receptors on the cell membrane triggering a cascade of biochemical signalling events that ultimately result in short- or long-term adaptation of the physiological system they are part of or interface with. The concept of translating a mechanical stimulus into a biochemical reaction is called "mechanotransduction" (Orr et al., 2006).

As described elsewhere, a significant facilitator of space physiology research has been the design and use of various terrestrial analogues to simulate immediate and long-term physiological effects of altered gravity environments under standardised laboratory conditions (see Figure 9.4). For terrestrial microgravity research, the widely accepted gold standard for studying physiological long-term effects is head-down tilt (HDT) bed rest (with a tilt angle of −6°; Sundblad et al., 2016; see Figure 9.3). Such a gold standard does not exist for hypogravity research, since every model has certain advantages and disadvantages (e.g. the temporal limitations during parabolic flights vs. its highest fidelity for microgravity/ hypogravity). Most importantly, thus far, space physiology researchers are still lacking a feasible and valid terrestrial simulation for long-term hypogravity studies on the ground, which render it necessary to select the model that is best suited to address a certain research question (De Martino et al., 2023), and to extrapolate immediate effects for predicting long-term adaptations to hypogravity (Richter et al., 2017).

In −6° head-down position in bed, gravitational loading in the G_z axis is removed whilst the head-down position leads to a redistribution of body fluids like what has been observed during actual microgravity exposure. HDT bed rest studies are also used to study the feasibility, tolerance, and effectiveness of proposed spaceflight countermeasures in the long term. These countermeasures can be nutritional countermeasures (Sandal et al., 2020), exercise countermeasures (Fiebig et al., 2019), passive countermeasures (e.g. centrifugation, lower body negative pressure; Holt et al., 2016) or a combination of different countermeasures.

The physiological long-term adaptation to hypogravity is an under-researched field. It is one of the greatest challenges for the international human spaceflight communities to design valid terrestrial analogues and to conduct human studies that are needed to safely send crew to the Moon and beyond for prolonged periods.

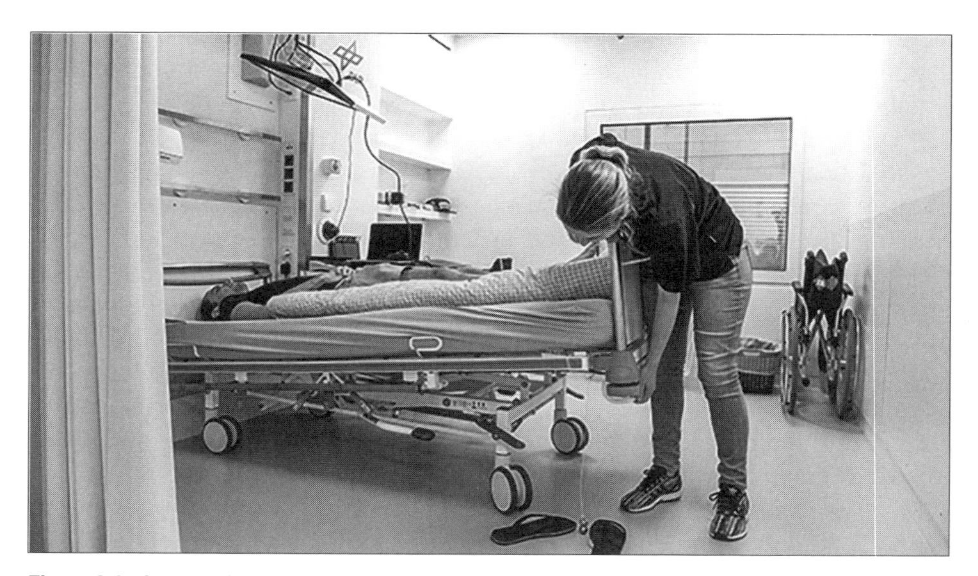

Figure 9.3. Set-up of head-down tilt bed rest study. © Deutsches Zentrum für Luft- und Raumfahrt (DLR). Reprinted with permission.

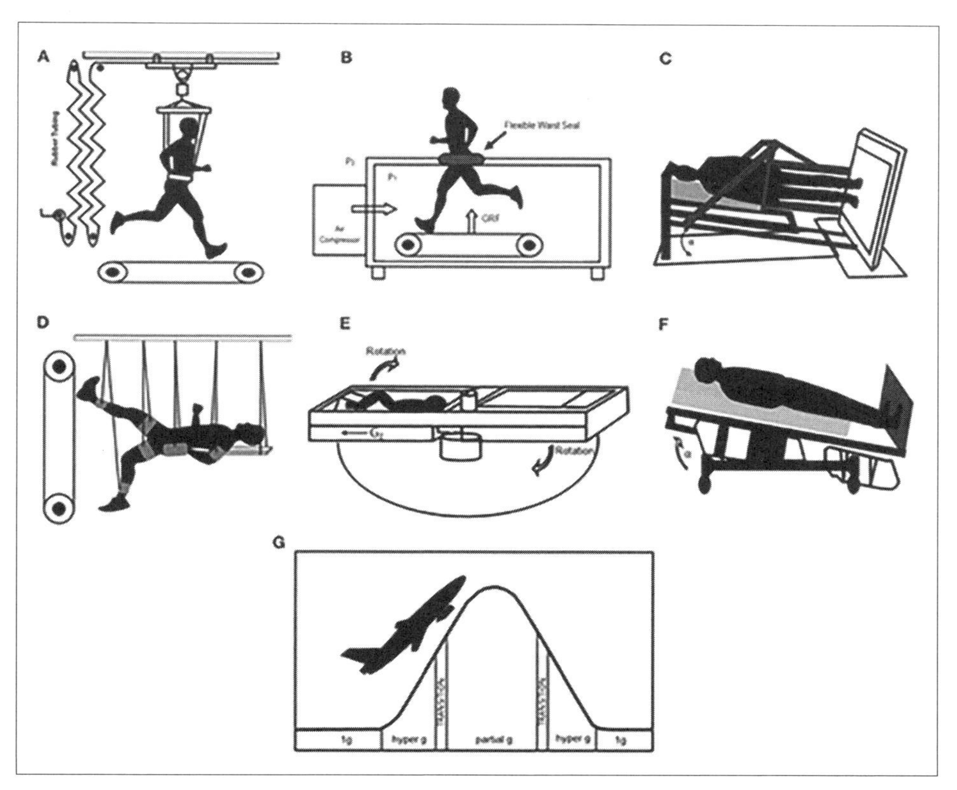

Figure 9.4. Experimental set-ups for the study of hypogravity. From "Human Biomechanical and Cardiopulmonary Responses to Partial Gravity – A Systematic Review," by C. Richter, B. Braunstein, A. Winnard, M. Nasser, and T. Weber, 2017, *Frontiers in Physiology*, *8*, Article 583, Figure 1 on p. 3. © The Author(s). CC BY license (http://creativecommons.org/licenses/by/4.0/)

Whilst the ISS offers a key platform for performing life science investigations in an actual microgravity environment, longer-duration missions allow for the collection of a substantial number of data, in addition to extensive investigations of the physiological and cognitive effects of microgravity. To evaluate the effectiveness of the exercise program used by crewmembers in preserving skeletal muscle size and function, one study has utilised pre- and post-mission magnetic resonance images as well as muscle biopsies and a torque velocity dynamometer to assess changes in calf muscle volume, microanatomy and performance, respectively (Trappe et al., 2009). Researchers carried out pre- and post-flight measurements to assess bone demineralisation in astronauts on long-duration missions to the ISS using dual-energy X-ray absorptiometry (DXA) to obtain the areal bone mineral density (BMD) of the hip and spine, as well as volumetric quantitative computed tomography to measure cortical, integral and trabecular volumetric BMD (Lang et al., 2004).

Additionally, studies have also investigated how altered gravity causes biomechanical differences in musculoskeletal tissue. Vico et al. (2017) employed a range of imaging techniques such as pre-and post-flight high-resolution peripheral quantitative computed tomography and DXA scans to assess cortical and trabecular bone microstructure. Fasting blood samples were also taken to assess biochemical markers of bone cellular activities including serum osteocyte markers. Sensorimotor changes associated with altered gravity environments have also been explored during extended stays in microgravity. For example, an investigation by Miller et al. (2018) studied balance and functional task performance in astronauts on the ISS as well as in individuals on terrestrial bed rest. Data were collected before and after the flight, and astronauts performed a variety of functional task test trials including *the seat egress and walk*, the *object translation* test, and *fall/stand* tests. Recorded metrics included the overall time for course completion and split times for each of the main sections. Researchers also used the Computerised Dynamic Posturography (CDP) test on crewmembers to assess balance (Miller et al., 2018).

Summary and Outlook

This chapter should be treated as an introduction to the topic of space human factors and the research methods used to study them. Many of the methods presented here overlap between the mentioned hazards. Probably the most distinctive methodological profiles can be observed in radiation and microgravity studies as they investigate specialised physiological adaptations. At the same time, isolation and confinement, distance from Earth, and enclosed and hostile environments impose many psychological challenges and the methods used to study them are often similar. For example, analogue space missions, surveys and interviews with astronauts or testing of supportive technologies are used. To summarise this abundance of knowledge, some take-home messages are presented.

- The five hazards of human space exploration listed by NASA (Lewis, 2022) are radiation, distance from Earth, isolation and confinement, altered gravitational fields, and hostile environments.
- Knowledge about the effect of radiation on humans is mostly based on scarce data from extraordinary events such as atomic bombs. Because of this, it must be complemented by molecular epidemiology and animal studies in Earth-based space-radiation analogues.
- Analogue space missions are used to study psychological, technological, and human factors related to space missions.
- Isolation studies, for example at Concordia Station in Antarctica, are used to study human response to isolation and confinement.

- Bed rest and dry immersion studies are analogues used to investigate the effects of microgravity on the human body.
- There is currently a lack of long-term hypogravity analogue studies to investigate the physiological effects of prolonged exposure to hypogravity.

Acknowledgements

We would like to thank Dr. Angelique Van Ombergen and Dr. Nathan Smith for their support and inspiring ideas.

References

Alcibiade, A., Schlacht, I. L., Finazzi, F., Di Capua, M., Ferrario, G., Musso, G., & Foing, B. (2020). Reliability in extreme isolation: A natural language processing tool for stress self-assessment. In *Advances in human factors and systems interaction: Proceedings of the AHFE 2020 virtual conference on human factors and systems interaction* (pp. 350–357). Springer. https://doi.org/10.1007/978-3-030-51369-6_47

Casario, K., Howard, K., Cordoza, M., Hermosillo, E., Ibrahim, L., Larson, O., Nasrini, J., & Basner, M. (2022). Acceptability of the Cognition Test Battery in astronaut and astronaut-surrogate populations. *Acta Astronautica, 190*, 14–23. https://doi.org/10.1016/j.actaastro.2021.09.035

Cermack, M. (2006). Monitoring and telemedicine support in remote environments and in human space flight. *British Journal of Anaesthesia, 97*(1), 107–114. https://doi.org/10.1093/bja/ael132

Childress, S. D., Williams, T. C., & Francisco, D. R. (2023). NASA space flight human-system standard: Enabling human spaceflight missions by supporting astronaut health, safety, and performance. *npj Microgravity, 9*(31), 1–7. https://doi.org/10.1038/s41526-023-00275-2

Crucian, B., Babiak-Vazquez, A., Johnston, S., Pierson, D. L., Ott, C. M., & Sams, C. (2016). Incidence of clinical symptoms during long-duration orbital spaceflight. *International Journal of General Medicine, 9*, 383–391. https://doi.org/10.2147/IJGM.S114188

Crucian, B., Johnston, S., Mehta, S., Stowe, R., Uchakin, P., Quiriarte, H., Pierson, D., Laudenslager, M. L., & Sams, C. (2016). A case of persistent skin rash and rhinitis with immune system dysregulation onboard the International Space Station. *Journal of Allergy and Clinical Immunology: In Practice, 4*(4), 759–762.e8. https://doi.org/10.1016/j.jaip.2015.12.021

De Martino, E., Green, D. A., Ciampi de Andrade, D., Weber, T., & Herssens, N. (2023). Human movement in simulated hypogravity – bridging the gap between space research and terrestrial rehabilitation. *Frontiers in Neurology, 14*, Article 1062349. https://doi.org/10.3389/fneur.2023.1062349

Durante, M. (2014). Space radiation protection: Destination Mars. *Life Sciences in Space Research, 1*(1), 2–9. https://doi.org/10.1016/J.LSSR.2014.01.002

Fiebig, L., Winnard, A. J., Nasser, M., Braunstein, B., Scott, J., Green, D., & Weber, T. (2019). Effectiveness of resistive exercise countermeasures in bed rest to maintain muscle strength and power – a systematic review. In J. Van Loon, M. Heer, C. Fuller, & M.-A. Custaud (Eds.), *Conference abstracts: 39th ISGP meeting & ESA life sciences meeting* (pp. 70–75). Frontiers Media. https://doi.org/10.3389/conf.fphys.2018.26.00020

Fischer, U., Mosier, K., Schmid, J., Smithsimmons, A., & Brougham, R. (2023). Braiding – a novel approach to supporting space/ ground communication under signal latency. *Acta Astronautica, 207*, 411–424. https://doi.org/10.1016/J.ACTAASTRO.2023.03.023

Flynn-Evans, E. E., Barger, L. K., Kubey, A. A., Sullivan, J. P., & Czeisler, C. A. (2016). Circadian misalignment affects sleep and medication use before and during spaceflight. *npj Microgravity, 2,* Article 15019. https://doi.org/10.1038/npjmgrav.2015.19

Fogtman, A., Baatout, S., Baselet, B., Berger, T., Hellweg, C. E., Jiggens, P., La Tessa, C., Narici, L., Nieminen, P., Sabatier, L., Santin, G., Schneider, U., Straube, U., Tabury, K., Tinganelli, W., Walsh, L., & Durante, M. (2023). Towards sustainable human space exploration – priorities for radiation research to quantify and mitigate radiation risks. *npj Microgravity 2023, 9*(8), 1–6. https://doi.org/10.1038/s41526-023-00262-7

Groemer, G., Gruber, V., Bishop, S., Peham, D., Wolf, L., & Högl, B. (2010). Human performance data in a high workload environment during the simulated Mars expedition "AustroMars". *Acta Astronautica, 66*(5), 780–787. https://doi.org/10.1016/j.actaastro.2009.08.017

Groemer, G., Losiak, A., Soucek, A., Plank, C., Zanardini, L., Sejkora, N., & Sams, S. (2016). The AMADEE-15 Mars simulation. *Acta Astronautica, 129*, 277–290. https://doi.org/10.1016/j.actaastro.2016.09.022

Groemer, G., Soucek, A., Frischauf, N., Stumptner, W., Ragonig, C., Sams, S., Bartenstein, T., Häuplik-Meusburger, S., Petrova, P., Evetts, S., Sivenesan, C., Bothe, C., Boyd, A., Dinkelaker, A., Dissertori, M., Fasching, D., Fischer, M., Föger, D., Foresta, L., ... Zanella-Kux, K. (2014). The MARS2013 Mars Analog Mission. *Astrobiology, 14*(5), 360–376. https://doi.org/10.1089/ast.2013.1062

Gruber, S., Groemer, G., Paternostro, S., & Larose, T. L. (2020). AMADEE-18 and the Analog Mission performance metrics analysis: A benchmarking tool for mission planning and evaluation. *Astrobiology, 20*(11), 1295–1302. https://doi.org/10.1089/ast.2019.2034

Hafner, L., Walsh, L., & Schneider, U. (2021). Cancer incidence risks above and below 1 Gy for radiation protection in space. *Life Sciences in Space Research, 28*, 41–56. https://doi.org/10.1016/J.LSSR.2020.09.001

Hawkey, A. (2003). The physical price of a ticket into space. *Journal of the British Interplanetary Society, 56*(5–6), 152–159. https://europepmc.org/article/med/14552355

Healey, B. (2020). White space: Applications of research and development derived from space flight analogues for developing solutions for global health worldwide. Concordia, Antarctica case study. *Studies in Space Policy, 22*, 165–172. https://doi.org/10.1007/978-3-030-21938-3_15

Holt, J. A., Macias, B. R., Schneider, S. M., Watenpaugh, D. E., Lee, S. M. C., Chang, D. G., & Hargens, A. R. (2016). WISE 2005: Aerobic and resistive countermeasures prevent paraspinal muscle deconditioning during 60-day bed rest in women. *Journal of Applied Physiology, 120*(10), 1215–1222. https://doi.org/10.1152/japplphysiol.00532.2015

Lewis, R. (2022, October, 4). *Human spaceflight hazards.* NASA. https://www.nasa.gov/feature/human-spaceflight-hazards

Jones, C. W., Basner, M., Mollicone, D. J., Mott, C. M., & Dinges, D. F. (2022). Sleep deficiency in spaceflight is associated with degraded neurobehavioral functions and elevated stress in astronauts on six-month missions aboard the International Space Station. *Sleep, 45*(3), Article zsac006. https://doi.org/10.1093/sleep/zsac006

Kanas, N. (2012). Psychological, psychiatric, and interpersonal aspects of long-duration space missions. *Journal of Spacecraft and Rockets, 27*(5), 457–463. https://doi.org/10.2514/3.26165

Kintz, N. M., & Palinkas, L. A. (2016). Communication delays impact behavior and performance aboard the International Space Station. *Aerospace Medicine and Human Performance, 87*(11), 940–946. https://doi.org/10.3357/AMHP.4626.2016

Lang, T., LeBlanc, A., Evans, H., Lu, Y., Genant, H., & Yu, A. (2004). Cortical and trabecular bone mineral loss from the spine and hip in long-duration spaceflight. *Journal of Bone and Mineral Research, 19*(6), 1006–1012. https://doi.org/10.1359/JBMR.040307

Leon, G. R. (2005). Men and women in space. *Aviation, Space, and Environmental Medicine, 76*(6 Suppl.), B84–B88

McMenamin, J., Allen, N. J., & Battler, M. (2020). Team processes and outcomes during the AMADEE-18 Mars Analog Mission. *Astrobiology, 20*(11), 1287–1294. https://doi.org/10.1089/ast.2019.2035

Miller, C. A., Kofman, I. S., Brady, R. R., May-Phillips, T. R., Batson, C. D., Lawrence, E. L., Taylor, C. L., Peters, B. T., Mulavara, A. P., Feiveson, A. H., Reschke, M. F., & Bloomberg, J. J. (2018). Functional task and balance performance in bed rest subjects and astronauts. *Aerospace Medicine and Human Performance, 89*(9), 805–815. https://doi.org/10.3357/AMHP.5039.2018

Mosier, K. L., & Fischer, U. M. (2021). Meeting the challenge of transmission delay: Communication protocols for space operations. *Human Factors, 65*(6), 1235–1250. https://doi.org/10.1177/00187208211047085

Musson, D. M., & Helmreich, R. L. (2005). Long-term personality data collection in support of spaceflight and analogue research. *Aviation Space and Environmental Medicine, 76*(6 Suppl.), B119–B125).

Neigut, J. (2015). *Overview of the Human Exploration Research Analog (HERA)*. https://ntrs.nasa.gov/citations/20150003017

Nicolas, M., Suedfeld, P., Weiss, K., & Gaudino, M. (2016). Affective, social, and cognitive outcomes during a 1-year wintering in Concordia. *Environment and Behavior, 48*(8), 1073–1091. https://doi.org/10.1177/0013916515583551

Orr, A. W., Helmke, B. P., Blackman, B. R., & Schwartz, M. A. (2006). Mechanisms of mechanotransduction. *Developmental Cell, 10*(1), 11–20. https://doi.org/10.1016/j.devcel.2005.12.006

Osborn, H. F. (1894). *From the Greeks to Darwin: An outline of the development of the evolution idea*. Macmillan and Co. https://doi.org/10.1037/12938-000

Posselt, B. N., Velho, R., O'Griofa, M., Shepanek, M., Golemis, A., & Gifford, S. E. (2021). Safety and healthcare provision in space analogs. *Acta Astronautica, 186*, 164–170. https://doi.org/10.1016/J.ACTAASTRO.2021.05.033

Richter, C., Braunstein, B., Winnard, A., Nasser, M., & Weber, T. (2017). Human biomechanical and cardiopulmonary responses to partial gravity – a systematic review. *Frontiers in Physiology, 8*, Article 270211. https://doi.org/10.3389/fphys.2017.00583

Sandal, P. H., Kim, D., Fiebig, L., Winnard, A., Caplan, N., Green, D. A., & Weber, T. (2020). Effectiveness of nutritional countermeasures in microgravity and its ground-based analogues to ameliorate musculoskeletal and cardiopulmonary deconditioning – a systematic review. *PLOS ONE, 15*(6), Article e0234412. https://doi.org/10.1371/journal.pone.0234412

Schuy, C., Weber, U., & Durante, M. (2020). Hybrid active-passive space radiation simulation concept for GSI and the future FAIR facility. *Frontiers in Physics, 8*, Article 566353. https://doi.org/10.3389/fphy.2020.00337

Shepanek, M. (2005). Human behavioral research in space: Quandaries for research subjects and researchers. *Aviation Space and Environmental Medicine, 76*(6 Suppl.), B25–B30.

Strangman, G. E., Sipes, W., & Beven, G. (2014). Human cognitive performance in spaceflight and analogue environments. *Aviation Space and Environmental Medicine, 85*(10), 1033–1048. https://doi.org/10.3357/ASEM.3961.2014

Strewe, C., Moser, D., Buchheim, J. I., Gunga, H. C., Stahn, A., Crucian, B. E., Fiedel, B., Bauer, H., Gössmann-Lang, P., Thieme, D., Kohlberg, E., Choukèr, A., & Feuerecker, M. (2019). Sex differences in stress and immune responses during confinement in Antarctica. *Biology of Sex Differences, 10*(1), 20. https://doi.org/10.1186/s13293-019-0231-0

Sundblad, P., Orlov, O., Angerer, O., Larina, I., & Cromwell, R. (2016). Standardization of bed rest studies in the spaceflight context. *Journal of Applied Physiology, 121*(1), 348–349. https://journals.physiology.org/doi/full/10.1152/japplphysiol.00089.2016 https://doi.org/10.1152/japplphysiol.00089.2016

Tafforin, C. (2015). Confinement vs. isolation as analogue environments for mars missions from a human ethology viewpoint. *Aerospace Medicine and Human Performance, 86*(2), 131–135. https://doi.org/10.3357/AMHP.4100.2015

Tafforin, C., Vinokhodova, A., & Gushin, V. (2021). Individual diversity and temporal stability during a 4-month confinement experiment (SIRIUS-19) for human space exploration. *Human Ethology, 36*, 36–48. https://doi.org/10.22330/he/36/036-048

Trappe, S., Costill, D., Gallagher, P., Creer, A., Peters, J. R., Evans, H., Riley, D. A., & Fitts, R. H. (2009). Exercise in space: Human skeletal muscle after 6 months aboard the International Space Station. *Journal of Applied Physiology, 106*(4), 1159–1168. https://doi.org/10.1152/japplphysiol.91578.2008

Ushakov, I. B., Vladimirovich, M. B., Bubeev, Y. A., Gushin, V. I., Vasil'eva, G. Y., Vinokhodova, A. G., & Shved, D. M. (2014). Main findings of psychophysiological studies in the Mars 500 experiment. *Herald of the Russian Academy of Sciences, 84*(2), 106–114. https://doi.org/10.1134/S1019331614020063

Van Puyvelde, M., Gijbels, D., Van Caelenberg, T., Smith, N., Bessone, L., Buckle-Charlesworth, S., & Pattyn, N. (2022). Living on the edge: How to prepare for it? *Frontiers in Neuroergonomics, 3*, Article 1007774. https://doi.org/10.3389/fnrgo.2022.1007774

Verheyden, B., Liu, J., Beckers, F., & Aubert, A. E. (2009). Adaptation of heart rate and blood pressure to short and long duration space missions. *Respiratory Physiology and Neurobiology, 169*(Suppl.), 13–16. https://doi.org/10.1016/j.resp.2009.03.008

Vernikos, J., Walter, N., Worms, J. C., & Blanc, S. (2016). THESEUS: The European research priorities for human exploration of space. *npj Microgravity, 2,* 16034, 1–3. https://doi.org/10.1038/npjmgrav.2016.34

Vico, L., van Rietbergen, B., Vilayphiou, N., Linossier, M. T., Locrelle, H., Normand, M., Zouch, M., Gerbaix, M., Bonnet, N., Novikov, V., Thomas, T., & Vassilieva, G. (2017). Cortical and trabecular bone microstructure did not recover at weight-bearing skeletal sites and progressively deteriorated at non-weight-bearing sites during the year following International Space Station missions. *Journal of Bone and Mineral Research, 32*(10), 2010–2021. https://doi.org/10.1002/jbmr.3188

Wotring, V. E. (2015, February 15). Medication use by U.S. crewmembers on the International Space Station. *The FASEB Journal, 29*(11), 4417–4423. https://doi.org/10.1096/FJ.14-264838

Zuur, A. F., Ieno, E. N., Walker, N. J., Saveliev, A. A., & Smith, G. M. (2009). Mixed effects modelling for nested data. In A. F. Zuur, E. N. Leno, N. J. Walker, A. A. Saveliev, & G. M. Smith. (Eds.). *Mixed effects models and extensions in ecology with R* (pp. 101–142). Springer. https://doi.org/10.1007/978-0-387-87458-6_5

Chapter 10

CIMON – The First Artificial Crew Assistant in Space

Till Eisenberg, Gerhard Reichert, Ralf Christe, Judith Irina Buchheim, Christian Karrasch, and Ioana V. Koglbauer

Abstract

The human spaceflight community is in preparation to expand their exploration beyond the current frontier, from the Moon to Mars. While we have already developed effective countermeasures for many challenges of space travel, the psychological impact of deep space missions needs to be further addressed. Influences of the space environment such as microgravity, radiation and isolation far away from Earth are challenging conditions for the human body and mind. The crew for these missions will be as small as possible to limit the mass and volume. Despite a very thorough preparation, unfavourable group dynamics might appear triggered by the specific stressors of this mission, such as isolation. Also, the "Earth-out-of-sight" effect will occur for the first time in human existence when the characteristic bluish-green colour of planet Earth disappears from their view on the way to Mars. In 2014/2015, Airbus developed the concept of an artificial crew mate for space missions that could provide technical services to the crew such as assistance in procedures, environmental monitoring, and mission handbook access. This crew mate could be part of the team, serve as a mentor, and detect, mirror or introduce new emotions to the users. In 2016, the German Space Agency at the German Aerospace Center initiated the program Crew Interactive Mobile Companion (CIMON) with the participation of multiple partners. The initial demonstration was accomplished 2 years later in the frame of the European Space Agency's Horizons Mission. This chapter describes innovative methods used for CIMON with the focus on human–machine interaction in the unique environment of space travel. In addition, the future testing of new features for CIMON is addressed. This project is important for the success of future deep space missions.

Keywords

artificial intelligence, assistance system, human space flight, International Space Station (ISS), strategic design

Strategy to Realise CIMON

The target mission and first user of CIMON were pre-defined in 2016. The European Space Agency's astronaut Alexander Gerst was chosen for the commissioning of the first interaction between a human being and the one-of-a-kind artificial intelligence (AI)-driven robotic crew mate.

The time frame between the kick-off and delivery was limited to about 2 years, whereas applying all standard processes, workflows and review durations of data packages would normally require a period of about 5 years. The highly motivated project team came up with – at that point in time and in that working environment – rather new tools, methods and strategies to give the project the chance of success. CIMON was classified as a technology demonstrator where functionality outweighed performance. The payload specification was defined accordingly and, in parallel, meaningful mission goals were agreed. Scientific requirements were also created and included into the specification. An agile approach was used to speed up the work. The focus was on finding solutions, not on problems.

Methods, Techniques and Technologies to Create CIMON

Strategic Design Development

By means of strategic design development, the development and design processes are coordinated with the overall strategy of a company/program. Industrial design was used in CIMON as a strategic tool to achieve the holistic development processes. The key aspects were:

1. Vision and Purposeful Design
   It all started with a vision of a helpful intelligent robot named "Simon" from a science fiction movie. Simon accompanied Captain Future in space and always stood by his side. The interdisciplinary development team first concentrated on core requirements such as mobility and interaction with the astronauts. With its unique design and functions, CIMON was created to improve the living conditions of astronauts, to support the human exploration of deep space and life on the International Space Station (ISS).
2. Holistic Approach
   For the development of CIMON, a multitude of aspects were considered such as functions, user experience, new technologies, specific environmental conditions (e.g., on the ISS), safety, aesthetics and other important factors.
3. Focus on User Needs
   In the strategic design process for CIMON, the focus was placed on the needs and requirements of the astronauts by integrating design thinking and other de-

velopment methods. This approach can also be transferred to other applications (e.g., in the medical field).

4. Innovation-Enabling
   Our strategic design development encouraged innovation and creativity by using design methods such as agile prototyping, co-creation and iterative processes. This approach enabled the CIMON team to generate innovative ideas and to promptly incorporate changing requirements, new technological possibilities and user feedback from the astronauts. The CIMON project won the “Best of What’s New” award in 2018 and the “German prize for innovation” in 2019.
5. Adaptability
   Strategic design takes into account the dynamics of constant environmental change. Therefore, CIMON was designed to be adaptable and expandable, to incorporate improvements and additional features as technology and mission requirements continue to evolve.
6. Integration With Corporate Strategy and the Long-Term Perspective
   Another aspect of strategic design development was the alignment with the Airbus corporate strategy. With the development of CIMON, Airbus could demonstrate its innovative strength and continue to be a leading company in the field of space technology. The experiences with CIMON on the ISS could also help to improve human–machine interaction in other systems and domains (e.g., medical use cases).
7. Trans- and Interdisciplinary Collaboration
   The holistic development of CIMON required the interdisciplinary collaboration of aerospace engineers, industrial designers, graphic designers, software developers and others. Companies and institutions such as Airbus Defence and Space GmbH, Reichert Design, APOGAEUM Design, Laboratory for Translational Research “Stress & Immunity”, the German Aerospace Center in Cologne, and the German Space Agency at the German Aerospace Center worked together in the design development process.

Figure 10.1. Stages of intermediate development. © Airbus (stages 2015, 2016 and 2018). © German Aerospace Center (stage 2017). Reprinted with permission.

As CIMON clearly illustrates, strategic design can be used to link technical development, creative design and strategic planning in order to develop innovative and future-proof solutions that have a positive impact on companies, organisations and society.

Form Follows Function

"Form follows function" is a fundamental principle of design (Sullivan, 1896). This well-known design principle was first used by architects and designers in the late 19th and early 20th centuries and was validated through countless applications in various fields (e.g., industrial design). This principle was also applied to CIMON's design. For example, the white colour makes CIMON quickly identifiable and recognizable in the space station. Its spherical shape allows for the best multi-directional positioning in zero gravity. The design of the robot follows its functions as an autonomous companion and helper for astronauts. In summary, by aligning the design to meet the various requirements and functions, meaningful and effective product solutions can be achieved.

Buy-and-Adapt

The conventional "make-or-buy" dilemma is a critical decision-making process in many industries. It revolves around the question of whether it is better to manufacture a particular product or component or to procure it from external suppliers. The decision usually hinges on factors such as quality, cost, time, expertise, performance and strategic importance. However, in some scenarios, none of these pure approaches is optimal. This brings us to a more nuanced strategy in a fast-track space project like CIMON: *buy-and-adapt*. Thus, instead of creating a component from scratch or purchasing a product that fits perfectly into a system, products that are close enough to what is needed are procured and modified to meet specific requirements. This strategy combined the advantages of both making and buying such as time efficiency, cost-effectiveness, flexibility and risk mitigation.

By choosing the buy-and-adapt strategy, we managed to leverage the strengths of available products on the market, while making them fit seamlessly into our unique operational context. This hybrid approach allowed us to achieve our project objectives more efficiently. One example is the display of CIMON. A circular or even spherical display as a user interface would have been ideal for our purposes. However, the development of such a display would have been out of our budget and time schedule. Hence, it was decided to buy a commercial rectangular display and to modify it.

CIMON's Design

Spherical Shape Harmony

CIMON's basic body is reduced to the essentials – a sphere. This shape was chosen to enable optimal freedom of movement as it has the smallest surface area for the given volume. It also has the best flight characteristics in weightlessness (e.g., ISS). This allows CIMON to efficiently navigate through tight spaces and assist astronauts in their tasks without risking to cause damage to the station's interior or even injuring astronauts (Figure 10.1).

The spherical shape protects from injuries. CIMON has a smooth surface without sharp corners and edges and, thus, offers the highest level of safety for the crew in the extremely turbulent environment of the ISS. The smooth, white surface gives a futuristic look, it prevents dust and dirt from sticking on it and facilitates cleaning. Furthermore, the spherical design conveys calm and harmony in contrast to the environment on the ISS that is complex and overloaded with experimental setups, computers, monitors and cables. Thus, the design meets both aesthetic and practical requirements.

White Colour

The neutral white colour of CIMON follows aesthetic and functional specifications (e.g., visibility, recognition). Because the ISS is crammed with technical devices and equipment, a bright white CIMON is easier for the astronaut to detect and locate. This is especially important when CIMON autonomously moves around the space station. White is also a colour that can be easily identified from afar.

Mobility and Autonomy

CIMON can manoeuvre by pushing air through internal air channels. For propulsion, CIMON is equipped with 14 propellers, which are connected to the surface of the housing with connecting tubes on the inside (Figure 10.2). It can quickly follow the astronauts or navigate independently to a desired destination in the space station.

To fulfil the safety requirements, no finger, hair, pen or other objects should get in contact with the internally located propellers. Thus, CIMON was equipped with a protection made of a six-corner honeycomb grid. The size of the grid harmoniously decreases towards the outside surface (Figure 10.3).

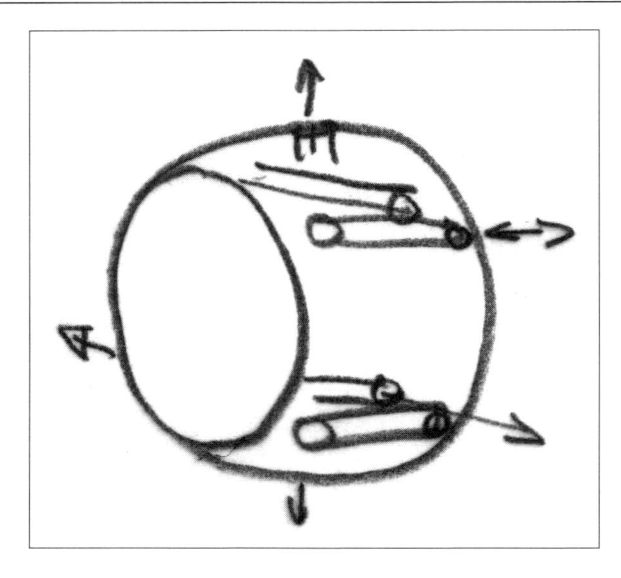

Figure 10.2
Sketch of air flow channels.
© Gerhard Reichert. Reprinted with permission.

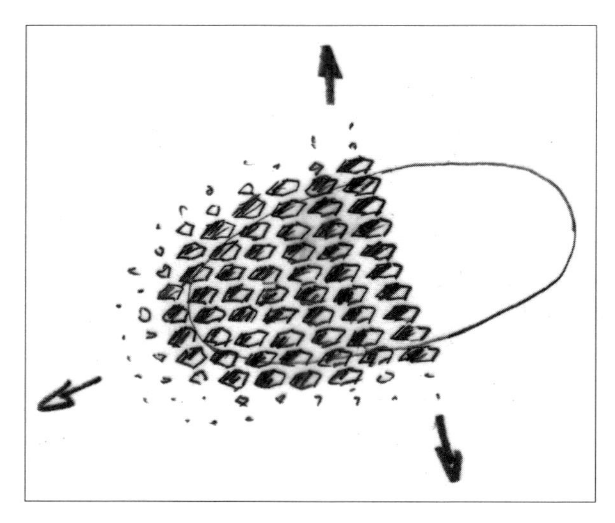

Figure 10.3
Draft design concept of a six-corner honeycomb grit.
© Gerhard Reichert. Reprinted with permission.

Display Showing a Human-Like Face

For better acceptance and communication with the astronauts, the face section of CIMON was designed in the size of an average human face (Figure 10.4). The interaction concept included facial recognition and response. Thus, CIMON's human-like interaction should facilitate a social bond between the robot and the astronauts. On future space missions away from family and friends for very long periods of time, CIMON will be given the important role of a buddy who can provide a kind of emotional support.

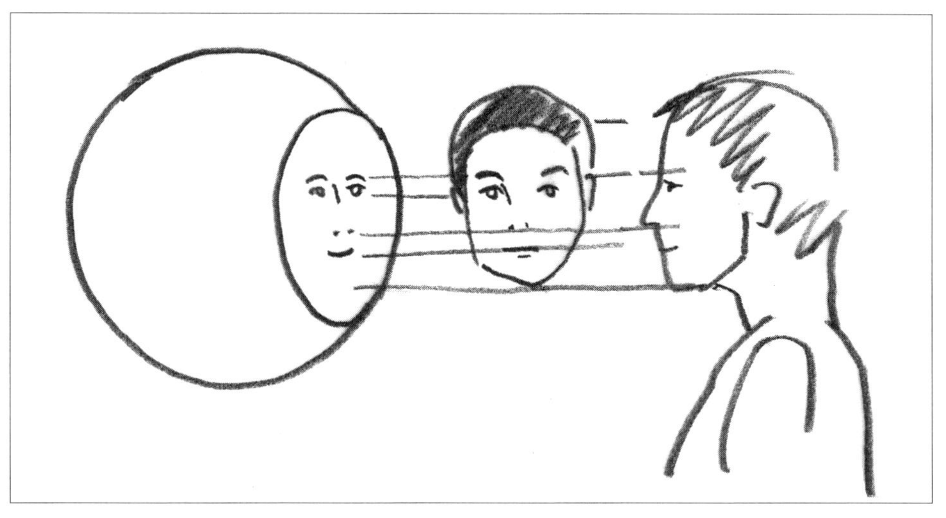

Figure 10.4 CIMON's size compared to a human's head. © Gerhard Reichert. Reprinted with permission.

CIMON's Facial Expression

To support the decision about the best fitting design for the use case of CIMON, the "uncanny valley" effect (Mori, 1970/2012) was considered. This describes the relationship between the increasing human-likeness of a robot and human's affinity for the robot, which increases until a certain point from which it abruptly decreases (Mori, 1970/2012). In the uncanny valley the human perceives the robot as eerie (Mori, 1970/2012). Thus, for CIMON, using a photorealistic face or an avatar was not considered a sensible approach.

The design of CIMON's face including eyes, eyebrows, nose and mouth was created in the field of tension between minimalism and maximalism. This means exploring solutions between a minimalist (e.g., semicolon face or smiley face) and a photorealistic depiction of a face. The fundamentally reduced graphic design of CIMON's face, with lines of equal thickness, served as the basis for different variants as shown in Figure 10.5. In addition, a gender-neutral design was selected for CIMON's face (Figure 10.6).

The animation of the eye area enables the representation of a human-like emotion. Blinking of the eyes and the movement of the eyebrows play an amplifying role in the depiction of emotions. The combination of gaze control, depiction of emotions and speech strengthen CIMON's communication with the human.

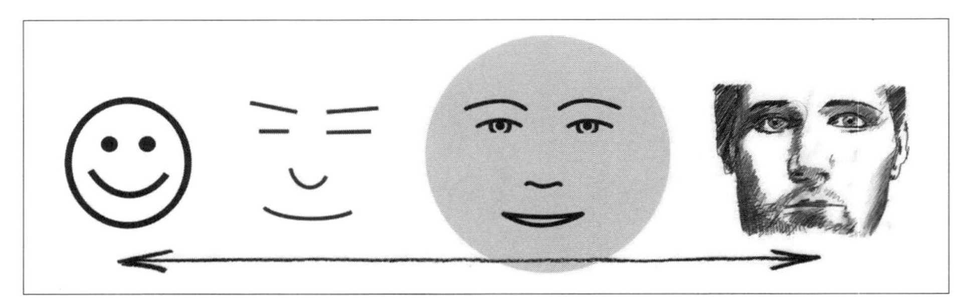

Figure 10.5 Design variants from abstract (left) to realistic (right) and final choice of the face of CIMON. © Gerhard Reichert. Reprinted with permission.

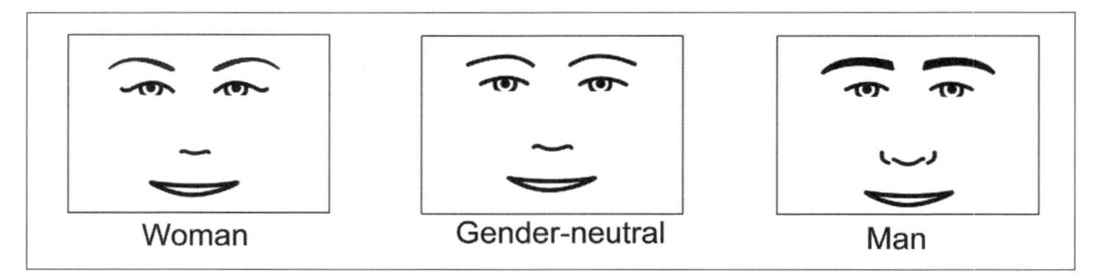

Figure 10.6 Female, gender-neutral and male variants of face design. © Ralf Christe. Reprinted with permission.

Animations of the mouth were developed to match the human mouth movement when speaking in the respective language (e.g., vowels, phonemes). Lip synchronisation was added because, at least for English native speakers who are not used to watching dubbed movies, this is an important element of communication (Figure 10.7).

CIMON's Voice

For the voice and assistance functionality, IBM Cloud services based on Watson were implemented into the overall CIMON ecosystem. In accordance with user feedback, one of the most human-like male voices was chosen for CIMON out of the various voice types available.

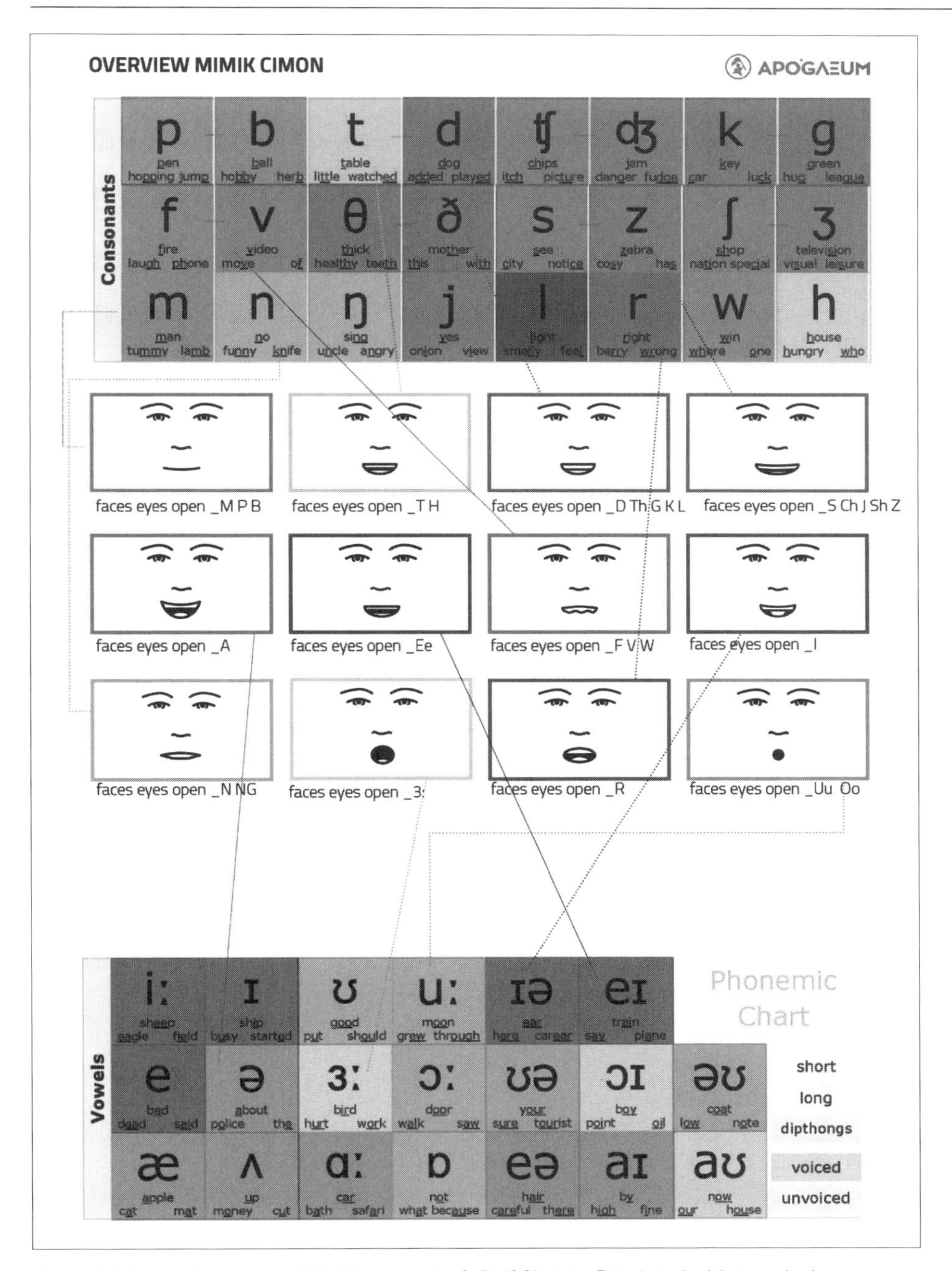

Figure 10.7. Animations of CIMON's mouth. © Ralf Christe. Reprinted with permission.

Artificial Intelligence Features

CIMON uses natural language processing to interact with the astronauts, to conduct scientific experiments and to act as a personal assistant. IBM has integrated elements of the IBM Cloud incorporating Watson Assistant technology into CIMON's artificial intelligence (AI) to enhance its capabilities and interaction with the astronauts. Using these technologies CIMON reacts to voice commands and guides the crew through pre-trained crew procedures. In addition, CIMON can take photos and videos for the documentation of experiments. It can also provide support for other features (e.g., putting on a favourite music playlist). In addition, CIMON uses a face detection and recognition software to orient towards a user and identify the user's name.

Touch Screen

There are several reasons why CIMON has no touch screen functionality. For example, in a weightless environment floating objects are pushed away at the slightest touch. Counteracting such unforeseeable forces would have triggered big efforts for the adjustment of CIMON's guidance, navigation and control system (GNC). Despite the lack of touch functionality, CIMON allows for voice control and frees both hands of the astronaut.

Minimalistic Hardware Design

The physical controls and interfaces have been reduced to the minimum to convey harmony and to support an intuitive use (Figure 10.8).

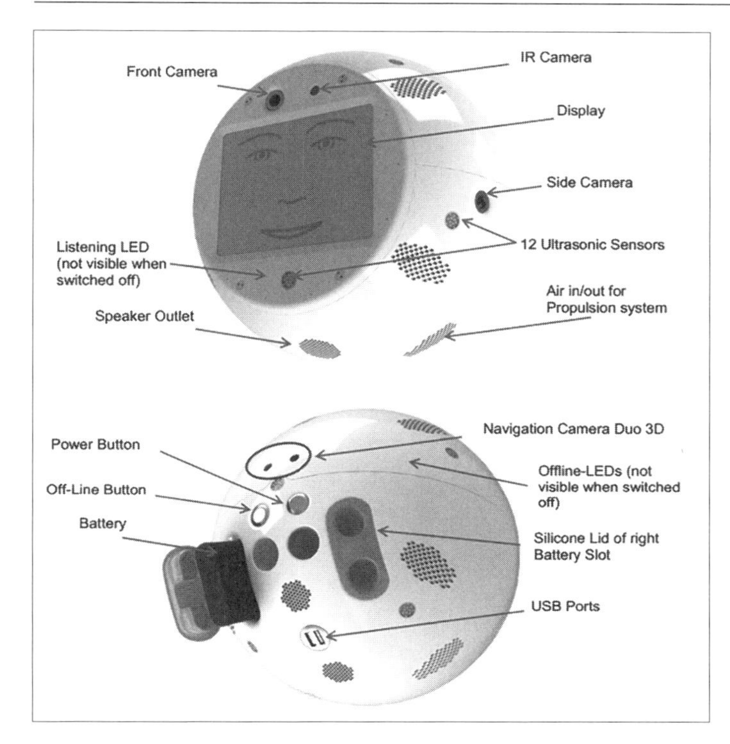

Figure 10.8. CIMON's external interfaces to the environment. © Airbus. Reprinted with permission.

Figure 10.9. CIMON with Alexander Gerst on the ISS 2018. © European Space Agency (ESA)/NASA. Reprinted with permission.

CIMON's Evaluations

Astronauts live and work under very unique conditions that can affect their well-being. They are affected by physical and psychological stressors such as microgravity, high workload, isolation and confinement, sleep disturbances or circadian rhythm disruption. Chronic stress triggers the release of stress hormones such as cortisol that affects immunity (Sorrells & Sapolsky, 2007). Concomitantly, despite a very good preparation and training, several health issues have been observed in long-duration space flight that can affect mission success such as immune dysregulation, wound healing problems or the reactivation of several dormant viruses that still cannot be prevented (Buchheim et al., 2019; Cialdai et al., 2022; B. Crucian et al., 2016; B. E. Crucian et al., 2018; Mehta et al., 2017; Monici et al., 2022). Chronic stress can also influence the efficiency of a team leading to the disruption of communication, high pressure towards uniformity or groupthink, wherein individual opinions are easily disregarded or suppressed (Dion, 2004; Turner & Pratkanis, 1998).

After an initial set of tests (Figure 10.9) different functionalities of CIMON were consolidated and further developed. Future research will assess whether CIMON's assistance and autonomy might be a countermeasure for these problems (e.g., reduce workload and stress) and also an effective means for maintaining the crew's health. Astronauts are highly trained, but the training has taken place some time ago on the ground. Each day, astronauts are supported by a large team of flight controllers to carry out tasks, experiments or maintenance work. Each flight controller has a support team of various people in the background that can be asked in the case of a problem. Also, the running of the ISS itself as well as the daily scheduling is done by people on the ground. In deep space exploration there will be a lower amount of earth-bound crew support available. Therefore, more crew autonomy and intelligent support technologies are needed (Nicogossian et al., 2001). Ideally, an intelligent support device will be capable of analysing complex or large numbers of data, and of providing the astronaut with an unbiased data read-out as a basis for decision-making. It should also support training and planning processes and prevent groupthink in which group members tend to switch to the expected, albeit incorrect, group opinion. Crew acceptance, transparency, autonomous mobility and voice recognition are expected to be critical factors for success.

A pilot study was conceptualised by the Laboratory for Translational Research, "Stress and Immunity", at the Department of Anaesthesiology of the LMU University Hospital of the Ludwig Maximilian University of Munich in Germany. These functionalities of CIMON are to be evaluated in 2024 for the first time on board the ISS. Since the technical set-up on the ISS is challenging, the initial read-out will be limited to the established functionality. Since the AI database is on Earth, access to the AI can currently only occur when a data connection to ground is available. The aim is to evaluate user satisfaction by means of a spoken questionnaire after the designated CIMON science sessions. The scenario will be an assisted multi-step activity wherein the astronaut needs to execute a certain task with the use of on-board hardware. The quality of the audio-visual instructions, the applicability of a mobile video documentation system compared to the common static camera as well as AI features such as voice recognition, speech-to-text and text-to-speech will be assessed. The user feedback will help to further develop CIMON's functions and to adapt the device to the wishes and needs of astronauts. This study will help to gain the first insights into whether an intelligent platform such as CIMON as well as its new features and connectivity set-up is feasible for on-board use. The success critically depends on user acceptance. To develop a safe and reliable AI system, the user should be involved in the development process early on to provide specific input to the developers (Taddeo & Floridi, 2018), which is exactly what we intended for the CIMON project. CIMON was not expected to be a final product but rather a first step to gain expertise and to better understand the challenges and opportunities that will allow this technology to evolve.

Summary and Outlook

Future human space exploration will make use of robotic assistance wherever feasible. With the return of humans to the Moon's surface in the frame of the Artemis III mission, this future is becoming more tangible. By moving further away from Earth, it is likely that the strain for astronauts will increase even more compared to the ISS (Manzey, 2004). Together with the construction of the Gateway in lunar orbit, the lunar habitat will be an excellent test platform for emerging and established robotic applications. A neutral, intelligent robot that will assist in daily routine work and can autonomously execute assigned tasks is likely to reduce stress and will be an important technology feature also for the Gateway. It is clear from the application of machine algorithms to the daily clinical work of physicians that support systems need to be designed together with the end user in order to achieve acceptance. Therefore, the user evaluation data and additional inputs by crew will provide indispensable insights that will allow us to adapt the functionalities accordingly in order to shape the intelligent assistant best suited for day-to-day support of humans living and working successfully in the extreme space environment.

The transfer of this expertise/technology could also benefit the people on Earth. This was another aim of the project. It might be possible that the experience gained with CIMON running in the specifically noisy ISS environment will help to provide better voice recognition and speech-to-text tools for the medical field. Patients could also benefit from a companion during their hospital stays that would help explain procedures or answer questions, even providing emotional support. The application of machine algorithms, especially in medicine, has gained momentum. However, the benefits are challenged by personal rights, ethical concerns and liability (Brady & Neri, 2020; Petkus et al., 2020). Nevertheless, using AI to process big data will help physicians to make better decisions, but it will also change the medical profession (Hardy & Harvey, 2020; Wartman & Combs, 2019). It is, however, imperative to explore the technology we are using and to better understand the benefits and disadvantages (Taddeo & Floridi, 2018). Only then will AI be implemented in a safe, transparent and trustworthy way.

Acknowledgments

The success of this project was only possible through the fruitful collaboration of various stakeholders and suppliers: The German Aerospace Center, Airbus Defence and Space GmbH, European Space Agency, Reichert Design, Apogaeum, International Business Machines (IBM), Biotesc and many more.

References

Brady, A. P., & Neri, E. (2020). Artificial intelligence in radiology – ethical considerations. *Diagnostics, 10*(4), Article 231. https://doi.org/10.3390/diagnostics10040231

Buchheim, J. I., Matzel, S., Rykova, M., Vassilieva, G., Ponomarev, S., Nichiporuk, I., Hörl, M., Moser, D., Biere, K., Feuerecker, M., Schelling, G., Thieme, D., Kaufmann, I., Thiel, M., & Choukèr, A. (2019). Stress related shift toward inflammaging in cosmonauts after long-duration space flight. *Frontiers in Physiology, 10*. https://doi.org/10.3389/fphys.2019.00085

Cialdai, F., Risaliti, C., & Monici, M. (2022). Role of fibroblasts in wound healing and tissue remodeling on Earth and in space. *Frontiers in Bioengineering and Biotechnology, 10*, Article 958381. https://doi.org/10.3389/fbioe.2022.958381

Crucian, B., Babiak-Vazquez, A., Johnston, S., Pierson, D. L., Ott, C. M., & Sams, C. (2016). Incidence of clinical symptoms during long-duration orbital spaceflight. *International Journal of General Medicine, 9*, 383–391. https://doi.org/10.2147/IJGM.S114188

Crucian, B. E., Choukèr, A., Simpson, R. J., Mehta, S., Marshall, G., Smith, S. M., Zwart, S. R., Heer, M., Ponomarev, S., Whitmire, A., Frippiat, J. P., Douglas, G. L., Lorenzi, H., Buchheim, J. I., Makedonas, G., Ginsburg, G. S., Ott, C. M., Pierson, D. L., Krieger, S. S., Baecker, N., & Sams, C. (2018). Immune system dysregulation during spaceflight: Potential countermeasures for deep space exploration missions. *Frontiers in Immunology, 9*, Article 1437. https://doi.org/10.3389/fimmu.2018.01437

Dion, K. L. (2004). Interpersonal and group processes in long-term spaceflight crews: Perspectives from social and organizational psychology. *Aviation, Space, and Environmental Medicine, 75*(7), C36–C43.

Hardy, M., & Harvey, H. (2020). Artificial intelligence in diagnostic imaging: impact on the radiography profession. *The British Journal of Radiology, 93*(1108), Article 20190840. https://doi.org/10.1259/bjr.20190840

Manzey, D. (2004). Human missions to Mars: New psychological challenges and research issues. *Acta Astronautica, 55*(3–9), 781–790. https://doi.org/10.1016/j.actaastro.2004.05.013

Mehta, S. K., Laudenslager, M. L., Stowe, R. P., Crucian, B. E., Feiveson, A. H., Sams, C. F., & Pierson, D. L. (2017). Latent virus reactivation in astronauts on the international space station. *npj Microgravity, 3*, Article 11. https://doi.org/10.1038/s41526-017-0015-y

Monici, M., van Loon, J., Chouker, A., & Iorio, C. S. (2022). Editorial: Wound management and healing in space. *Frontiers in Bioengineering and Biotechnology, 10*, Article 1078986. https://doi.org/10.3389/fbioe.2022.1078986

Mori, M. (2012). The uncanny valley [from the field] (K. F. MacDorman & N. Kageki, Trans.). *IEEE Robotics & Automation Magazine, 19*(2), 98–100. https://doi.org/10.1109/MRA.2012.2192811. (Original work published 1970)

Nicogossian, A. E., Pober, D. F., & Roy, S. A. (2001). Evolution of telemedicine in the space program and earth applications. *Telemedicine Journal and E-health, 7*(1), 1–15. https://doi.org/10.1089/153056201300093813

Petkus, H., Hoogewerf, J., & Wyatt, J. C. (2020). What do senior physicians think about AI and clinical decision support systems: Quantitative and qualitative analysis of data from specialty societies. *Clinical Medicine, 20*(3), 324–328. https://doi.org/10.7861/clinmed.2019-0317

Sorrells, S. F., & Sapolsky, R. M. (2007). An inflammatory review of glucocorticoid actions in the CNS. *Brain, Behavior, and Immunity, 21*(3), 259–272. https://doi.org/10.1016/j.bbi.2006.11.006

Sullivan, L. (1896). The tall office building artistically considered. *The Architectural Record, 12*, 403–50.

Taddeo, M., & Floridi, L. (2018). How AI can be a force for good. *Science, 361*(6404), 751–752. https://doi.org/10.1126/science.aat5991

Turner, M. E., & Pratkanis, A. R. (1998). Twenty-five years of groupthink theory and research: Lessons from the evaluation of a theory. *Organizational Behavior and Human Decision Processes, 73*(2), 105–115. https://doi.org/10.1006/obhd.1998.2756

Wartman, S. A., & Combs, C. D. (2019). Reimagining medical education in the age of AI. *AMA Journal of Ethics, 21*(2), E146–E152. https://doi.org/10.1001/amajethics.2019.146

Chapter 11

Applied User Research in Virtual Reality

Tools, Methods, and Challenges

Leonie Bensch, Andrea Casini, Aidan Cowley, Florian Dufresne, Enrico Guerra, Paul de Medeiros, Tommy Nilsson, Flavie Rometsch, Andreas Treuer, and Anna Vock

Abstract

This chapter explores the practice of conducting user research studies and design assessments in virtual reality (VR). An overview of key VR hardware and software tools is provided, including game engines, such as Unity and Unreal Engine. Qualitative and quantitative research methods, along with their various synergies with VR, are likewise discussed, and some of the challenges associated with VR, such as limited sensory stimulation, are reflected upon. VR is proving particularly useful in the context of space systems development, where its utilisation offers a cost-effective and secure method for simulating extraterrestrial environments, allowing for rapid prototyping and evaluation of innovative concepts under representative operational conditions. To illustrate this, we present a case study detailing the application of VR to aid aerospace engineers testing their ideas with end users and stakeholders during early design stages of the European Space Agency's (ESA) prospective Argonaut lunar lander. This case study demonstrates the effectiveness of VR simulations in gathering important feedback concerning the operability of the Argonaut lander in poor lighting conditions as well as surfacing relevant ergonomics considerations and constraints. The chapter concludes by discussing the strengths and weaknesses associated with VR-based user studies and proposes future research directions, emphasising the necessity for novel VR interfaces to overcome existing technical limitations.

Keywords

virtual reality, user research, space systems design, European Space Agency, Argonaut

Introduction

Interactive simulations in virtual reality (VR) provide the means for rapid assessment and iterative development of novel design concepts without incurring many of the financial, logistical or safety constraints typically associated with real-world studies of working prototypes. This has made VR into a particularly useful tool in the field of space systems design, due to its capacity to simulate relevant extraterrestrial environmental conditions that cannot be replicated through other means in the real world, whilst also helping to reduce some of the massive costs normally entailed in development of novel space systems.

Such VR simulations enable aerospace engineers to visualise and continuously test relevant ideas with a wide range of end users and expert stakeholders from the earliest stages of a design process. The reflections, behavioural measurements and observations acquired through such user studies can subsequently help steer concept optimisation and further decision-making, whilst also allowing engineers and designers to gain a better understanding of the underlying problems in order to adjust or redefine their goals as needed. This, in turn, allows for rapid iteration of solutions during early design stages, thereby reducing the risk of having to implement expensive changes further down the development process.

Following this line of reasoning, the European Space Agency (ESA) has been drawing on VR to help guide the ongoing development of the prospective Argonaut lunar lander. In this chapter we offer a first-hand account of this endeavour, detailing ESA's approach and its outcomes. By reflecting on the technical limitations inherent in conventional VR interfaces, such as insufficient haptic feedback, we elaborate not just the opportunities, but also the challenges involved in using VR to conduct valid user studies. In doing so, we provide lessons learned from ESA's work that may help future practitioners responsibly harness some of the considerable potential offered by VR in space engineering and beyond.

Tools

VR technology has evolved significantly over the past few decades, becoming more widely available and accessible. Early advancements date back to the 1960s when Ivan Sutherland developed one of the forerunners of modern VR headsets, featuring head-tracking and the display of computer-generated images (Sutherland, 1998). NASA's Virtual Interface Environment Workstation (VIEW) introduced in 1988 further pushed the boundaries of VR with its advanced capabilities for immersive and highly interactive simulations (Fisher et al., 1988). The potential offered by this technology quickly became apparent when it played a pivotal role in training the Hubble space telescope flight team for a critical repair mission in 1993 (Loftin & Kenney, 1995).

Aside from head-mounted displays (HMDs), another noteworthy VR approach is the cave automatic virtual environment (CAVE) technology. CAVEs surround the user with screens projecting virtual images onto walls and the floor, eliminating the need for discomforting headsets. Additionally, participants in multi-user applications can physically see each other, enhancing social interaction (Manjrekar et al., 2014). However, due to the complexity and cost associated with setting up CAVE systems, their widespread adoption has been limited. This has resulted in the prevalence of modern HMDs in most domains, including research, industry, and the private sector.

Consumer-grade VR headsets, such as the Meta Quest 2 (Meta, 2020), HTC Vive (HTC, 2023), and Pico 4 (PicoXR, 2024), have gained popularity due to their affordability, high visual fidelity (up to 2448 × 2448 pixels per eye; Vive, 2023), and ease of use. Standalone headsets, exemplified by the Meta Quest 2, do not require an external PC or tracking devices due to their inside-out tracking capabilities. They can be easily used by simply powering them on and placing them on the head (connection to a PC for performance-consuming applications is still possible with most models). On the other hand, PC-supported headsets such as the HTC Vive Pro utilise lighthouse tracking, an infrared-based technology developed by Valve. These headsets require installation of external tracking devices before use.

Interacting with VR applications commonly involves the use of controllers, which resemble game console controllers with various input keys assigned to specific functions and actions. Alternative input methods, such as haptic gloves (Perret & Vander Poorten, 2018) or hand tracking and gesture control (Buckingham, 2021), also exist. However, the controller-based interaction scheme remains prevalent in the majority of VR applications. While VR hardware is crucial, the significance of software and virtual experiences cannot be overstated. Game engines such as Unity (Unity, 2023) and Unreal Engine (Engine, 2023) play a vital role in creating applications for modern VR headsets. These engines offer comprehensive development environments that enable the creation of complex three-dimensional simulations and immersive experiences. Unity and Unreal Engine, in particular, stand out due to their extensive VR functionalities, support for common VR platforms, and large developer communities. These engines offer comprehensive toolsets that simplify otherwise complex technical processes such as performance optimisation, interaction design, and realistic physics simulation. Additionally, their asset import pipelines facilitate seamless integration with various digital creation tools, including 3D modelling and texturing software. As a result, they enable developers to quickly and effectively model hypothetical design solutions and relevant scenarios, which can subsequently be explored by relevant stakeholders and assessed through user research methods.

Methods

Much like real-world studies, user research conducted in VR can be broadly categorised into qualitative and quantitative approaches. Qualitative approaches primarily strive to generate a comprehensive in-depth understanding and interpretation of users' needs, actions, behaviours, beliefs, and emotions, unearthing the processes and meanings that drive them. As such, they offer insights into the "why" of user behaviours or beliefs (Hennink et al., 2020). This is typically achieved through the application of descriptive research methods, including, but not limited to, interviews and focus groups. Such studies, particularly in the highly specialised context of space systems design, typically involve a relatively small, handpicked sample of participants, such as domain experts. By leveraging the informed perspectives of these participants, researchers gain a unique opportunity to develop an understanding of relevant design issues from the viewpoint of potential users and other key stakeholders. The immersive nature of VR enabling participants to engage in hypothetical situations while being put in the shoes of prospective users, (e.g. future astronauts) can serve as a fertiliser for such qualitative enquiries. Indeed, its capacity to foster empathy and provide an in-depth understanding of diverse user perspectives has led to VR being dubbed the "ultimate empathy machine" (Herrera et al., 2018).

Similarly, the ability to convey rich immersive settings allows researchers to explore the contextual factors influencing users' behaviours and perspectives. This is particularly useful in projects concerned with environments that are difficult to access in the real world, such as space systems design, where frequent evaluations of concepts in remote space environments (e.g. the Moon, Mars, International Space Station [ISS]) are necessary. In doing so, VR effectively endows design assessments with elements of contextual inquiry (Duda et al., 2020), putting the spotlight on synergies and frictions arising between the depicted design solutions, users and environments (Nilsson et al., 2023). Methodologies employed to elicit such reflections typically pivot around reflections gathered through the think-aloud protocol as well as interviews and collaborative design sessions conducted with the user immersed in the VR environment.

In contrast to qualitative research, quantitative research aims to interpret phenomena through statistical patterns. This approach involves a larger sample size and seeks to derive findings that can be generalised to a broader population. By collecting measurable numerical data and employing statistical techniques, quantitative methods equip researchers with the means to address questions such as "how much", "what," and "where" (Apuke, 2017).

The requisite numerical data are primarily collected through experiments or user surveys, which are then scrutinised using statistical methods, including correlation or regression analysis. This process enables the clear affirmation or rejection of pre-formulated hypotheses based on the statistical findings. In the realm of space system design, quantitative research in VR is particularly useful for conducting comparative studies of multiple potential solutions (A/B testing) or evaluating users' physiological responses. Common techniques employed in this domain include task performance measurements (such as completion times and error rates), analysing self-reported survey responses (e.g. using Likert scales to measure workload via the NASA TLX scale; Hart, 2006), eye-tracking (for attention, information processing, user engagement), motion capturing (mocap), and biometric measurements (e.g. heart rate or brain activity; Becker et al., 2022).

However, it is worth noting that qualitative and quantitative methods are not mutually exclusive. The most comprehensive insights often emerge from a hybrid, or *mixed-methods*, approach (Sandelowski, 2000). For example, qualitative user reflections might be gathered through a think-aloud protocol and semi-structured interviews during immersive VR sessions in order to better understand the domain experts' perspectives. Simultaneously, quantitative data may be collected to gain insights into their physiological responses (e.g. heart rate) to various stimuli present in the VR environment.

Challenges

Despite the numerous valuable applications of VR simulations, several challenges persist that hinder the utilisation of VR to its full potential in the field of space systems design. While VR technology has made significant advancements in visual fidelity, there is a prevalent concern about its limited ability to stimulate other sensory modalities, which may undermine the validity of observations made in VR environments. Major discrepancies between human responses in VR environments and their real-world equivalents have been linked to such technical limitations. Notably, the so-called super soldier syndrome causes users to exhibit game-like attitudes, such as careless behaviour and unrealistic risk-taking, thus potentially invalidating behavioural user data collected during such studies (Barlow & Morrison, 2005).

Technical difficulties associated with providing accurate haptic feedback, in particular, have been shown to impede the perceived realism of VR environments, thus posing a considerable challenge for user research activities. Given the important role of altered sensory experiences in space, such as the sensation of hypogravity, movement restrictions incurred by bulky space suits, inertia, and other haptic or tactile sensations, the space systems design domain may be particularly affected by such limitations, potentially restricting the applicability of VR-based research approaches in ongoing design and development endeavours. Instead, relevant conditions, such as hypogravity, have traditionally been simulated using real-world neutral buoyancy facilities or testbeds on the seafloor (Koutnik et al., 2021) as well as parabolic flight campaigns (De Martino et al., 2020). Such practices come with their own limitations: Neutral buoyancy facilities require equipment to be specifically modified for the right buoyancy in order to simulate appropriate gravity levels, and the hypogravity conditions experienced during parabolic flights typically only last for approximately 20–30 s. Moreover, such approaches likewise entail oftentimes prohibitively high costs. There is a need, then, for novel and efficient interfaces capable of compensating for some of the limitations associated with VR. One technology in this vein that has shown promise is a gravity offload system. NASA's ARGOS, for instance has been proven capable of recreating the effects of desired gravity levels by offloading VR users by a desirable factor, thereby more reliably simulating environmental conditions on the Moon, Mars or the ISS (Orr et al., 2022).

Similarly, whilst spacesuits can be visually simulated in a VR environment, the movement restrictions associated with an actual suit cannot be replicated digitally. Given the limited access to actual spacesuits due to their considerable weight and cost, it is vital for future studies to identify adaptable and efficient simulation methods. Potential solutions include the development of cheaper spacesuit mock-ups that realistically recreate some of the relevant movement constraints. Here, systems such as the Austrian Space Forum Aouda.X space suit simulator (Groemer et al., 2012) or the Extravehicular Activity Space Suit Simulator (EVA S3) developed at the Massachusetts Institute of Technology (Meyen, 2013) could prove useful to provide realistic and cost-effective means to simulate movement restrictions in virtual environments. To simulate haptic sensations, haptic gloves (Zhu et al., 2020) or suits can be used to replicate forces and tactile sensations. Another approach involves combining physical mock-ups with virtual visual content, as demonstrated in NASA's hybrid reality lab (Delgado & Noyes, 2017).

The Argonaut Case Study

The notion of reviving crewed missions to the Moon and establishing a permanent human presence on its surface is rapidly gaining prominence among private entities and governmental space agencies alike. The fruition of this ambition will hinge on the development and implementation of robust logistics systems capable of ensuring a dependable delivery of vital supplies and cargo to the lunar surface in support of future human expeditions. Against this backdrop, the ESA is currently engaged in a feasibility study aimed at laying the foundation for future development of the Argonaut lunar lander. The primary objective of this autonomous lunar landing vehicle will be the transportation of diverse crew supply payloads and scientific experiments to the lunar surface. With an anticipated initial deployment in 2029, Argonaut is poised to play a key role in Europe's pursuit of sustainable human exploration and colonisation of the Moon.

Primary aims of the early stages of this project include elaborating known design challenges, uncovering novel ones, and validating potential design solutions.

Given the nascent nature of this endeavour, with no physical prototype of the Argonaut having been constructed so far, VR has been deemed a fitting tool for fulfilling these objectives. Consequently, our team was tasked with constructing a virtual prototype of the Argonaut lander that could then be subjected to assessments using interactive VR. Drawing on input from the Argonaut project management and responsible engineering teams, we produced a representative configuration of the lander (Nilsson et al., 2022). The result was an approximately 2.8-m-tall virtual lander featuring an octagonal cargo deck of approximately 14 m^2 on top. In order to foster comprehensive discussions encompassing the broader operational context and to facilitate the examination and evaluation of aspects concerning usability and human factors challenges and limitations, we strived to incorporate an additional level of detail into the model, such as communication antennas and radiators, as well as a set of four cargo containers located at the cargo deck. A ladder and transportation cart were likewise included. The VR users were also embodied in a 3D model of the Exploration Extravehicular Mobility Unit (xEMU) EVA suit. For a schematic drawing of the lander mock-up and accompanying hardware, see Figure 11.1.

In addition to creating the virtual Argonaut mock-up and relevant hardware, we also needed to construct a representative lunar environment wherein the lander's operations could be simulated and evaluated. To accomplish this, we utilised topographic scans of the lunar surface captured by NASA's Lunar Reconnaissance Orbiter (LRO; D. E. Smith et al., 2017). Drawing on these scans, we reconstructed digitally an 8 × 8-km area in close proximity to the Shackleton crater situated at the lunar south pole (89.9°S, 0.0°E). We selected this particular location due to it

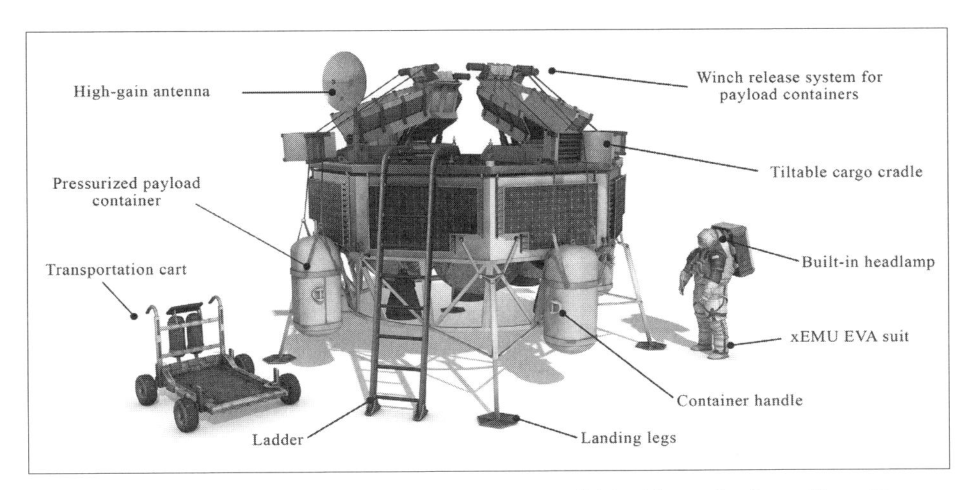

Figure 11.1. The virtual Argonaut lander mock-up. From "Using Virtual Reality to Shape Humanity's Return to the Moon: Key Takeaways From a Design Study", by T. Nilsson et al., 2023, Figure 2. © The Owner/Author(s). CC BY license (http://creativecommons.org/licenses/by/4.0/)

having been identified as one of the potential landing sites for the initial Artemis human landing (M. Smith et al., 2020), thereby providing a credible setting for our Argonaut assessment.

The LRO topographic scans had a relatively low resolution, with each pixel representing an area of 100 × 100 m on the real lunar surface. Consequently, finer surface details (e.g. boulders or smaller craters) had to be recreated manually using conventional 3D modelling software, such as Cinema 4D. Additionally, we developed a custom terrain shader to enhance the visual realism of the virtual lunar surface. Throughout this process, we collaborated closely with an experienced lunar geologist to ensure the utmost accuracy and authenticity of our virtual lunar landscape.

Accurate replication of the unique lighting conditions on the lunar south pole was another crucial aspect of our work. The virtual sun (i.e. a directional light in the VR environment) was oriented towards the north, set at an angle of 1.5° above the horizon, and its intensity was adjusted to 1.37 kW/m^2 (Ross et al., 2023). To faithfully reproduce the absence of a lunar atmosphere, all forms of indirect lighting and light scattering were deactivated, resulting in the creation of deep, dark shadows and blinding highlights that typify the lunar environment.

Once the development of the virtual lunar landscape was completed, we positioned the Argonaut mock-up at its centre. Finally, to bring the entire VR experience to life, we utilised the Unreal Engine 4 game engine. By combining the realistic lunar environment, the detailed Argonaut mock-up, and the immersive capabilities of the game engine, we thus provided an authentic and engaging VR simulation for evaluation and assessment purposes. To facilitate such an evaluation, we hand-picked and invited a number of experts from the field of human spaceflight to experience our VR simulation of the Argonaut mock-up. Notably, our study included two active astronauts, both of whom have extensive experience conducting EVA operations outside the ISS. Other participants included engineers responsible for providing technical support for ISS, including the maintenance of several of its modules. Multiple participants had similar experience with monitoring of astronaut missions through the European Spacecraft Communicator (EUROCOM) – an international flight control centre responsible for direct communication with the ISS crew.

Each participant experienced the VR simulation individually. Upon being briefed about the study purpose, the participant was asked to complete a questionnaire regarding their demographics and previous experience with VR. After a quick demonstration of the VR controllers, the participant was then given freedom to navigate the virtual environment and explore the Argonaut lander at leisure. Following the think-aloud protocol (Ericsson & Simon, 1980), participants were encouraged to verbalise their thoughts and reasoning while completing this examination. Once done examining the lander, participants were asked to answer a set of semi-structured interview questions. These questions were open-ended and aimed to elicit participants' reflections and comments on their experience. Specific topics of inquiry included various features of the lunar lander, such as the cargo unloading mechanism, antenna placement, design of the cargo containers, ladder, and transportation cart. Participants were also asked to identify potential safety hazards and suggest design improvements. The study did not impose any time limits, allowing participants to take as much time as needed to explore the lander and respond to our questions. As a result, the total duration of the sessions varied between 40 and 80 min.

Participant responses were documented through notes, audio recordings, and questionnaire responses. The dataset was then independently coded by three researchers, and any inconsistencies were resolved through discussion. The data were synthesised using a qualitative thematic analysis (Braun & Clarke, 2006). A recurring theme throughout the study was the concerns raised by our participants about the lighting conditions on the Moon and its impact on various design aspects of the Argonaut lander. The Argonaut's landing legs, for instance were seen as potential tripping hazards in poor lighting. Similarly, locating misplaced payload containers and equipment was likewise seen as problematic in shadowed areas. Suggestions included equipping mobile hardware with artificial light sources and making greater use of LED bars or light strips.

Conversely, in other areas, the blinding sunlight, exacerbated by the absence of the Moon's atmosphere, posed additional challenges. Participants expressed concerns, for instance about being blinded and experiencing momentary vision impairment while performing EVA procedures. Reflections from the Argonaut's aluminium and metallic components were also problematic, with anti-glare coatings and sun visors proposed as potential remedies. Solar power generation and the ideal placement of solar panels on the lander also frequently sparked discussion. Participants considered scenarios where the lander inadvertently lands in a shadow, rendering solar power generation impossible, and proposed countermeasures, such as mounting solar panels on tethered transportation carts for repositioning. Matters concerning temperature management and solar power generation in the design of future landers and other lunar infrastructure were likewise frequently brought up.

Another frequent topic of reflection was the dimensions and ergonomic suitability of design elements, particularly the cargo containers. Our study participants, for example found that the container handles were too slim to enable efficient manipulation during extravehicular activities, given the bulky gloves that astronauts would likely be wearing. Limited visibility while manually carrying the large containers and potential difficulties in maintaining balance were also identified as key design considerations. Redesigning container handles by, for instance adding attachment mechanisms or wheels, and using rectangular containers (to improve tileability) were suggested as possible solutions. The notion of astronauts having to climb up and operate on top of the Argonaut cargo deck likewise attracted numerous comments pertaining to potential safety and visibility issues. Recommendations included adjusting ladder angles, using side-view mirrors, and implementing protective measures for sensitive instruments.

As expected, the study also exposed technical limitations of VR that may have influenced some of the participants' comments. The lack of haptic feedback hindered accurate assessment of weight and object manipulation. The absence of movement constraints imposed by bulky spacesuits also affected the evaluations. When prompted to elaborate on such limitations, our participants generally argued the VR simulation was more suitable for providing situational awareness rather than for assessing physical mechanics. In particular, participants appreciated the contextual experience and found VR useful for planning EVA operations and visualising robotic actions. Despite its limitations, VR was thus recognised as a valuable design tool with strengths to aid future design activities.

Take-Aways and Future Directions

As demonstrated in the case study outlined here, VR can be purposefully applied to facilitate user studies during early design stages. Two of its qualities make it particularly suitable for space systems design – the capacity to simulate any chosen environment, including extreme environments, in a visually compelling and highly accessible manner, as well as its suitability for integration with other digital systems. More specifically, intelligibly conveying adverse conditions in a cost-effective, flexible, and, above all, safe way can provide invaluable insights into early-stage designs and operational concepts. Such adverse conditions may include disruptive lighting, rough terrain, scale- and distance-gauging difficulties as well as low-gravity conditions and their associated consequences, to give some examples. The suitability of VR for integration with broader operational ecosystems, on the other hand, enables model-based systems engineering project approaches, allowing for correct evaluation of, and interaction with, digital twins of spaceflight systems. Owing to the accessible digital interfaceability of VR, real-time integration of live data such as satellite observations, spacecraft telemetry or other forms of state information from ongoing missions becomes significantly more practical for such applications.

While effectively allowing for varied visual inputs, the lack of multisensory – particularly physical – interactions remains a problem. Although an astronaut on the lunar surface may be able to lift objects weighing upwards of 100 kg, the non-trivial dynamics involved in this situation cannot be intuitively conveyed through purely visual cues as offered by conventional VR interfaces. Yet, understanding effects resulting from the object's inertia, the astronaut's own reduced weight as well as motion constraints imposed by an EVA suit are essential for planning an operational concept including such activities. Similarly, pure VR studies have limitations in terms of evaluating the ergonomics of spaceflight systems beyond non-contact aspects, such as accessibility or lighting assessments. In the future, approaches combining VR simulations with additional physical interfaces might serve to close that gap. For this purpose, we are investigating the use of various haptic devices, as well as other technology capable of introducing physical interactions into VR. As an example, ESA and the German Aerospace Center (DLR) are currently developing and building a novel lunar analogue facility named LUNA in Cologne, Germany (Casini et al., 2023). This facility will allow us to replicate some of the crucial harsh environmental conditions on the Moon such as reduced gravity conditions via a gravity offloading system, a dusty surface using a regolith simulant and the challenging lighting conditions using a Sun simulator. LUNA is meant to serve as a training and technology centre where hardware, protocols and concepts of operations can be tested in a realistic lunar set-up.

Furthermore, extended reality (XR) simulations are also envisioned to further enhance the immersion levels and to advance VR-based technologies for design and simulation studies. For smaller-scale and more flexible ad hoc campaigns, solutions employing stand-alone haptic devices, such as exoskeletal gloves, tactile feedback systems, as well as effective use of spatially tracked props (weight simulators, generic contact surfaces and physical interfaces) are currently under investigation. These can be applied in a more general and varied way to VR scenarios, and their integration adapted on the fly. Additionally, cutting-edge manufacturing processes and technologies, such as additive manufacturing and laser cutting, can be leveraged to achieve prop interface specificity as required, without incurring substantial financial and labour costs. Moreover, they allow for fast iteration on a level closely approaching that of VR.

Summary and Outlook

In this chapter, we have provided a broad overview of VR-based user research practices, with a particular focus on the space systems design domain. The Argonaut study presented here constitutes one of the first cases of VR technology being used to ameliorate the overall design process of a future lunar system. Reflecting on the study outcomes, we have explored not only the efficacy of VR to elicit relevant feedback, but also some of the limitations stemming from the predominantly audiovisual nature of conventional VR interfaces. Overcoming these challenges and responsibly integrating VR into future design and engineering processes could have a transformative impact. Beyond simply reducing some of the associated costs, the highly accessible nature of VR makes it well positioned to help make space systems design activities accessible to professionals who may have previously been excluded from such projects. By enabling a wider range of individuals to contribute to the generation of ideas, designs, and solutions that will shape our future beyond Earth, VR thus exemplifies a disruptive technology that has the potential to foster a more inclusive, innovative, and dynamic space industry.

Authorship

All authors have contributed equally to this work.

References

Apuke, O. D. (2017). Quantitative research methods: A synopsis approach. *Kuwait Chapter of Arabian Journal of Business and Management Review, 33*(5471), 1–8.

Barlow, M., & Morrison, P. (2005). Challenging the super soldier syndrome in 1st person simulations. In *Proceedings of SimTecT 2005 Conference*. Simulation Industry Association of Australia.

Becker, L., Nilsson, T., & Cowley, A. (2022). *Electroencephalography (EEG), electromyography (EMG) and eye-tracking for astronaut training and space exploration*. arXiv. https://doi.org/10.48550/arXiv.2212.06139

Braun, V., & Clarke, V. (2006). Using thematic analysis in psychology. *Qualitative Research in Psychology, 3*(2), 77–101. https://doi.org/10.1191/1478088706qp063oa

Buckingham, G. (2021). Hand tracking for immersive virtual reality: Opportunities and challenges. *Frontiers in Virtual Reality, 2*, Article 728461. https://doi.org/10.3389/frvir.2021.728461

Casini, A. E. M., Mittler, P., Uhlig, T., Rometsch, F., Schlutz, J., Ferra, L., Cowley, A., & Fischer, B. (2023, March 6–10). *Crew training and concept of operations testing for future lunar mission at the ESADLR lunar facility* [Presentation]. 17th International Conference on Space Operations, Dubai, UAE.

De Martino, E., Salomoni, S. E., Winnard, A., McCarty, K., Lindsay, K., Riazati, S., Weber, T., Scott, J., Green, D. A., Hides, J., Debuse, D., Hodges, P. W., van Dieën, J. H., & Caplan, N. (2020). Hypogravity reduces trunk admittance and lumbar muscle activation in response to external perturbations. *Journal of Applied Physiology, 128*(4), 1044–1055.

Delgado, F. J., & Noyes, M. (2017, May 8–11). *NASA's hybrid reality lab: One giant leap for full dive* [Presentation]. GPU Technology Conference (GTC), San Jose, CA, USA.

Duda, S., Warburton, C. & Black, N. (2020). Contextual research. In M. Kurosu (Ed.), *Human-computer interaction. Design and user experience* (pp. 33–49). Springer International Publishing. https://doi.org/10.1007/978-3-030-49059-1_3

Engine, U. (2023). *The most powerful real-time 3d creation tool.* [Apparatus and software]. https://www.unrealengine.com/en-US/

Ericsson, K. A., & Simon, H. A. (1980). Verbal reports as data. *Psychological Review, 87*(3), 215–251. https://doi.org/10.1037/0033-295X.87.3.215

Fisher, S. S., Wenzel, E. M., Coler, C., & McGreevy, M. W. (1988). Virtual interface environment workstations. *Proceedings of the Human Factors Society Annual Meeting, 32*(2), 91–95. https://doi.org/10.1177/154193128803200219

Groemer, G. E., Hauth, S., Luger, U., Bickert, B., Sattler, B., Hauth, E., Föger, D., Schildhammmer, D., Agerer, C., Ragonig, C., Sams, S., Kaineder, F., & Knoflach, M. (2012). The Aouda. X space suit simulator and its applications to astrobiology. *Astrobiology, 12*(2), 125–134. https://doi.org/10.1089/ast.2011.0673

Hart, S. G. (2006). NASA-task load index (NASA-TLX); 20 years later. *Proceedings of the human factors and ergonomics society annual meeting, 50*(9), 904–908. https://doi.org/10.1177/154193120605000909

Hennink, M., Hutter, I., & Bailey, A. (2020). *Qualitative research methods* (2nd rev. ed.). Sage. https://doi.org/10.1007/s11135-023-01660-5

Herrera, F., Bailenson, J., Weisz, E., Ogle, E., & Zaki, J. (2018). Building long-term empathy: A large-scale comparison of traditional and virtual reality perspective-taking. *PloS One, 13*(10), Article e0204494. https://doi.org/10.1371/journal.pone.0204494

HTC. (2023). *Finde die richtige high-end-vr für dich* [Find the right high-end vr for you] [Apparatus and software]. https://www.vive.com/de/product/

Koutnik, A. P., Favre, M. E., Noboa, K., Sanchez-Gonzalez, M. A., Moss, S. E., Goubran, B., Ari, C., Poff, A. M., Rogers, C. Q., DeBlasi, J. M., Samy, B., Moussa, M., Serrador, J. M., & D'Agostino, D. P. (2021). Human adaptations to multiday saturation on NASA NEEMO. *Frontiers in Physiology, 11*, Article 610000. https://doi.org/10.3389/fphys.2020.610000

Loftin, R. B., & Kenney, P. (1995). Training the Hubble space telescope flight team. *IEEE Computer Graphics and Applications, 15*(5), 31–37. https://doi.org/10.1109/38.403825

Manjrekar, S., Sandilya, S., Bhosale, D., Kanchi, S., Pitkar, A., & Gondhalekar, M. (2014). Cave: An emerging immersive technology – a review. *2014 UKSIM-AMSS 16th International Conference on Computer Modelling and Simulation*, 131–136. https://doi.org/10.1109/UKSim.2014.20

Meyen, F. E. (2013). *Engineering a robotic exoskeleton for space suit simulation* [Unpublished doctoral dissertation]. Massachusetts Institute of Technology.

Meta. (2020). *Meta quest 2: Immersives all-in-one VR-headset* [Apparatus and software]. https://www.meta.com/de/quest/products/quest-2/

Nilsson, T., Rometsch, F., Casini, A.E.M., Guerra, E., Becker, L., Treuer, A., de Medeiros, P., Schnellbaecher, H., Vock, A. & Cowley, A. (2022). Using virtual reality to design and evaluate a lunar lander: The EL3 case study. In S. Barbosa, C. Lampe, C. Appert, & D. A. Shamma (Eds.), *Extended abstracts of the 2022 CHI Conference on Human Factors in Computing Systems* (Article 305). Association for Computing Machinery. https://doi.org/10.1145/3491101.3519775

Nilsson, T., Rometsch, F., Becker, L., Dufresne, F., Demedeiros, P., Guerra, E., Casini, A. E. M., Vock, A., Gaeremynck, F., & Cowley, A. (2023). Using virtual reality to shape humanity's return to the moon: Key takeaways from a design study. In A. Schmidt, K. Väänänen, T. Goyal, P. O. Kristensson, A. Peters, S. Mueller, J. R. Williamson, & M. L. Wilson (Eds.), *Proceedings of the 2023 CHI Conference on Human Factors in Computing Systems* (Article 305). Association for Computing Machinery. https://doi.org/10.1145/3544548.3580718

Orr, S., Casler, J., Rhoades, J., & de León, P. (2022). Effects of walking, running, and skipping under simulated reduced gravity using the NASA active response gravity offload system (ARGOS). *Acta Astronautica, 197*, 115–125. https://doi.org/10.1016/j.actaastro.2022.05.014

Perret, J., & Vander Poorten, E. (2018, June 25–27). *Touching virtual reality: A review of haptic gloves* [Presentation]. ACTUATOR 2018, 16th International Conference on New Actuators, Bremen, Germany.

PicoXR. (2024). *PICO, a virtual reality company offering immersive interactive VR experiences with our all-in-one VR headsets* [Apparatus and software]. https://www.picoxr.com/de/products/pico4

Ross, A. K., Ruppert, S., Gläser, P., & Elvis, M. (2023). Preliminary quantification of the available solar power near the lunar South Pole. *Acta Astronautica, 211*, 616–630. https://doi.org/10.1016/j.actaastro.2023.06.040

Sandelowski, M. (2000). Combining qualitative and quantitative sampling, data collection, and analysis techniques in mixed-method studies. *Research in Nursing & Health, 23*(3), 246–255. https://doi.org/10.1002/1098-240X(200006)23:3<246::AID-NUR9>3.0.CO;2-H

Smith, D. E., Zuber, M. T., Neumann, G. A., Mazarico, E., Lemoine, F. G., Head III, J. W., Lucey, P. G., Aharonson, O., Robinson, M. S., Sun, X., Torrence, M. H., Barker, M. K., Oberst, J., Duxbury, T. C., Mao, D., Barnouin, O. S., Jha, K., Rowlands, D. D., Goossens, S., ... McClanahan, T. (2017). Summary of the results from the lunar orbiter laser altimeter after seven years in lunar orbit. *Icarus, 283*, 70–91.

Smith, M., Craig, D., Herrmann, N., Mahoney, E., Krezel, J., McIntyre, N., & Goodliff, K. (2020, March 7–10). *The Artemis program: An overview of NASA's activities to return humans to the moon* [Presentation]. IEEE Aerospace Conference, Big Sky, MT, USA. https://doi.org/10.1109/AERO47225.2020.9172323

Sutherland, I. E. (1998). A head-mounted three dimensional display. In R. Wolfe (Ed.), *Seminal graphics: Pioneering efforts that shaped the field* (pp. 295–302, Volume 1). Association for Computing Machinery. https://doi.org/10.1145/280811.281016

Unity. (2023). *Unity real-time development platform | 3D, 2D, VR AR engine* [Apparatus and software]. https://unity.com/

Vive. (2023). *Vive pro 2 full kit specs.* https://www.vive.com/de/product/vive-pro2-full-kit/specs/

Zhu, M., Sun, Z., Zhang, Z., Shi, Q., He, T., Liu, H., Chen, T. & Lee, C. (2020). Haptic-feedback smart glove as a creative human-machine interface (HMI) for virtual/augmented reality applications. *Science Advances, 6*(19), Article eaaz8693. https://doi.org/10.1126/sciadv.aaz8693

Chapter 12

Past, Present, and Future Trends in Aviation for the Usage of Extended Reality

Christophe Hurter, Mickaël Causse, and Maxime Cordeil

Abstract

Extended, virtual, mixed and augmented reality (XR, VR, MR, AR) devices allow users to interact with computer-generated environments that replicate or enhance the perception of the real world. AR superimposes digital information on the real world, while VR immerses the user in a virtual environment. MR combines elements of both to seamlessly blend the virtual and real worlds. XR technology provides immersive experiences, creating a sense of presence in the virtual realm. XR applications span various domains, such as training, education, medicine, digital health, and aviation, and can benefit pilots and air traffic controllers (ATCOs). In this chapter, we share our experiences using XR technology, detailing findings and the process of efficient implementation for end users such as pilots and ATCOs. We present examples of improving the paper strip by projecting dynamic and updated information, ensuring smoother monitoring for controllers. Our remote tower project incorporates remote sensing with sound spatialisation, enhancing situation awareness during airfield management. We also discuss developing an immersive analytic tool using XR devices for investigating recorded aircraft trajectories and its potential replication through dedicated application programming interface (API). We explore prospects for studying human behaviour using VR devices under more immersive conditions, manipulating mental workload, prototyping, evaluating complex interfaces (e.g. cockpit design), and understanding conflict-resolution algorithms through explainable AI (XAI). Finally, we address future challenges, such as improving interaction capabilities with tangibility in VR, and the opportunities XR technology could bring to aviation activities.

Keywords

immersive analytics, air traffic control, pilot, aeronautical data processing, extended reality

Introduction

Air traffic controllers (ATCOs) and pilots have to analyse large amounts of information in order to support efficient decision-making in a constrained time frame. Information can be structured (e.g. aircraft data such as position, altitude, call sign) or unstructured and complex, and may include sources such as audio streams. While the goal of ATCOs is to maintain a safe distance between aircraft, today it is almost as critical to optimise the traffic flow in terms of flight duration and fuel consumption. Research and development in this domain are active with the study of new systems and techniques and seek to improve traffic management. ATCOs use dedicated interactive visualisations to build a mental model by arranging *four-dimensional (4D, space + time)* trajectories. While recent technological improvements in terms of visualisation and interaction techniques arise, most existing systems display 2D representations of these 4D data. New opportunities emerge with head-mounted display (HMD) and other types of immersive environments such as cave automatic virtual environment (CAVE)-style rooms to display and interact with 4D data. Evolving with technology, visualisations have become more accurate with high-resolution displays, and now provide a sufficient number of pixels to display fine-grained information.

Immersive technology also supports more natural and immersive interactions (e.g. the use of HMDs with multi-modal input through the head and gesture tracking). Extended reality (XR) technology provides immersive and engaging experiences that make users feel more present in a virtual world. XR includes technologies such as virtual reality (VR), augmented reality (AR) and mixed reality (MR) devices. These tools allow users to interact with computer-generated simulations of real or plausible environments while enhancing their perception of the world. AR overlays digital information on the user's view of the real world, providing an augmented experience. VR, on the other hand, immerses the user in a virtual environment and isolates them from the real world. MR combines elements of AR and VR, seamlessly blending virtual and real-world components. XR has applications in a variety of fields, including training, education, medical and digital health, and aviation, such as for pilots and ATCOs. In addition, XR plays a critical role in research and development disciplines, enabling scientists and engineers to interact with and analyse complex data in innovative ways. For these reasons, this chapter focuses on recent investigations involving XR technology in the aviation domain, exploring its potential and benefits in this specific field.

In this chapter, we first provide a high-level description of how ATCOs work today to manage *the position of aircraft in the sky in real time*. We identify what is challenging with currently used technology and introduce the *augmented paper strips*. We then introduce another aspect of the controller domain where immersive analytics can potentially play an important role: the new remote tower concept. We provide a detailed analysis of this activity through design requirements and a scenario of usage.

In the following sections we also present the problem of large data analysis (e.g. large quantity of aircraft trajectories) involved in air traffic management (ATM). We review the previous tools used for this application and we demonstrate a tool that shows how immersive analytics, thanks to the use of HMDs, can be useful for performing collaborative air traffic data analysis. We then conclude this chapter with a discussion and the research agenda where all these efforts are in line with the evolution of future traffic (EUROCONTROL, 2013) in which every model predicts continual traffic increase.

Air Traffic Control and Novel Technologies

ATCOs optimise traffic with the following considerations: reducing fuel consumption and noise, traffic management, preventing waiting loops, avoiding crowded areas, and coordinating aircraft between air spaces. While controllers monitor aircraft trajectories to prevent ground and air collisions, their activity also involves cooperation with aeronautical practitioners. From a high-level perspective, traffic management can be carried out either in real time or in a delayed time frame. With direct commands to aircraft, real-time traffic management is equivalent to actual aircraft management. Data analytics is preferably used in deferred time management to study traffic in order to enable long-term traffic optimisation.

Working procedures for controllers are continuously improved while preserving and sustaining safety considerations. Every modification must be evaluated and validated, which slows down but secures the process. By offering new methods to capitalise on the operations of ATCOs, visual analytics suggests valuable potential. Particularly, 360-degree immersive displays surround people to address the constraints of conventional screens. The user can select to view the required information by turning their head. MR, as offered by the Microsoft Hololens, is a new and developing technology that offers additional interesting capabilities. Holograms are shown and anchored in the space as if they were a natural part of the user's surroundings, but the user is not entirely submerged in a VR. With the metaphor of a very big transparent display (360° of field of vision), this new technology offers tangible opportunities for ATCOs. In this sense, Willett et al. (2017) discussed opportunities to embed data representation in immersive reality or MR environments and they also provide a possible research agenda in this area.

VR is also booming in the field of piloting, making it possible, for example to train new procedures or checklists or to learn piloting without relying on complex and expensive flight simulators. For research purposes, it provides an easy way to duplicate the complex pilot environments through simulated scenarios (Labedan et al., 2021; Peysakhovich et al., 2020). Oberhauser et al. (2018) conducted a study to investigate the functional accuracy of a VR flight simulator compared to a traditional screen-based flight simulator. The study aimed to assess different aspects of the pilot's flight performance, such as heading, altitude, flight path deviations, and delays in control operations. Their results showed that deviations in flight performance were significantly greater in VR than in conventional flight simulation. As such, VR remains an environment where interactions can be more challenging than in mixed reality.

The Augmented Paper Strip

A paper strip is a physical strip used by ATCOs to record flight information. It contains details about an aircraft's call sign, altitude, speed, heading, etc. Controllers use it to track and coordinate aircraft movements for safe and efficient air traffic flow. Several earlier works have examined ATM and air traffic control (ATC) from various angles (Hurter et al., 2014a). While Mackay (1999) specifically investigated the usage of paper strips to learn more about ATC activity, Letondal et al. (2013) explored mental models for ATCOs with tangibility support, and Hurter et al. (2010) looked into augmented paper strips. Numerous real-time control interfaces, interactions, analytical tools, and techniques have been developed as a result of research in this application space. In this section, we go into further depth on the idea of *augmented paper strips* to help decision-making.

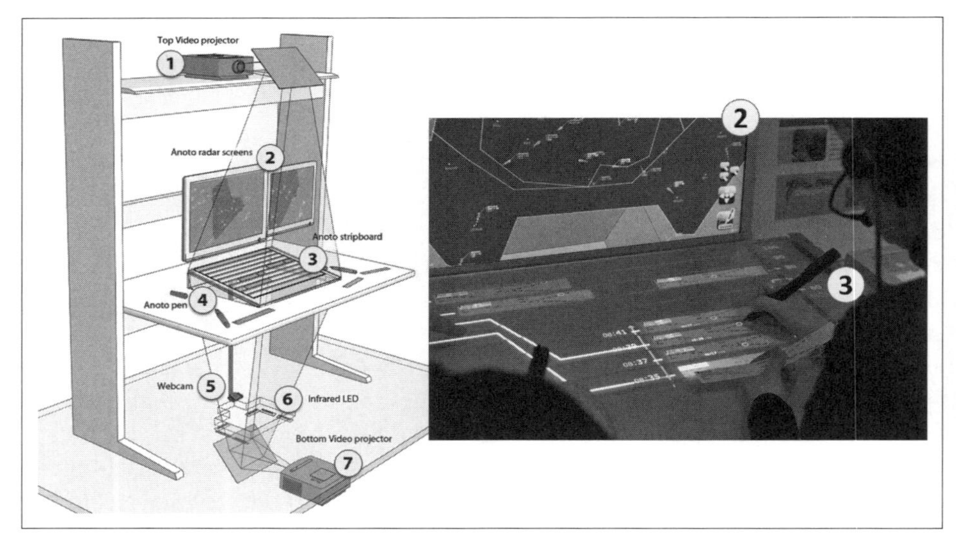

Figure 12.1. Strip'TIC (Stripping Tangible Interface for Controllers) is an augmented and tangible environment for control. This prototype consists of seven parts: an upper projector (1), a lower projector (7), an aircraft situation display (radar screen) (2), a strip board (3), digital pens (4), a webcam (5) and an infrared lighting system (6). © ENAC, University of Toulouse. Reprinted with permission.

Strip'TIC System Description

Once printed, the paper strip cannot be updated, nor can the information reported on the strip be used as input into the system. This strong limitation hinders the development of more responsive systems with more accurate future aircraft locations, complex conflict detection and resolutions. Taking into account this limitation, Strip'TIC (Stripping Tangible Interface for Controllers) has been developed (Hurter, Lesbordes, et al., 2012).

This prototype (see Figure 12.1) combines augmented paper and digital pens on a multi-touch glass strip board, using computer-based tracking and augmented rear and front projections. The paper strips, the strip board and the aircraft situation display are all covered with Anoto Digital Pen Patterns (DP patterns), which are small patterns used by the pen's infrared digital camera to calculate the pen position on the dotted surface. The controller's actions with the digital pen are sent to the software in real time. They can annotate strips of paper, and the resulting strokes are both physical (pen ink) and digital (projected ink). Controllers can use the pen to point at symbols on the main display or at names on the paper strips. The strip board, which helps to structure the set of strips, is semi-opaque. This allows for both a bottom projection on the strip board and strip tracking using AR patterns printed on the back of the strips. A second projector, located above the device, displays graphics on top of the strip board and the strips (e.g. projected ink, aircraft labels). The user can manipulate the paper strips as tangible objects and use gestures to fulfil ATM (Savery et al., 2013).

In the following scenario, we describe how controllers handle time and organise sequences of actions with paper strips for aircraft landings. When approaching an airport with dense traffic or bad weather, controllers can delay aircraft landings and place them in standby holding stacks. The stacks are organised vertically by flight altitude, and each aircraft entering the stack is placed at the highest altitude, flying in a hippodrome at this level. The controller gradually asks the aircraft to descend to the next available altitude. When a plane leaves the stack, a fixed time elapses until it lands on the runway. For example if aircraft are landing east on Runway 06 at Paris Orly airport, the time to reach the runway is 9 min from the ODRAN stack and 17 min from the MOLEK stack. In this case, the ATCO must mentally calculate when each aircraft must leave the stack, sequencing the aircraft from the two stacks and maintaining a 90-s interval between each landing. Figure 12.1 shows the arrangement of the strips on the strip board to help the ATCO form a mental picture of each aircraft in the stack. The strips represent the aircraft in the ODRAN and MOLEK stacks, as well as the planes allowed to leave the stacks in a given landing sequence.

Thanks to an immersive environment with augmented items (paper strips) and multi-modalities (pen, touch, tangible objects), Strip'TIC demonstrates how to support ATC operations (Hurter, Lesbordes, et al., 2012; Letondal et al., 2013).

Immersive Environment for Remote Towers

ATC service providers never cease innovating with new methods and facilities in order to maintain or increase safety levels and efficiency. One of the most recent inventions is called the "remote tower" (RT) concept (Arico et al., 2019). Aeronautical services can be performed using RT from a place other than the original control tower (Figure 12.2). To do this, an airfield mast with numerous sensors (cameras, radio antennas, etc.) is installed. Thanks to network connection systems, the video streams and additional data are sent to a remote location. Many nations have begun working on establishing RTs to support aviation services in aerodromes with little traffic or challenging access and to reduce development expenses compared to actual towers.

In the following, we present scenarios where RTs have shown to be effective in terms of supporting decision-making and safety level (Arico et al., 2019; Reynal et al., 2019).

Figure 12.2. Typical remote tower room. Image generated with stable diffusion model 1.4 (left). The immersive environment for a remote tower concept (right). © ENAC, University of Toulouse. Reprinted with permission.

Sound Location

At 2 p.m., the airspace is busy as usual with various small visual flight rules (VFR) and large instrument flight rules (IFR) aircraft. A new aircraft enters and contacts the tower, but the controller cannot establish visual contact due to other aircraft in the same direction. One potential solution is a sound location system with a surround set-up in the RT. It accurately reproduces airport sounds at their actual locations. For instance, if an aircraft is on the left of the controller, its sound will come from the left, helping the controller locate it. The system also alters the speech of pilots from further distances, indicating their distance from the airport. This technology enables controllers to identify an aircraft's position and establish visual contact more efficiently. It provides additional information, enhancing working speed. Even in good weather, spotting aircraft from the tower can be challenging, but the surround system helps identify them faster and with greater precision.

Haptic Feedback

At 5 p.m. the airport experiences high traffic density with numerous ground movements, particularly at taxiway crossings. Currently, two aircraft are taxiing towards one of these crossing points. The controller receives an alarm indicating the situation, but it takes time to pinpoint the exact location due to the high volume of ground movements. One potential solution is to use a haptic seat with SpeechCon and graphic prediction. The ground controller receives feedback via vibrating lines in the backrest, indicating the location of the possible danger. The vibration tempo correlates with the distance between the involved aircraft. The controller also hears an alarm saying, "Crossing! Crossing!" through the SpeechCon system. Additionally, a visual prediction of the path is displayed on the ground radar. These indicators draw the controller's attention to the dangerous situation, enabling a quick resolution. The SpeechCon system allows for customised messages for specific situations, aiding faster problem identification. The haptic seat helps emphasise the significance of the situation.

Immersive Environment for Pilots

Numerous aspects related to aircraft piloting can be addressed through XR. The development of virtual flight simulators (Valentino et al., 2017) enabled the creation of realistic situations in which individuals can train for piloting or understand the functioning of flight instruments. The space and financial gains are substantial, and it becomes feasible to easily design new systems and assess their usability during virtual in-flight conditions. Additionally, it is possible to create immersive environments simulating piloting scenarios with the aim of challenging pilots or operators and evaluating, for instance their capacity to manage high mental workload. A study by Gozzi et al. (2022) aimed to develop a virtual version of the well-known MATB-II task (Comstock Jr & Arnegard, 1992). The purpose of this study was to assess the potential benefits offered by an immersive version of the

MATB-II (Figure 12.3). Indeed, this task is widely used in the field of piloting to evaluate, for example the ability to manage multiple tasks simultaneously under high workload. However, the original version implemented on a simple 2D computer screen provides limited immersion. Furthermore, it does not account for the spatial distribution of the different cockpit tasks simulated by MATB-II. For instance the tracking task of the MATB-II replicates the control of the aircraft's attitude, and this instrument is located in front of the pilots, whereas the occurrence of visual alerts, simulated by the system monitoring task, may occur at distant locations within the cockpit.

The result, obtained with 31 participants using "MATB-II VR", showed that mental and physical effort was higher in the virtual version compared to the classic 2D MATB-II. Heart rate, measured throughout the task, was also higher with the VR version. These findings confirmed that immersive environments can be more challenging than aeronautical tasks performed on a classic 2D computer screen. However, the authors highlighted that interaction within virtual environments must be improved before complex operational tasks, such as piloting, can be effectively performed.

XR Usages for Aeronautical Data Analytics

Controllers and pilots deal with 4D real-time information: managing and controlling aircraft positions in the sky over time. However, most aeronautical interfaces rely on 2D radar screens, leading to inherent perceptual challenges with 3D+time data and constantly burdening the mental load of ATCOs. Researchers in the domain have explored solutions using immersive technology such as stereo displays, CAVEs, and HMDs. Mazzucchelli and Montelone (2008) introduced the AD4 project, which utilises immersive technology (auto-stereoscopic monitors) and 3D visualisations of air traffic data. However, they did not study users' performance with this set-up and assumed that immersive technology might not necessarily reduce the mental load, but could offer new decision-making opportunities for controllers. Lange et al. (2003) proposed a multi-modal immersive control system using 3D stereo screens and voice commands. They assumed that replacing voice commands for object selection could reduce the need for gestures and hand-based interactions. However, navigation, flight selection, and way-point manipulation still rely on a "virtual reality wand" (i.e. a 6DOF tracked pointer). The visualisation system integrates 3D aircraft representations, flight paths, waypoints, weather, and high-fidelity geography. The authors outlined a plan for evaluating the system, but no actual evaluation was provided. Little work has been done to assess the effectiveness of VR environments in supporting controller work.

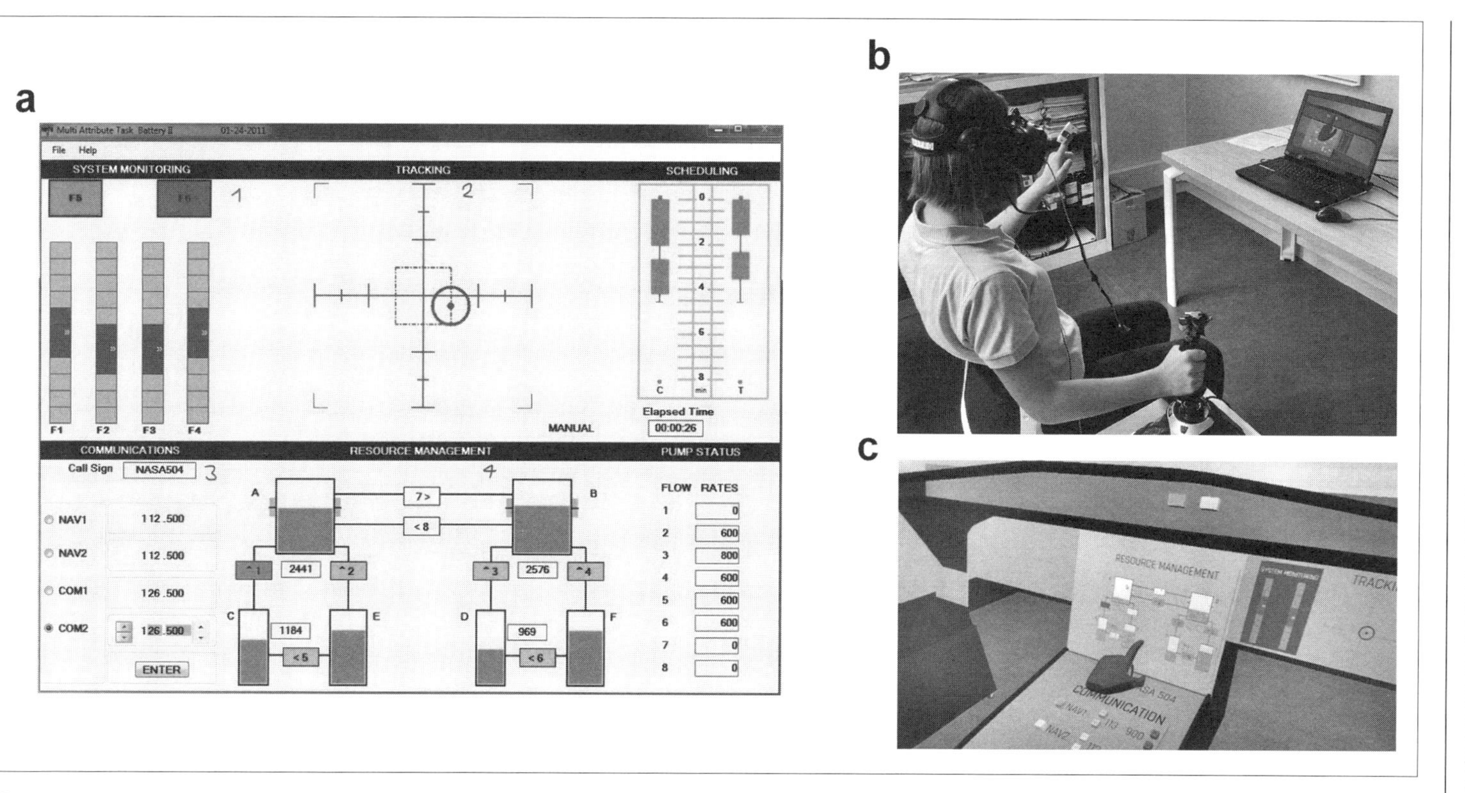

Figure 12.3. The classic MATB-II task developed by NASA (a). A participant performing the MATB-II VR task (b). The MATB-II VR task as seen by the participant (c). The different tasks of the classic MATB-II are now located in different areas of the virtual cockpit (e.g. tracking task, resource management task). Adapted from Gozzi et al., 2022. © ISAE–SUPAREO, University of Toulouse. Reprinted with permission.

Immersive Analysis of Air Traffic Data

Air traffic analysts study vast records of air traffic data. For instance, a single day's recorded traffic over France contains around 500,000 multidimensional data points. Understanding how traffic is reconfigured in non-nominal situations is crucial for ensuring fluidity and safety. Air traffic analysis involves various tools for extracting knowledge from real-time or recorded aircraft trajectory data. These tools offer different metrics and interaction techniques to understand the traffic structure and evolution (Hurter et al., 2014b). One approach is analysing traffic from a flow perspective, enabling users to compare trajectories and examine temporal changes (Scheepens et al., 2016). Simplification techniques such as Edge Bundling aid in visually exploring flow patterns (Hurter, Ersoy, et al., 2012). Dynamics of flows can be analysed using dynamic network and schematising techniques (Hurter, Ersoy, et al., 2014; Hurter et al., 2010). Visual analytic tools also help extract flow dynamics (Andrienko et al., 2013), sometimes revealing unexpected information such as meteorological parameters (Hurter, Alligier, et al., 2014). Interacting with large aircraft trajectory datasets is challenging, but pixel-based techniques (Hurter, 2015) offer solutions. Previous research in this field enhances our understanding of air traffic data management and its unique characteristics.

FiberClay (Hurter et al., 2019) is a tool that employs a bi-manual interaction metaphor, simulating the manipulation of clay to carve out content and create data queries from dense 3D dataset. The tool facilitates the analyst's tasks of filtering and selecting interesting 3D curve patterns by directly interacting with the 3D content using hand controllers. For trajectory analysis (Figure 12.4), users can select and filter subsets of trajectories through Boolean operations (e.g. selections,

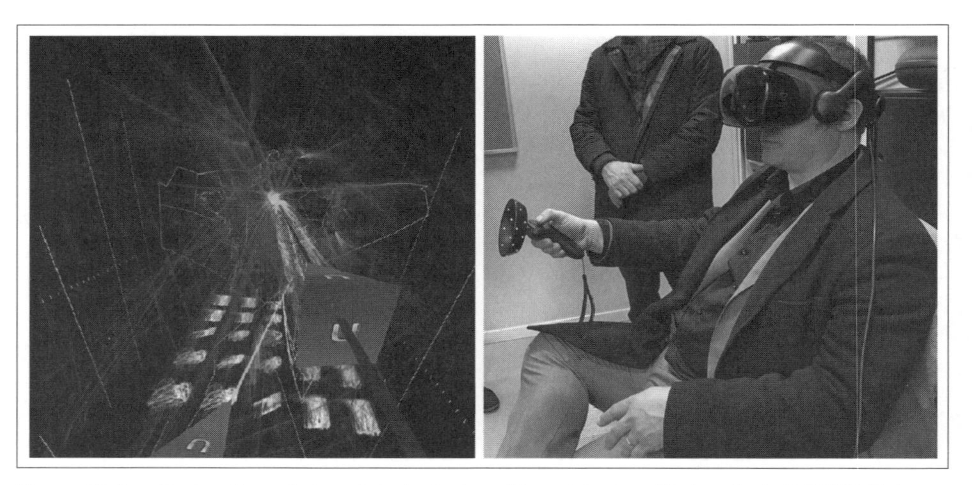

Figure 12.4. Exploration of aircraft trajectories with a virtual reality headset. © ENAC, University of Toulouse. Reprinted with permission.

intersections, exclusions) similar to a selection brush tool in photo editing software. This design aims to reduce cognitive load and task difficulty when dealing with complex 3D scenes, which can be challenging in traditional 2D desktop set-ups. Initial evaluations with domain experts show promising results, with users finding the tool easy to use and helpful in discovering and analysing curve patterns.

XR Usage for Collaborative Trajectories Visualisation

The previous section, Immersive Analysis of Air Traffic Data, calls for collaborative and immersive aircraft trajectory visualisation to facilitate common understanding, shared expertise, and reporting from distant locations. Currently, this is achieved using 2D visualisation tools displaying 3D aircraft trajectories, along with video conferencing systems for sharing insights. However, this set-up is not optimal for viewers when presenters navigate the visualisation (e.g. pan and zoom operations) and describe 3D data using 2D visualisations. Affordable immersive display and input technology, such as the Oculus Rift HMD and the Leapmotion hand-tracking device, enables the creation of efficient tools for remote collaboration in trajectory analysis. We developed a platform that facilitates collaborative aircraft trajectory visualisation, allowing multiple users to analyse large datasets together using affordable HMDs and hand-tracking devices.

As a collaborative platform for immersive big data visualisation, we have identified the following system needs: immersive visualisation, navigation, land-mark, user gaze visualisation, and data filtering. Cordeil et al. (2016) created a networked Unity application that connects two users from remote sites using Oculus Rift headsets for immersive visualisation (Figure 12.5, top). The users' hands and fingers are tracked with Leapmotion, providing high-fidelity feedback within the Oculus Rift. Navigation is achieved through head-tracking, leaning, and rotating the head, along with gaze direction and a game pad. Rendezvous interaction allows users to send location cues in the virtual environment, while basic filtering operations enable traffic replay and opacity adjustments for trajectory visualisation. Communication is supported with VOIP clients such as Skype or Zoom. This platform is an initial step toward enhancing communication between remote experts.

Explainable AI (XAI) Opportunities

In ATM, the decision-making process is already integrated with artificial intelligence (AI), aiming to assist controllers in their daily tasks. However, AI systems still face challenges regarding acceptability. Recent surveys, such as the one by Degas et al. (2022), highlight the current use of XAI (explainable AI) and its potential application in future aeronautical systems. Currently, AI/machine

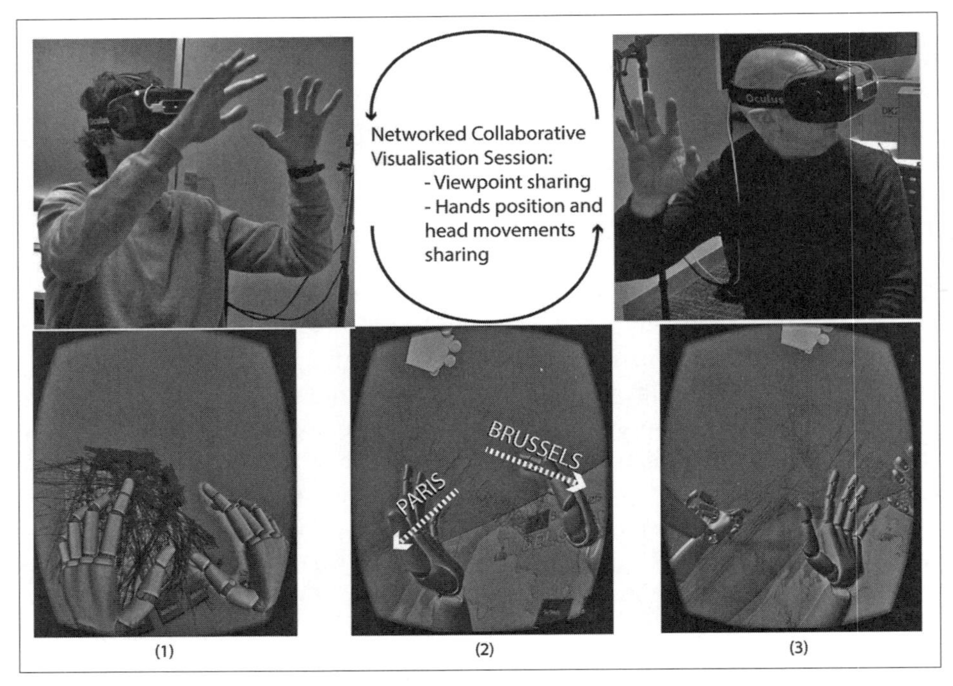

Figure 12.5. Top: Remote users in the collaborative visualisation platform wearing Oculus Rift DK2 head-mounted displays with Leapmotion for hand tracking. Bottom: Visual representation of the remote collaborative session in the Unity client. The session begins with an overview of the dataset (1), showcasing a day of air traffic data over France. An expert sends a rendezvous position to the other viewer and explains how the stack was formed and handled (2). Both participants share the same virtual space, communicate orally through VoIP (Voice over Internet Protocol), and can directly describe and display specific spatial arrangements of trajectories using their hands (green and blue hands) (3). © ENAC, University of Toulouse. Reprinted with permission.

learning automation systems lack transparency as they only provide results without additional information to support explanations. The goal is to transform the decision-making process into a "white box" delivering understandable outcomes through transparent processes. The project ARTIMATION (n.d.) focuses on achieving this by using data visualisation, data-driven storytelling, and immersive analytics to create a transparent and explainable AI model (Hurter et al., 2022). This project leverages human perceptual capabilities to enhance our understanding of AI algorithms through appropriate data visualisation, facilitating XAI. Immersive analytics is also explored as a means to present information effectively.

A VR proof of concept was developed by adapting the existing tool FiberClay (Hurter et al., 2019). This prototype explores trajectory optimisation possibilities,

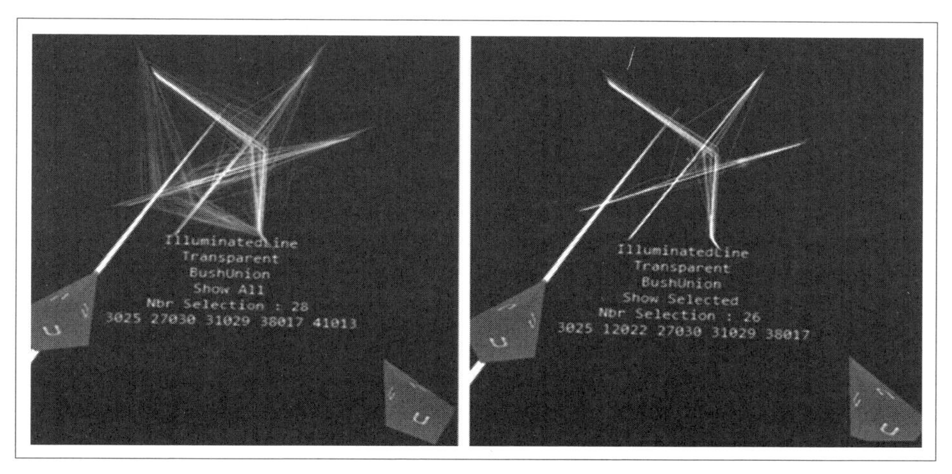

Figure 12.6. Display of a typical traffic situation with stacked trajectories showcasing potential optimised traffic flow. © ENAC, University of Toulouse. Reprinted with permission.

utilising the third dimension to differentiate between candidate solutions (Figure 12.6). The virtual reality environment aids in explaining traffic optimisation behaviour and offers various dataset presentations and filtering options. This interface proves beneficial to developers for analysing and certifying behaviours, especially in post-operations.

Summary and Outlook

Aeronautical activities are cautious about technical changes due to safety concerns. The working methods of ATCOs and pilots and the technological environment cannot change rapidly. Modifications must be carefully assessed for safety. While immersive environments are readily available, widespread adoption will take time. However, as a long-term plan, they offer valuable opportunities for integration with aeronautical practitioners. Deferred data analysis using immersive environments supports effective decision-making. Immersive analytics has remarkable data visualisation and interaction capabilities, but faces challenges with emerging technologies. Long-term effects on users' physiology are not well understood. Introducing this technology may affect the work duration of ATCOs and pilots and lead to different types of fatigue. New media or devices must undergo validation for data presentation quality, security, and resilience. While immersive environments have a long history, the recent interest in immersive analytics lacks scientific work on its support for aeronautical activities. Investigation and potential applications in future developments are promising.

References

Andrienko, G., Andrienko, N., Hurter, C., Rinzivillo, S., & Wrobel, S. (2013). Scalable analysis of movement data for extracting and exploring significant places. *IEEE Transactions on Visualization and Computer Graphics, 19*(7), 1078–1094. https://doi.org/10.1109/TVCG.2012.311

Arico, P., Reynal, M., Di Flumeri, G., Borghini, G., Sciaraffa, N., Imbert, J.-P., Hurter, C., Terenzi, M., Ferreira, A., Pozzi, S., Betti, V., Marucci, M., Telea, A. C., & Babiloni, F. (2019). How neurophysiological measures can be used to enhance the evaluation of remote tower solutions. *Frontiers in Human Neuroscience, 13*, Article 303. https://doi.org/10.3389/fnhum.2019.00303

ARTIMATION. (n.d.). https://www.artimation.eu/

Comstock, J. R., Jr., & Arnegard, R. J. (1992). *The multi-attribute task battery for human operator workload and strategic behaviour research* [Technical report No. NAS 1.15: 104174].

Cordeil, M., Dwyer, T., & Hurter, C. (2016). Immersive solutions for future air traffic control and management. In *Proceedings of the 2016 ACM companion on interactive surfaces and spaces*, (pp. 25–31). ACM. https://doi.org/10.1145/3009939.3009944

Degas, A., Islam, M. R., Hurter, C., Barua, S., Rahman, H., Poudel, M., Ruscio, D., Ahmed, M. U., Begum, S., Rahman, M. A., Bonelli, S., Cartocci, G., Di Flumeri, G., Borghini, G., Babiloni, F., & Arico, P. (2022). A survey on artificial intelligence AI and explainable AI in air traffic management: Current trends and development with future research trajectory. *Applied Sciences, 12*(3), Article 1295. https://doi.org/10.3390/app12031295

EUROCONTROL. (2013). *Challenges of growth 2013*. https://www.eurocontrol.int/publication/challenges-growth-2013

Gozzi, Z., Peysakhovich, V., Cantu, A., & Causse, M. (2022). Behavioral and physiological assessment of a virtual reality version of the MATB-II task. In *Proceedings of the 17th International Joint Conference on Computer Vision, Imaging and Computer Graphics Theory and Applications – Volume 2: VISIGRAPP* (pp. 77–87). https://doi.org/10.5220/0010912100003124

Hurter, C. (2015). *Image-based visualization: Interactive multi- dimensional data exploration*. Morgan & Claypool Publishers. https://doi.org/10.1007/978-3-031-02601-0

Hurter, C., Alligier, R., Gianazza, D., Puechmorel, S., Andrienko, G., & Andrienko, N. (2014). Wind parameters extraction from aircraft trajectories. *Computers, Environment and Urban Systems, 47*, 28–43. https://doi.org/10.1016/j.compenvurbsys.2014.01.005

Hurter, C., Conversy, S., Gianazza, D., & Telea, A. (2014a). Interactive image-based information visualization for aircraft trajectory analysis. *Transportation Research Part C: Emerging Technologies, 47*, 207–227. https://doi.org/10.1016/j.trc.2014.03.005

Hurter, C., Conversy, S., Gianazza, D., & Telea, A. (2014b). Interactive image-based information visualization for aircraft trajectory analysis. *Transportation Research Part C: Emerging Technologies, 47*(Pt. 2), 207–227. https://doi.org/10.1016/j.trc.2014.03.005

Hurter, C., Degas, A., Guibert, A., Durand, N., Ferreira, A., Cavagnetto, N., Islam, M. R., Barua, S., Ahmed, M. U., Begum, S., Bonelli, S., Cartocci, G., Flumeri, G. D., Borghini, G., Babiloni, F., & Aricio, P. (2022). Usage of more transparent and explainable conflict resolution algorithm: Air traffic controller feedback. *Transportation Research Procedia, 66*, 270–278.

Hurter, C., Ersoy, O., Fabrikant, S. I., Klein, T. R., & Telea, A. C. (2014). Bundled visualization of dynamic graph and trail data. *IEEE Transactions on Visualization and Computer Graphics, 20*(8), 1141–1157. https://doi.org/10.1109/TVCG.2013.246

Hurter, C., Ersoy, O., & Telea, A. (2012). Graph bundling by kernel density estimation. *Computer Graphics Forum, 31*(3pt1), 865–874. https://doi.org/10.1111/j.1467-8659.2012.03079.x

Hurter, C., Lesbordes, R., Letondal, C., Vinot, J.-L., & Conversy, S. (2012). Strip'tic: Exploring augmented paper strips for air traffic controllers. In *Proceedings of the International Working Conference on Advanced Visual Interfaces* (pp. 225–232). ACM. https://doi.org/10.1145/2254556.2254598

Hurter, C., Riche, N. H., Drucker, S. M., Cordeil, M., Alligier, R., & Vuillemot, R. (2019). Fiberclay: Sculpting three dimensional trajectories to reveal structural insights. *IEEE Transactions on Visualization and Computer Graphics, 25*(1), 704–714. https://doi.org/10.1109/TVCG.2018.2865191

Hurter, C., Serrurier, M., Alonso, R., Tabart, G., & Vinot, J.-L. (2010). An automatic generation of schematic maps to display flight routes for air traffic controllers: Structure and color optimization. In *Proceedings of the International Conference on Advanced Visual Interfaces* (pp. 233–240). ACM. https://doi.org/10.1145/1842993.1843034

Labedan, P., Darodes-de Tailly, N., Dehais, F., & Peysakhovich, V. (2021). Virtual reality for pilot training: Study of cardiac activity. In *Proceedings of the 16th International Joint Conference on Computer Vision, Imaging and Computer Graphics Theory and Applications -- Volume 2: VISIGRAPP* (pp. 81–88). https://doi.org/10.5220/0010296700810088

Lange, M., Hjalmarsson, J., Cooper, M., Ynnerman, A., & Duong, V. (2003). 3D visualization and 3D and voice interaction in air traffic management. In *The annual Swedish Chapter of Eurographics (SIGRAD) conference. Special theme-real-time simulations. Conference proceedings from SIGRAD2003, No. 10* (pp. 17–22). Linkoping University Electronic Press.

Letondal, C., Hurter, C., Lesbordes, R., Vinot, J.-L., & Conversy, S. (2013). Flights in my hands: Coherence concerns in designing strip'tic, a tangible space for air traffic controllers. In *Proceedings of the SIGCHI Conference on Human Factors in Computing Systems* (pp. 2175–2184). ACM. https://doi.org/10.1145/2470654.2481300

MacKay, W. E. (1999). Is paper safer? the role of paper flight strips in air traffic control. *ACM Transactions on Computer-Human Interactions, 6*(4), 311–340. https://doi.org/10.1145/331490.331491

Mazzucchelli, L., & Monteleone, A. (2008). A 4D immersive virtual reality system for air traffic control. *Proceedings of the 7th EUROCONTROL innovative research workshop & exhibition*. EUROCONTROL Experimental Centre.

Oberhauser, M., Dreyer, D., Braunstingl, R., & Koglbauer, I. (2018). What's real about virtual reality flight simulation? *Aviation Psychology and Applied Human Factors, 8*(1), 22–34. https://doi.org/10.1027/2192-0923/a000134

Peysakhovich, V., Monnier, L., Gornet, M., & Juaneda, S. (2020). Virtual reality vs. real-life training to learn checklists for light aircraft. In *Eye-tracking in aviation. Proceedings of the 1st International Workshop (ETAVI 2020)* (pp. 47–53). ISAE-SUPAERO, Universite de Toulouse, Institute of Cartography and Geoinformation (IKG).

Reynal, M., Imbert, J.-P., Arico, P., Toupillier, J., Borghini, G., & Hurter, C. (2019). Audio focus: Interactive spatial sound coupled with haptics to improve sound source location in poor visibility. *International Journal of Human-Computer Studies, 129*, 116–128. https://doi.org/10.1016/j.ijhcs.2019.04.001

Savery, C., Hurter, C., Lesbordes, R., Cordeil, M., & Graham, T. C. N. (2013). When paper meets multi-touch: A study of multi-modal interactions in air traffic control. In P. Kotzé, G. Marsden, G. Lindgaard, J. Wesson, & M. Winckler (Eds.), *Human-Computer Interaction – INTERACT 2013* (pp. 196–213). Springer. https://doi.org/10.1007/978-3-642-40477-1_12

Scheepens, R., Hurter, C., Wetering, H. V. D., & Wijk,J. J. V. (2016). Visualization, selection, and analysis of traffic flows. *IEEE Transactions on Visualization and Computer Graphics, 22*(1), 379–388. https://doi.org/10.1109/TVCG.2015.2467112

Valentino, K., Christian, K., & Joelianto, E. (2017). Virtual reality flight simulator. *Internetworking Indonesia Journal, 9*(1), 21–25.

Willett, W., Jansen, Y., & Dragicevic, P. (2017). Embedded data representations. *IEEE Transactions on Visualization and Computer Graphics, 23*(1), 61–470. https://doi.org/10.1109/TVCG.2016.2598608

Chapter 13

A Human Factors Approach for Evaluating Virtual Reality Training in the Aeronautics Domain

Cédric Bach, Stéphane Drouot, Nawel Khenak, Anne-Claire Collet, Federico Nemmi, and Florence Buratto

Abstract

In light of the lack of scientific studies and official guidelines, there is a need for the development of a standard testing method to assess virtual reality (VR) training devices. In this chapter we propose a method for the validation of VR training devices in a pilots' training context. This method is based on three pillars: reducing the risk of cybersickness, assessing effectiveness (crucial learning acquisition), and assessing efficiency (better learning/time ratio). The three-pillar approach we propose appears as a valid means to integrate VR in professional training while ensuring that critical learning is comparable to existing means. It allows for a systematic validation of the training devices and lays a foundation for the required change in the management of safe and efficient VR integration. At each of the three steps of this approach, it enables users to iterate on the devices and to adjust technical or usage parameters that can enhance the user experience.

Keywords

virtual reality training, cybersickness, human factors method, learning effectiveness and efficiency

Introduction

The rapid development of immersive technologies such as virtual reality (VR) in the training environment raises questions about the physiological effects and pedagogical benefits of using VR as a training device technology. For example, VR training devices must be shown to be at least as effective for learning critical tasks for pilots or cabin crew as the existing means of training. On the educational side, VR training devices tend to show advantages for motor and visuospatial task learning over theoretical knowledge acquisition (Renganayagalu et al., 2021). However, many studies available in the literature point out the potential adverse effects of

VR technology on human physiology and cognition such as the so-called cybersickness. This phenomenon is increasingly documented also in the international standards (International Organization for Standardization, 2020a; NATO Science and Technology Office, 2021). Thus, a user-centered approach is required to prevent "savage" insertions missing testing and iteration of new products and their usage. Specifically, integrating a VR solution to train pilots should address some crucial human factors (HF) dimensions.

Due to the lack of scientific studies and official guidelines, we developed a standard testing method to assess VR training devices using a research action approach. This method is based on three pillars: (1) reducing the risk of cybersickness, (2) assessing effectiveness – crucial learning acquisition, and (3) assessing efficiency – better learning/time ratio. After describing cybersickness and its impact on VR training in the next section, this chapter exposes the methods and the key findings for the three pillars.

Cybersickness and Its Impact on VR Training

Main Symptoms of Cybersickness

The most general definition of cybersickness has been proposed by Nesbitt and Nalivaiko (2018, p. 1) as "an uncomfortable side effect experienced by users of immersive interfaces commonly used for Virtual Reality." Cybersickness is thought to severely impact up to 5% of VR users (Stanney et al., 2003; Lawson, 2014), forcing them to terminate the exposure prematurely. However, milder symptoms and discomfort, which may not force exposure termination, but can impact performance and learning, are thought to be experienced by up to one third of VR users (Stanney et al., 2020). For particularly aversive environments (e.g., roller-coaster simulator) or tasks (e.g., requiring ample and frequent head movements), cybersickness symptoms might be experienced by the totality of users (Davis et al., 2015). However, cybersickness symptoms can emerge not only during or immediately after exposure, but also until 24 hr after exposure termination. The most frequently reported symptoms of cybersickness are visual fatigue, nausea, headaches, disorientation, tiredness, and postural instability (Bouchard et al., 2011; Bruck & Watters, 2011). These symptoms are among those measured by the Simulator Sickness Questionnaire (SSQ; Kennedy et al., 1993), a questionnaire that is also widely used for measuring cybersickness.

Visual Fatigue

Visual fatigue or eyestrain refers to an array of conditions or symptoms experienced by individuals who perform tasks requiring extended visual concentration or focus. *Blurred vision*, *headache*, *dry eyes*, *difficulty in maintaining visual focus*, *eye pain*, and *tiredness* are all facets of visual fatigue (Sheedy, 2014). The symptomatology of visual fatigue also includes extraocular disorders such as *headache*, *neck pain*, *back and shoulder pain*, *decreased performance in mental activities*, and *loss of concentration* (Blehm et al., 2005; Emoto et al., 2005). The last two symptoms are quite important to consider regarding the training activities, as they could negatively impact training efficiency. On top of these subjective symptoms, objective symptoms have also been reported after VR exposure, such as *exophoria*, that is, lack of eyes coordination with one eye closed (Morse & Jiang, 1999), and an increase in the near point of accommodation and convergence (Yoon et al., 2020).

There is a high incidence of visual fatigue in VR users, with up to 60% reporting visual fatigue following VR immersion using a head-mounted display (HMD; Mon-Williams et al., 1993), mainly due to the vergence-accommodation conflict. The systematic presence of visual fatigue after VR exposure with HMD has been confirmed by a recent meta-analysis (Yuan et al., 2018). Although the scientific literature suggests that objective symptoms categorized under visual fatigue are transient and disappear rapidly (in approximately less than 1 hr), hazard still exists. Indeed, the presence of persistent visual fatigue and objectively measured change in vergence and accommodation after VR exposure suggest extreme caution when VR is used in vocational contexts that require driving, using heavy machinery, or generally where vision-related problems could lead to accidents. Therefore, people exposed to VR should be encouraged to wait at least 1 hr before undertaking tasks that are demanding on the visual system.

Disorientation

Disorientation can refer to two distinct but somewhat related concepts: the loss of one's sense of direction or one's position in relation to the surroundings, and a more general state of confusion with regard to time and place. Specific symptoms pertaining to disorientation are fullness of the head, dizziness, and vertigo. It has been suggested that a predominance of disorientation symptoms over oculo-motor and nausea-related symptoms is what differentiates cybersickness from simulator sickness and motion sickness, respectively (Kim et al., 2018). A recent meta-analysis of VR studies using HMD reports that the disorientation subscale of the SSQ has the highest average value for post-exposure assessments (Saredakis et al., 2020). A mismatch between proprioceptive feedback (e.g., balance organs in the inner ear) and perceptual visual feedback is thought to be one of the underpinning causes of disorientation.

Because disorientation is at the core of cybersickness (Kim et al., 2018) and shows the highest pre-post exposure increase (Saredakis et al., 2020), particular attention should be paid to reducing/monitoring the effects of this symptom. For example, if disorientation persists after VR exposure, exposed individuals should be warned not to drive, use heavy machinery, and generally perform dangerous tasks. However, it should be noted that there is a lack of studies focusing on the time course of disorientation and visual fatigue aftereffects. While 1 hr seems a reasonable estimate on average, these aftereffects could last longer than that. Therefore, individuals should be informed of possible longer-term effects.

Nausea

Nausea is defined as a diffuse sensation of unease and discomfort, often perceived as an urge to vomit. Several symptoms classified under the umbrella of nausea or gastrointestinal symptoms are included in the SSQ (Kennedy et al., 1993) and Motion Sickness Assessment Questionnaire (MSAQ; Gianaros et al., 2001) such as increased salivation, sweating, stomach awareness, burping, feeling sick to stomach, or feeling queasy. Although nausea and gastrointestinal-related symptoms are among the most cited in cybersickness studies (LaViola, 2000), precise estimates of the incidence of this family of symptoms are hard to find. A recent study using state-of-the-art technology and three off-the-shelf VR games (with a gameplay time between 9 and 19 min) showed that 64 out of 195 individuals (33%) somewhat or strongly agreed with the sentence "the experience was nauseating" (with 5% strongly agreeing; Stone, 2017). Other authors report a similar estimate of 25% of individuals reporting mild or severe nausea symptoms, using HMD and a static-environment visual stimulation (Moss & Muth, 2011). A somewhat similar proportion of 20% of participants reported having nausea-related symptoms in one of the first studies concerning HMD VR (Mon-William et al., 1993). Another study, using two intentionally provocative roller-coaster rides simulation, found that all participants (12 out of 12) reported being at least mildly nauseated, with eight participants quitting the immersion across the two rides (Davis et al., 2015). Of note, nausea symptoms are also associated with age, with older individuals more prone to developing symptoms than younger ones, and with gender, with women more prone to developing symptoms than men (Kemeny et al., 2020).

As for disorientation, a sensory mismatch between proprioception and visual feedback could be the root of nausea symptoms. It has been suggested that nausea-related symptoms are the least frequently reported for cybersickness (Kim et al., 2018). Moreover, Saredakis et al. (2020) reported in their meta-analysis that the nausea subscale of the SSQ shows the smallest average difference between pre- and post-exposure. However, nausea is still a commonly reported symptom, observed even for relatively static environments where the only movements are head movements. Moreover, the incidence of nausea is thought to be much higher than the one reported "on average" if the VR environment or task presented to the participants is particularly aversive, especially with the presence of a sensory-cue conflict (e.g., VR presents a velocity movement while the user is still static).

Impact of Cybersickness on VR Training Efficiency

So far, there are not yet clear links between the cybersickness level and its consequences on pilot training efficiency. Research indicates contrasting results, showing both the presence (Souchet, 2020) and the absence (Souchet et al., 2019) of cybersickness effects on the learning rate (Souchet et al., 2022). Nevertheless, visual fatigue is promising for explaining the impact of cybersickness on VR training, as the symptomatology includes extraocular symptoms such as *decreased performance in mental activities* and *loss of concentration*. Indeed, these cognitive activities are central to training. Besides, there is some evidence that exposure to 3D contents can cause visual fatigue and adverse cognitive effects. For example, Szpak et al. (2019) have shown that 30 min of VR immersion led to adverse cognitive effects including *changes in decision times* that may be related to *alertness* and *attention*. According to Sepich et al. (2022), the overall task performance in VR was negatively correlated with cybersickness, emphasizing the importance of reducing cybersickness for demanding tasks such as flying an aircraft.

Therefore, it is important to consider the risk of cybersickness before introducing VR training in operations with pilots. The first step in assessing the risk of cybersickness could be an *expert inspection*. This type of method is common in usability engineering and is also relevant for identifying the potential risk of cybersickness that could jeopardize the use of VR (Bach & Scapin, 2010). An inspection of a VR device should follow four main steps: (a) the familiarization with the VR environment, (b) the cybersickness risk assessment itself, (c) the quotation of the risk among cybersickness relative dimensions, and (d) the analysis and recommendations. While these steps are standard in expert inspection, there is a specific difficulty to inscribe them within a structure based on known cybersickness symptoms, due to the lack of studies in this field. Therefore, to provide a guide for organizing an inspection process, we defined seven HF dimensions that may have an impact on cybersickness risk (Khenak, Bach, & Buratto, 2023):

1. User Characteristics
   This dimension consists of knowing more about the targeted population of users and their *inter- and intra-individual susceptibility* to cybersickness. Many factors can influence this susceptibility. Thus, the risk of experiencing cybersickness is lower if the targeted population is selected with a low level of susceptibility (e.g., pilots and military). Conversely, the general population is statistically more exposed to the cybersickness risk.
2. Technology Properties
   VR technology is characterized by technical and ergonomic properties that can have a significant impact on cybersickness. Technical properties mostly include visual factors such as the field of view, the visual clutter, and the latency. Regarding ergonomic properties of HMDs, one should pay attention to the weight and the weight distribution, which can increase the inertia on the neck causing muscle fatigue. Besides, the inter-pupillary distance (IPD) settings on some

headsets do not fit with the required IPD adjustment, potentially leading to visual fatigue.

3. Task to Perform
   The type of tasks and how users achieve them, that is, the interaction methods, can impact VR users' comfort. For example, observation tasks could lead to translational or rotational head movements that may increase the motion parallax of objects, which is an important predictor of cybersickness.
4. VR Content/Application
   The content represents the interface in/with which users interact. Many content-related factors affect cybersickness such as motion characteristics (e.g., acceleration) and visual characteristics (e.g., level of realism). For example, it has been observed that contents with high dynamic motion features (e.g., multiple moving objects) are prone to causing more cybersickness than static features.
5. Physical Space
   The physical space refers to the room where the VR interaction takes place. Space-related factors can negatively impact the safety of use beyond the cybersickness phenomenon itself. Besides, user comfort can be impacted by environmental conditions such as high temperature, the presence of noise, and variation of luminosity and brightness, which may also affect the tracking system.
6. Exposure Duration
   The duration of VR exposure is pointed out as one of the main contributors to cybersickness. Indeed, different studies demonstrated that cybersickness increases with exposure time (Dużmańska et al., 2018). The repetition of VR exposures and the rest time between exposures may also influence cybersickness risk.
7. Cybersickness Risk Knowledge and Prevention
   The level of cybersickness knowledge and the risk associated with VR has a direct impact on the detection of cybersickness symptoms. First, it can positively impact the awareness of users who would perform an auto-assessment of their fitness. Moreover, it could raise the consciousness of designers, product owners, and experimenters who would pay more attention to the negative impacts of cybersickness on users.

These dimensions are based on the lessons we have learned and recommendations from the literature. As cybersickness is a complex phenomenon, the reader should bear in mind that the dimensions are interrelated and interdependent. It is also important to note that expert inspection should be accompanied by user testing to gather direct feedback on how users experience VR and to confirm expert judgment and/or identify potential cybersickness issues that experts may have overlooked (see test examples in the next sections).

Pillar 1: Reducing the Risk of Cybersickness

Inspecting a VR Training Device

Description of the VR Training Device Under Study

As a case study, we present the example of a VR device developed for training in the aeronautical domain. The environment is designed to be used by commercial airline pilots to practice procedural tasks, and consists of an off-the-shelf hardware paired with a software fully designed according to the pilot's training needs. The software was developed and modified on an iterative experimental basis. It is a high-fidelity 3D immersive environment of an Airbus A320 cockpit on a one-to-one scale. The user can interact with the cockpit, but the aircraft itself is static and there is no movement in the external environment (i.e., the user cannot perceive any motion of the aircraft). In addition, the software includes various user interfaces for setting up and selecting lessons, helping users, and monitoring their performance. Each lesson can be performed with different levels of assistance according to a specific combination of audio, textual, and visual aids. The user can take on the role of any member of the flight crew (i.e., pilot flying or pilot monitoring), from any seat in the cockpit. This means that while one user plays the selected role in a lesson, the role of the other pilot is played automatically by an avatar in a chronologically coordinated way.

This virtual environment is displayed in an HP reverb G2 HMD with a resolution of 4320 × 2160 (combined). The headset, which weighs 550 g, is composed of a Dual LCD 2.89'' diagonal screen and two front-facing and two side-facing cameras for tracking. More importantly, it provides a 90-Hz refresh rate and a 114° field of view to users and enables adjustment of the IPD within a range of 60–68 mm using a hardware slider. The headset is connected via a cable to a VR-capable computer with the following characteristics: Intel i7+ 11th Gen CPU, Nvidia RTX 3080 GPU, 32 Gb RAM, 500 GB NVME SSD, 800 W+ PSU, running on Windows 10. The interaction with the virtual cockpit is performed through two HP reverb G2 controllers (167 g each) connected to the HMD via Bluetooth, which provide haptic feedback to users. Besides, audio is rendered through the HMD headphone. Finally, to replicate as much as possible the seating position in the real aircraft, pilots were seated on a chair with small armrests.

Results of the Inspection

The inspection did not identify serious risks regarding the dimensions of the *user characteristics*, the *technology properties*, the *physical space*, the *exposure duration*, and the *cybersickness risk knowledge and prevention*. The users were certified pilots with general good health and regular medical checks, especially regarding vision and balance perception. The VR technologies had no obvious defect in this area. The physical space was configured for VR use. The exposure was around 20 min per session, up to 1 hr per day including three 20-min sessions with breaks (i.e., in line with the main recommendations). Users, designers, and training teams had good knowledge about the cybersickness risks. The main initial concerns were about *tasks to perform* and *VR content/application*. Some procedures in the VR training include actions on the overhead panel of the cockpit and quite intense head movements with potential motion parallax. These procedures need to be carefully monitored, as they can expose users to cybersickness risks. Moreover, VR training took place in a cockpit where most of the actions to perform were at a harmful distance with no way to change that. This distance was not comfortable for the vergence/accommodation conflict and could be a source of visual fatigue (which can be compensated for by the use of an adapted exposure duration planning). In short, the initial screening of the cybersickness risk based on expert inspection identified some issues regarding the potential visual fatigue and a risk of disorientation.

User Testing

Beyond expert inspection, each training application should be tested for adverse effects, focusing on the specificity of the target population. In the following, we present an example of such a test. During this test, two characteristic elements of the target population were of particular interest:

- Commercial pilots have specific sight requirements in order to maintain their medical certificate. Thus, ascertaining that the VR exposure has no long-lasting effects on pilots' vision is of the utmost importance for the adoption of this technology.
- The great majority of published VR studies were conducted with university students usually in the age range of 18–30 years. However, many experienced pilots and instructors are older.

Thus, 26 commercial pilots (25 male pilots and one female pilot) aged between 43 and 57 years old ($M = 50.5$, $SD = 4.6$) were recruited. The participants were included in two waves (Group 1 = 14 pilots, and Group 2 = 12 pilots). The pilots were exposed to the VR training device for three consecutive days. On the first day, they were exposed to a single 20-min VR session. On the second day, they attended two 20-min sessions and on the third day three 20-min sessions. The training consisted of several pre-flight routines. At the beginning and at the end of each day, several variables were measured: the near point of accommodation, the near point of convergence, and the cybersickness intensity (measured using the SSQ). Moreover, pilots were asked to report any aftereffects they may have experienced between days and after the last day and up to 72 hr.

Results

The results did not show any significant effect of VR exposure on sight for near accommodation or convergence distance before and after each singular VR exposure. Moreover, no effect of the day was observed, except for the near point of accommodation recovery. However, the difference between before and after measures *decreased* along the days (Figure 13.1). From a quantitative viewpoint, a small and not statistically significant increase (around 1 cm) for the near accommodation and convergence distance was observed, but this disappeared 30 min after exposure for all pilots.

We observed a significant increase in cybersickness symptoms after VR exposure for the second and third days in Group 1. Indeed, all pilots included in this group reported severe dizziness and nausea during a specific lesson on Day 2. This lesson required them to perform repetitive and ample head and torso movements that were recognized to be the cause of the *disorientation* symptoms. In addition, these movements were considered as inappropriate because they were not required from a normal pilot-seating position and they might also cause falls. Consequently, it was decided to slightly modify the protocol en route for Group 2 by removing

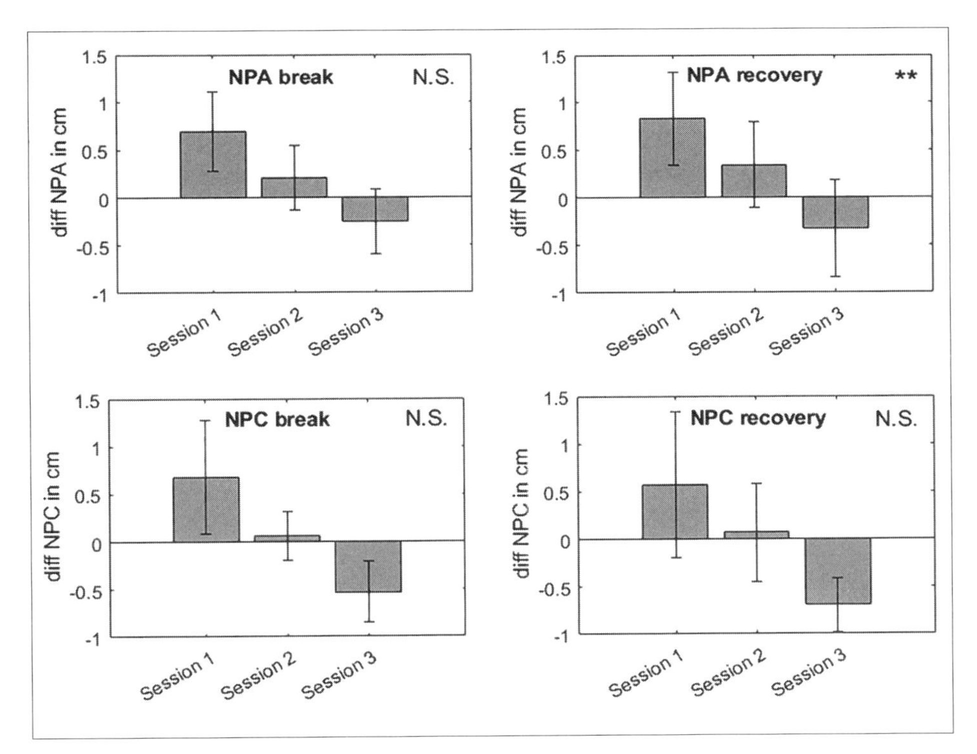

Figure 13.1. Before and after differences in near point of accommodation (NPA, upper row) and near point of convergence (NPC, lower row). N.S. = not significant. ** $p < .01$.

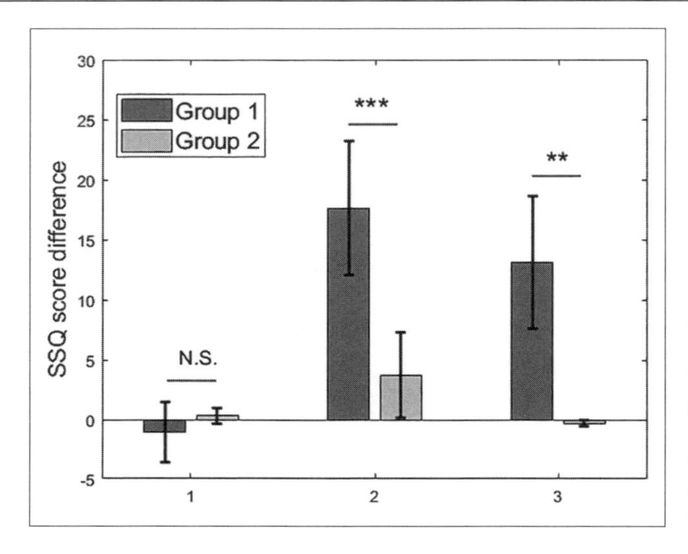

Figure 13.2.
SSQ score differences before and after VR exposure. Error bars represent the standard error to the mean. N.S. = not significant; SSQ = Simulator Sickness Questionnaire.
$**p < .01$. $***p < .001$.

the aversive lesson on Day 2 and doubling one of the other lessons to keep the time of exposure identical between Group 1 and Group 2. With the protocol thus modified, we observed no significant increase in cybersickness for Group 2, for each day (Figure 13.2).

As for aftereffects, quantitatively, cybersickness symptoms were less frequently reported by Group 2 than by Group 1 (15 pilots reporting aftereffects for Group 1 vs. two pilots reporting aftereffects for Group 2). Interestingly, there were significant correlations between the severity of cybersickness symptoms on Days 2 and 3 and the number of reported aftereffects. The most frequently reported aftereffects were *unusual fatigue*, *eyestrain*, and *headache*, usually lasting until the evening of the day of exposure.

Human Factors Outcomes

Overall, the results showed that repeated exposures to 20-min training sessions in a virtual environment are safe in terms of the sight requirements to be declared fit for duty as a commercial pilot. Specifically, on average, the training environment did not lead to any significant change in the near point of convergence or accommodation distance. Moreover, from a quantitative and single-subject perspective, all observed changes in the near point of accommodation and convergence distances disappeared at least 30 min after the end of the exposure. In conclusion, the test demonstrated that the selected VR training device is safe for the eyesight of a population with selected characteristics (i.e., sight requirements to be declared fit for duty; an age higher than that of participants usually included in VR study). Also, the VR device can be used without eliciting cybersickness symptoms and aftereffects, when avoiding specific movements during training. Overall, the methods presented for Pillar 1 such as the risk expert inspection and controlled experiments can be taken as a footprint for future studies of cybersickness related to VR as a training environment.

Assessing Pillar 2: Evaluating the Effectiveness of VR Training

Assessing Knowledge Retention and Skill Acquisition

The next main objective was to determine whether the VR training device enables users to effectively learn a set of standard operating procedures (SOPs). Thus, in a second test phase, the effectiveness of the VR device was assessed in comparison with a non-immersive training device with similar learning goals. Effects of the devices on the learning effectiveness were defined in accordance with the International Organization for Standardization (2020b). This section reports on the main results of the comparative test. For detailed information, please refer to the whole study (Khenak, Bach, Drouot, et al., 2023). The test was conducted with 12 pilots (non-commercial profile and no experience with the SOP of an Airbus aircraft). There were three women and nine men with ages ranging from 24 to 50 years ($M = 38.8$, $SD = 7.1$). Five participants reported having no experience with VR, while the others reported having used VR at least once. The task consisted of learning two preflight normal procedures containing theoretical knowledge and practical skills. Knowledge acquired during the task was assessed using a theoretical post-questionnaire, while skill acquisition was assessed by a post-check on an Airbus simulated procedure trainer (APT) device. In addition, the SSQ was administered before and after the VR exposure, and the usability of both devices was assessed using the System Usability Scale (SUS; Brooke, 1996).

Results

The knowledge retention score obtained after using VR was 47%, whereas it was 83% after using the non-immersive device. Also, the skill acquisition score was lower with the immersive device (79%) than the non-immersive device (89%). However, this score was achieved with less time spent using the VR device (−43%), indicating greater efficiency of the latter in terms of skill acquisition. The average cybersickness score was 6.54 immediately post-exposure, which is associated with low physiological impacts (Stanney et al., 2020). In addition, no statistically significant effect of VR exposure on the SSQ score was found (Wilcoxon signed-rank test: $p = .28$, Cohen's $d = .39$). Finally, the SUS scores indicated that the VR device was rated as better (80.6) than the non-immersive device (73.75) in terms of usability. Also, both devices provided a "good" perceived usability according to Bangor's scale (Bangor et al., 2008).

Human Factors Outcomes

The user test tended to show the effectiveness of the VR training device to initiate skill acquisition of a procedure application. On the other hand, results indicated that the VR device is less effective for knowledge retention, which is consistent with previous research reporting that VR may not be superior to conventional non-immersive devices for acquiring certain intellectual knowledge (Babu et al., 2018). Thus, further studies are required to better assess the effectiveness of such a VR training device as support for pilot training.

Pillar 3: Evaluating the Efficiency of VR Training

Assessing Completion Time and Performance

The third test phase described in this section aimed to evaluate the learning efficiency of the VR device within the framework of a qualification training program with airline pilots. Specifically, the choice was made to implement the use of the VR device in a training program allowing the acquisition of an aircraft type rating on Airbus A320. This program includes several stages designed to progressively develop the different pilot competencies through simple to complex training scenarios, combining the acquisition of systems knowledge and the practice of the normal and abnormal operational procedures of the aircraft. The integration of VR lessons into the training program took into account the HF recommendations identified during the two previous tests in order to guarantee the success of this training course. Thus, a gradual increase in VR exposure time was set up. In terms of session duration, 20 min per session (with a tolerance of 3 min to allow users to complete a lesson) and 60 min per day were retained. Besides, 10-min breaks were planned between VR sessions. This allowed us to define an experimental 6-hr program over 7 days as follows:

- Day 1: one lesson including one VR session of 20–23 min
- Day 2: one lesson including two VR sessions of 20–23 min with one 10-min break
- Days 3–7: one lesson per day, including three VR sessions of 20–23 min with two 10-min breaks

Training lessons with the VR device were implemented at the start of the current program targeting existing training objectives provided by a non-immersive training device[1]. Moreover, the device was used in individual mode (although it could be used in collective mode) in order to compare pilot performance obtained after VR training without the potential influence of collective training. Three airline crews (six pilots) took part in the test. The sample was composed entirely of men with ages ranging from 27 to 46 years (M = 36.2). Their IPDs were between 60 and 67 mm (M = 63.4), which are values covered by the HMD. Three pilots had experience with VR through games and three had no experience at all with VR systems. Their goal was to complete the learning program with the VR device. The degree of cybersickness was measured by administering the SSQ before and after each VR session. As done in the second study, the usability of the VR device was measured using the SUS. In addition, after training with the VR device, the learning performance of the pilots was assessed in an APT device by measuring the execution or non-execution of the learned procedural tasks. Also, the time spent using the VR device was recorded.

Results

The completion time and performance observed during the assessment session were compared with a defined reference time and with the performance expected during the first procedural training session after training with a similar but non-immersive training device. The results showed that, compared to the referential training, completion duration increased by 22%. However, the average performance increased by 85%, indicating that the pilots improved their execution of the procedures learned. Consequently, the learning efficiency (= performance / duration) increased by 51%. The average cybersickness score was 2.23 immediately post-exposure, which is associated with low physiological impacts (Stanney et al., 2020). Finally, the training device was rated as 85.42 out of 100 points on SUS, which is associated with the adjective "best imaginable" on Bangor's scale (Bangor et al., 2008).

Human Factors Outcomes

In this test, the HF recommendations identified in Pillars 1 and 2 were applied. Results demonstrated appropriate mitigation of the risks associated with cybersickness. In addition, results tend to demonstrate the efficiency of the VR technology in the memorization and execution of procedures containing visuospatial tasks. In sum, VR used as a training device for commercial pilots appears well suited to training in the operational procedures of an airliner. This test was the third and final pillar of the method to verify the conditions for safe and efficient use of the VR

1 The original training program uses the non-immersive device only after aircraft systems lessons. Yet, progressive exposure to the VR device over the same period implied that the VR lessons had to be programmed on the same day as the aircraft systems lessons. The latter was therefore scheduled in the morning and the VR exposures in the afternoon.

device for pilot training. In a regulated aeronautical context, this method could serve as a basis for argumentation and demonstration to the relevant authorities when applying for authorization / certification to use this technology in qualifying training programs.

Next Step: VR Training With Motion

The addition of external motion (i.e., motion not directly related to users' movements) to VR training could increase the risk of cybersickness. Therefore, many open questions should be explored such as the use (or not) of a motion platform, the velocity of motion, and the number of tasks involving both head/torso movement and aircraft motion in different flight phases (e.g., training sessions about visual approach patterns, taxiing, go-around) to name a few. Mainly, different HF dimensions presented in the section Impact of Cybersickness on VR Training Efficiency could be impacted by the addition of motion. Regarding the technology properties, the use of a motion platform could complexify the question of delays/synchronization of the different interaction modalities between inputs and outputs. A well-designed technological set-up could match HF requirements, but this would need to be evaluated. The tasks to perform could be complex to design for ensuring an acceptable cybersickness risk, especially regarding the sensorial mismatch that could be a source of nausea symptoms. The VR content/application will evolve into a moving landscape that could increase the visual complexity. The physical space, especially with a motion platform, will have to be carefully considered to ensure the safety of interaction. Regarding exposure duration, as the use of motion will increase the complexity of interaction, 20-min sessions could be too long and might be rescheduled with a different distribution of breaks. The cybersickness risk knowledge and prevention could evolve to take into account unknown effects of motion on the overall training program and/or on the design of VR training. In any case, this evolution must be supported by an iterative and user-centered design/evaluation process including the different stakeholders.

Summary and Outlook

VR devices are very promising tools to improve training. This chapter provided a three-pillar approach to address the benefits and limitations of VR in pilot training, and to advance VR applications and related technologies. Although specific VR devices can provide a more efficient training for pilots as compared to conventional non-immersive devices, their integration into an existing training program should be progressive and iterative in order to facilitate the transition between actual and future (full-scale) VR training. The lessons learned shared in this chapter are the first step toward finding a way to facilitate this transition. Specifically, the user-centered approach we followed allowed us to identify, through inspections and user testing, a number of issues associated with cybersickness that might otherwise have been overlooked. Thus, the three-pillar method presented in this chapter appears to be an appropriate means for integrating VR into the professional training environment and ensuring that the way VR is used does not involve a regression in learning performance in critical domains compared with existing means. This method enables a systematic validation of the training devices and lays the foundation for the change management required for safe and efficient VR integration.

References

Babu, S. K., Krishna, S., Unnikrishnan, R., & Bhavani, R. R. (2018). Virtual reality learning environments for vocational education: A comparison study with conventional instructional media on knowledge retention. In *2018 IEEE 18th international conference on advanced learning technologies* (pp. 385–389). IEEE. https://doi.org/10.1109/ICALT.2018.00094

Bach, C., & Scapin, D. L. (2010). Comparing inspections and user testing for the evaluation of virtual environments. *International Journal of Human-Computer Interaction, 26*(8), 786–824. https://doi.org/10.1080/10447318.2010.487195

Bangor, A., Kortum, P. T., & Miller, J. T. (2008). An empirical evaluation of the system usability scale. *International Journal of Human-Computer Interaction, 24*(6), 574–594. https://doi.org/10.1080/10447310802205776

Blehm, C., Vishnu, S., Khattak, A., Mitra, S., & Yee, R. W. (2005). Computer vision syndrome: A review. *Survey of Ophthalmology, 50*(3), 253–262. https://doi.org/10.1016/j.survophthal.2005.02.008

Bouchard, S., Robillard, G., Renaud, P., & Bernier, F. (2011). Exploring new dimensions in the assessment of virtual reality induced side effects. *Journal of Computer and Information Technology, 1*(3), 20–32.

Brooke, J. (1996). SUS: A "quick" and "dirty" usability scale. In P. W. Jordan, B. Thomas, B. A. Weerdmeester, & I. L. McClelland (Eds.), *Usability evaluation in industry* (pp. 189–194). Taylor and Francis.

Bruck, S., & Watters, P. A. (2011). The factor structure of cybersickness. *Displays, 32*(4), 153–158. https://doi.org/10.1016/j.displa.2011.07.002

Davis, S., Nesbitt, K., & Nalivaiko, E. (2015). Comparing the onset of cybersickness using the Oculus Rift and two virtual roller coasters. In Y. Pisan, K. Nesbitt, & K. Blackmore (Eds.), *Proceedings of the 11th Australasian conference on interactive entertainment* (Vol. 27, p. 30). CRPIT. https://crpit.scem.westernsydney.edu.au/confpapers/CRPITV167Davis.pdf

Dużmańska, N., Strojny, P., & Strojny, A. (2018). Can simulator sickness be avoided? A review on temporal aspects of simulator sickness. *Frontiers in Psychology, 9*, Article 2132. https://doi.org/10.3389/fpsyg.2018.02132

Emoto, M., Niida, T., & Okano, F. (2005). Repeated vergence adaptation causes the decline of visual functions in watching stereoscopic television. *Journal of Display Technology, 1*(2), 328–340. https://doi.org/10.1109/JDT.2005.858938

Gianaros, P. J., Muth, E. R., Mordkoff, J. T., Levine, M. E., & Stern, R. M. (2001). A questionnaire for the assessment of the multiple dimensions of motion sickness. *Aviation, Space, and Environmental Medicine, 72*(2), 115–119.

International Organization for Standardization. (2020a). *Ergonomics of human-system interaction. Part 394: Ergonomic requirements for reducing undesirable biomedical effects of visually induced motion sickness during watching electronic images (ISO Standard No. 9241-394:2020)*. https://www.iso.org/standard/73227.html

International Organization for Standardization. (2020b). *Ergonomics of human-system interaction. Part 110: Interaction Principles (ISO Standard No. 9241-110:2020)*. https://www.iso.org/standard/75258.html

Kemeny, A., Chardonnet, J. R., & Colombet, F. (2020). Getting rid of cybersickness. *Virtual reality, augmented reality, and simulators*. Springer. https://doi.org/10.1007/978-3-030-59342-1

Kennedy, R. S., Lane, N. E., Berbaum, K. S., & Lilienthal, M. G. (1993). Simulator sickness questionnaire: An enhanced method for quantifying simulator sickness. *The International Journal of Aviation Psychology, 3*(3), 203–220. https://doi.org/10.1207/s15327108ijap0303_3

Khenak, N., Bach, C., & Buratto, F. (2023). Understanding the relationship between cybersickness and usability through the human factors dimensions: A comparison of a pilot and an assembly VR training. In *Ergo'IA '23: Proceedings of the 18th "Ergonomie et Informatique Avancée" Conference* (pp. 1–10). ACM. https://doi.org/10.1145/3624323.3624342

Khenak, N., Bach, C., Drouot, S., & Buratto, F. (2023). Evaluation of virtual reality training: Effectiveness on pilots' learning. In *IHM'23-34e Conférence Internationale Francophone sur l'Interaction Humain-Machine*. HAL. https://hal.science/hal-04046414/document

Kim, H. K., Park, J., Choi, Y., & Choe, M. (2018). Virtual reality sickness questionnaire (VRSQ): Motion sickness measurement index in a virtual reality environment. *Applied Ergonomics, 69*, 66–73. https://doi.org/10.1016/j.apergo.2017.12.016

LaViola, J. J. (2000). A discussion of cybersickness in virtual environments. *ACM Sigchi Bulletin, 32*(1), 47–56. https://doi.org/10.1145/333329.333344

Lawson, B. D. (2014). Motion sickness symptomatology and origins. In K. S. Hale & K. M. Stanney (Eds.), *Handbook of virtual environments: Design, implementation, and applications* (pp. 531–600). CRC Press.

Mon-Williams, M., Warm, J. P., & Rushton, S. (1993). Binocular vision in a virtual world: Visual deficits following the wearing of a head-mounted display. *Ophthalmic and Physiological Optics, 13*(4), 387–391. https://doi.org/10.1111/j.1475-1313.1993.tb00496.x

Morse, S. E., & Jiang, B. C. (1999). Oculomotor function after virtual reality use differentiates symptomatic from asymptomatic individuals. *Optometry and Vision Science, 76*(9), 637–642. https://doi.org/10.1097/00006324-199909000-00021

Moss, J. D., & Muth, E. R. (2011). Characteristics of head-mounted displays and their effects on simulator sickness. *Human Factors, 53*(3), 308–319. https://doi.org/10.1177/0018720811405196

NATO Science and Technology Office. (2021). *Guidelines for mitigating cybersickness in virtual reality systems. Peer-reviewed* (NATO STO-TR-HFM-MSG-323). https://www.sto.nato.int/publications/STO%20Technical%20Reports/STO-TR-HFM-MSG-323/$$TR-HFM-MSG-323-ALL.pdf

Nesbitt, K., & Nalivaiko, E. (2018). Cybersickness. In N. Lee (Ed.), *Encyclopedia of computer graphics and games* (pp. 1–6). Springer International Publishing. https://doi.org/10.1007/978-3-319-08234-9_252-1

Renganayagalu, S. K., Mallam, S. C., & Nazir, S. (2021). Effectiveness of VR head mounted displays in professional training: A systematic review. *Technology, Knowledge and Learning, 26*(4), 999–1041. https://doi.org/10.1007/s10758-020-09489-9

Saredakis, D., Szpak, A., Birckhead, B., Keage, H. A., Rizzo, A., & Loetscher, T. (2020). Factors associated with virtual reality sickness in head-mounted displays: A systematic review and meta-analysis. *Frontiers in Human Neuroscience, 14*, Article 96. https://doi.org/10.3389/fnhum.2020.00096

Sepich, N. C., Jasper, A., Fieffer, S., Gilbert, S. B., Dorneich, M. C., & Kelly, J. W. (2022). The impact of task workload on cybersickness. *Frontiers in Virtual Reality, 3*, Article 943409. https://doi.org/10.3389/frvir.2022.943409

Sheedy, J. (2014). Visual fatigue. *Points de Vue Magazine, N°70*. https://www.optiqueduvillard.ch/images/blog/pdf/Visual-fatigue-in-near-vision.pdf

Souchet, A., Philippe, S., Ober, F., Léveque, A., & Leroy, L. (2019). Investigating cyclical stereoscopy effects over visual discomfort and fatigue in virtual reality while learning. In *2019 IEEE International Symposium on Mixed and Augmented Reality* (pp. 328–338). IEEE. https://doi.org/10.1109/ISMAR.2019.00031

Souchet, A. (2020). *Impacts de la fatigue visuelle sur l'apprentissage via serious game en réalité virtuelle* [Impact of visual fatigue on learning via virtual reality serious game] [Doctoral dissertation, Paris 8 University]. https://www.theses.fr/2020PA080002

Souchet, A. D., Philippe, S., Lourdeaux, D., & Leroy, L. (2022). Measuring visual fatigue and cognitive load via eye tracking while learning with virtual reality head-mounted displays: A review. *International Journal of Human–Computer Interaction, 38*(9), 801–824. https://doi.org/10.1080/10447318.2021.1976509

Stanney, K. M., Hale, K. S., Nahmens, I., & Kennedy, R. S. (2003). What to expect from immersive virtual environment exposure: Influences of gender, body mass index, and past experience. *Human Factors, 45*(3), 504–520. https://doi.org/10.1518/hfes.45.3.504.27254

Stanney, K., Lawson, B. D., Rokers, B., Dennison, M., Fidopiastis, C., Stoffregen, T., Weech, S., & Fulvio, J. M. (2020). Identifying causes of and solutions for cybersickness in immersive technology: Reformulation of a research and development agenda. *International Journal of Human–Computer Interaction, 36*(19), 1783–1803. https://doi.org/10.1080/10447318.2020.1828535

Stone, W. B. (2017). *Psychometric evaluation of the Simulator Sickness Questionnaire as a measure of cybersickness* [Doctoral dissertation, Iowa State University]. https://doi.org/10.31274/etd-180810-5050

Szpak, A., Michalski, S. C., Saredakis, D., Chen, C. S., & Loetscher, T. (2019). Beyond feeling sick: The visual and cognitive aftereffects of virtual reality. *IEEE Access, 7*, 130883–130892. https://doi.org/10.1109/ACCESS.2019.2940073

Yoon, H. J., Kim, J., Park, S. W., & Heo, H. (2020). Influence of virtual reality on visual parameters: Immersive versus non-immersive mode. *BMC Ophthalmology, 20*, Article 200. https://doi.org/10.1186/s12886-020-01471-4

Yuan, J., Mansouri, B., Pettey, J. H., Ahmed, S. F., & Khaderi, S. K. (2018). The visual effects associated with head-mounted displays. *International Journal of Ophthalmology and Clinical Research, 5*(2), Article 085. https://doi.org/10.23937/2378-346x/1410085

Chapter 14

Investigating Augmented-Reality-Supported Flight Training

Ioana V. Koglbauer, Wolfgang Vorraber, Birgit Moesl, Harald Schaffernak, and Reinhard Braunstingl

Abstract

This chapter gives an overview of several studies on augmented reality (AR) applications for flight training. The studies reported here are rooted in the use-case technology-mapping framework that addresses innovations in the process and business model and merges the human-centred and process-driven perspectives with the technology-driven approach. AR use cases for type rating (TR) training were defined iteratively in workshops for opportunity scouting that used various data sources. The learning content, learning conditions and potential application areas for AR were evaluated in surveys with flight instructors and pilots. The AR technologies were mapped with promising use cases to ultimately create the process and business model innovations. A number of AR use cases have been identified, implemented and evaluated. Evaluation methods ranged from use case studies to controlled experiments that compared AR applications with the conventional training means. In addition, the basic structures and the value exchange relations of the ecosystem were analysed to determine their relevance for introducing the AR-based training into the TR syllabus. This analysis enables the re-design of the future system and helps to anticipate potential problems when deploying the AR-based training service. The results indicate the suitability of AR technologies to address diverse user needs and preferences, including gender preferences, as well as aspects of social and economic sustainability. The value, limitations and complementarity of the methods, as well as the involvement of various stakeholders in the assessment of AR-based pilot training, are discussed.

Keywords

augmented reality, flight training, human-centred, ecosystem, business model innovation, sustainability

Applications of AR Technology in Education

The terms "augmented reality" (AR) and "mixed reality" (MR) have been structured in the context of the reality–virtuality continuum (Milgram et al., 1995). AR is a form of MR in which augmented cues (e.g., holograms) are introduced in the real world. Multiple studies show that AR can improve the learning performance and experience (Wu et al., 2013). A famous example is the augmentation of written information in a book featuring pop-ups including three-dimensional (3D) or animated AR content (Billinghurst et al., 2001). Other AR features that have been studied include the provision of stereo-sound assisted guidance (Feng et al., 2023), remote guidance (Huang et al., 2022) and collaboration features (Billinghurst & Kato, 2002). The educational benefits of AR have been described in various domains, ranging from science laboratories (Vergara et al., 2020) to aviation training (Brown, 2019; Moesl et al., 2021; Vlasblom et al., 2019).

In a literature survey Billinghurst et al. (2015) found a surprisingly low proportion of user studies in AR research. Furthermore, only a few studies on AR used statistical methods to analyse training effects. In a meta-analysis of 25 studies from various domains, Kaplan et al. (2020) investigated the training effects of virtual reality (VR), AR and MR. The results of this study showed that VR, AR and MR were as effective as traditional training means. The main benefit of VR-, AR- and MR-based training is that it offers the opportunity for practicing in specific contexts where, for reasons such as danger or costs, traditional means cannot be applied. Thus, by using MR, trainees can prepare for future situations that are not yet a reality, are not easily accessible in reality, or are even dangerous for humans, such as the prospective missions to Mars (Hancock, 2017).

A study of AR features that could improve pilots' situational awareness was investigated by Blundell and Harris (2023). Based on focus group discussions with 11 ATPL pilots from the commercial aviation sector, these authors identified applications of AR for the purposes of airspace visualization (e.g., weather and restricted airspace information in planning the flight path), flight path simulation (e.g., future state navigation information for high-workload scenarios), critical controls and displays (e.g., highlighting key elements, particularly in emergency scenarios), traffic and terrain augmentation (Blundell & Harris, 2023). They also identified several potential risks of using AR in the cockpit, such as an increased information processing burden due to display clutter and the increased complexity of some tasks (Blundell & Harris, 2023).

A naturalistic study carried out in the context of naval aviation training by McCoy-Fisher et al. (2019) found that the use of MR for part-task training was associated with improvements in trainees' performance in the aircraft. In their study, all participants were enrolled in a naval aviation course, had free access to various VR/MR training devices and could choose whether to use these or not. Their analysis was based on self-reports of usage and performance data. For the MR part-task training, an AR device connected to a T-45C hardware simulator cockpit was used. The trainees who used AR for training on the ground performed better afterwards during the real flight as shown by fewer marginal flights, fewer poor performance events and fewer repetitions in the formation sessions as compared to the trainees who did not train with AR. In addition, the trainees who trained with AR more rapidly reached the minimum required score for advancement in the instruments part of the syllabus (McCoy-Fisher et al., 2019). Finally, these study results show that the AR application was considered helpful for learning to build a sight picture, prepare for the next event, perform remediation and learn new content (McCoy-Fisher et al., 2019).

In summary, the research performed to date has shown that AR-supported training can improve learning performance and experience and has a potential to support innovation in the highly traditional and standardized domain of flight training, at a time when the global aviation industry is striving to attract a new generation of talented and diverse pilots and to rethink their training procedures.

Type Rating Training

Pilots must take and successfully complete a type rating (TR) course that teaches them the type-specific elements of a certain aircraft they will operate in commercial air transport. The TR trainees typically hold at least a commercial pilot licence. Some trainees already have held various TRs, some do their first TR as part of a flight school curriculum or after finishing flight school. The TR course consists of theoretical and practical parts that are designed in accordance with standards and regulations (see European Union Aviation Safety Agency, 2023). Substantial costs (i.e., up to USD 130,000 per pilot) and effort are required to complete about 20–45 days of intensive learning. Currently, the TR courses rely on synchronous learning and on-site training methods. Thus, especially in business aviation, the trainees and instructors usually travel over long distances to take these courses, because certified full-flight simulators are expensive and, for some types of aircraft, also rare.

In this chapter we report on our studies of how the TR course contents to be supported by AR were prioritized (Moesl et al., 2022; Schaffernak, Moesl, Vorraber, Holy, et al., 2020; Schaffernak, Moesl, Vorraber, & Koglbauer, 2022). Thus, in an endeavour to improve the TR training methods in a sustainable manner, various impacts on environmental, economic and social sustainability were studied. Special attention was given to gender equality (United Nations, n.d.) – which is an aspect of social sustainability – in the use of AR as an enabling technology. Thus, we also designed AR-based TR training content and assessed them in a gender-sensitive manner by implementing features that match various gender preferences and by including a diverse team of participants in the process (Moesl et al., 2023a, 2023b). This is important because aviation is a male-dominated domain and women are still underrepresented in the pilot population worldwide (Ferla & Graham, 2019; Mitchell et al., 2005). Flying an aircraft has stereotypically been seen as a male profession. However, the demand for talented women in the aviation industry is growing (Woods, 2022), and it is important to address gender diversity aspects when shaping the future of the next-generation TR training in a socially sustainable manner.

The UCTM Framework for Addressing AR-Based Innovations in Flight Training

The use-case technology-mapping (UCTM) framework (Vorraber, Neubacher, et al., 2019) is a business model innovation concept for complex sociotechnical systems that has been applied to support business executives and experts from the fields of information technology and human factors, helping them to integrate new technologies into their organisations, processes and business models in various domains (e.g., health care, public safety, automotive production). UCTM merges technology-driven with human-centred and process-driven approaches. More guidance on the application of UCTM is provided by Vorraber, Neubacher, et al. (2019).

Figure 14.1 shows the UCTM application for creating innovative AR-based artefacts for TR training (Schaffernak et al., 2022). The human-centred and process-driven part of the framework identified use cases for the AR technology (Schaffernak et al., 2022). The process and potential use case analysis (the left side of Figure 14.1) was conducted iteratively and included results from several workshops, surveys and assessments (Schaffernak et al., 2022). Thus, the scoping of AR-based innovations addressed potential benefits for the trainees and improvements in the training quality that could be offered in an economic sustainable manner (Schaffernak et al., 2022).

In the rest of this chapter, these methods used for investigating AR-supported flight training in human- and process-centred approaches are presented. Their benefits and limitations, as well as an outlook for future developments, are discussed.

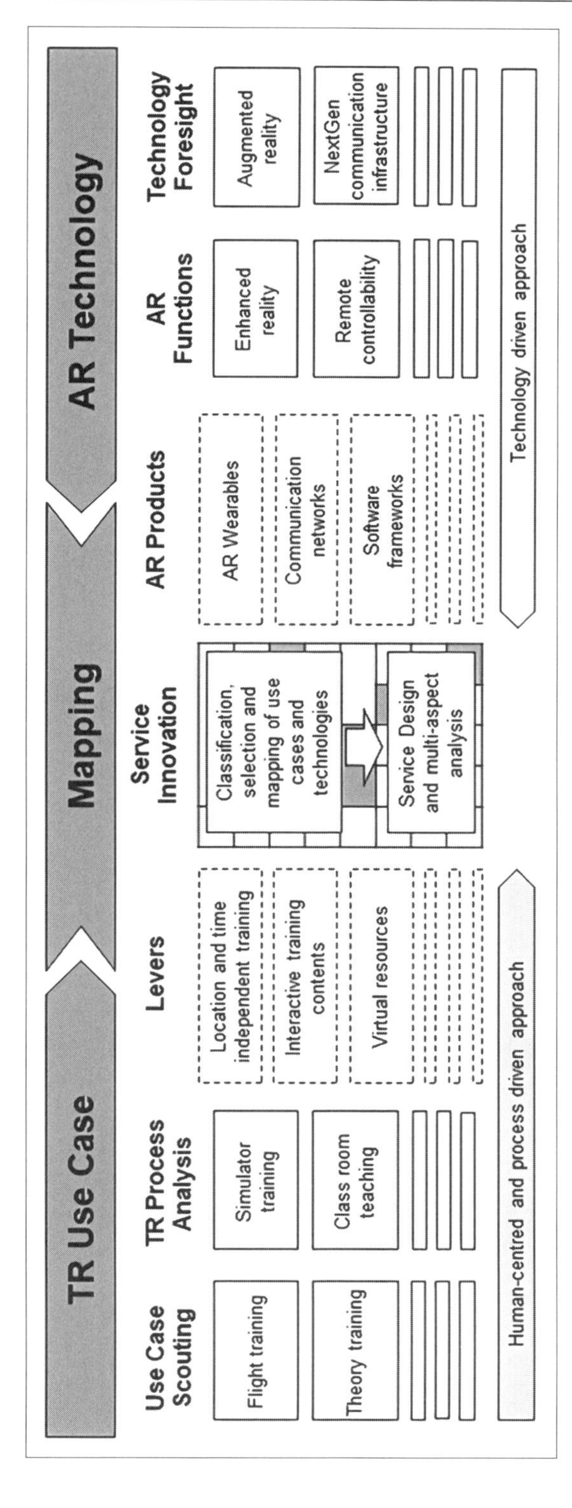

Figure 14.1. Application of the UCTM framework for implementing AR technology in TR training. AR = augmented reality; TR = type rating; UCTM = use-case technology-mapping. Adapted from Figure 2 in Schaffernak et al., 2022. (Changed to highlight the AR-type of technology addressed and the complementarity of the approaches.)

Initial High-Level Workshops

At the beginning of the study, workshops were conducted with representatives from different approved training organisations (ATOs): two senior managers (two men, one with a business degree, and the other with an engineering degree who is also an active airline pilot) and two middle managers (one woman with a degree in business education and one man with an engineering degree who is also an active pilot and instructor; Schaffernak et al., 2022). An academic team facilitated the workshops and the application of the methods (Schaffernak et al., 2022). To familiarise the participants with the AR technology, the HoloLens AR device (Microsoft, n.d.), its functionality and its generic industrial applications were demonstrated (Schaffernak et al., 2022). In addition, the process models of the ATOs were discussed and analysed (Schaffernak et al., 2022). A total of 11 initial AR use cases were identified and prioritised in these workshops, such as interactive theory training, outside check and procedure training (for more details, see Schaffernak et al., 2022).

Surveys With Pilots and Flight Instructors

Online surveys were conducted to obtain the perspectives of pilots and flight instructors (Schaffernak et al., 2020). A group of 60 pilots and flight instructors including 12 women evaluated potential application areas for AR-supported pilot training based on video presentations of typical AR use cases from other domains (for more details, see Schaffernak et al., 2020). Their results show that AR can be used in flight training for content such as theoretical instruction, pre-flight aircraft inspection and procedure training (Schaffernak et al., 2020). In addition, they assessed a number of gaming concepts that could be implemented in AR to make learning more interesting and engaging (Schaffernak et al., 2020). Although many similarities between the opinions of female and male pilots were identified, this online survey also identified gender-specific preferences for gaming concepts (Schaffernak et al., 2020).

Furthermore, the learning conditions and the TR course content were evaluated in a survey with 31 TR pilots (seven female pilots and 24 male pilots) and 22 TR male instructors (for details, see Moesl et al., 2022). Thus, the perspectives of instructors and former TR trainees could be merged (Moesl et al., 2022). Since the group sizes of female and male pilots and of TR instructors differed, the descriptive data analysis used percentages and weighted relative frequencies for comparisons (Moesl et al., 2022). Although this study identified many similarities between the identified gender groups, gender-specific preferences and difficulties related to the TR content, as well as the training and assessment conditions, were also recognised (Moesl et al., 2022). Simulator training was unanimously rated as the most enjoyable part of the course (Moesl et al., 2022). The procedures related to missed approach, landing, flight management system, abnormal and emergency flight procedures and instrument flight procedures were rated in this study as the most difficult TR course content items (Moesl et al., 2022). In addition, the TR pilots and instructors also identified a variety of course content that could be easier to learn if AR technology was used, such as emergency procedures, aircraft structure and equipment (Moesl et al., 2022). The women rated AR as most beneficial for training related to special requirements, the flight management system and the instrument approaches (Moesl et al., 2022). The men considered that learning the type-specific flight preparation content could be easier if AR was used (Moesl et al., 2022). The instructors suggested that also aircraft performance as well as abnormal and emergency flight procedures could be easier to learn if AR was used (Moesl et al., 2022).

Besides content, also difficult training conditions were identified in this study, such as the demand to assimilate a "huge quantity" of complex information during a relatively short duration of the course (Moesl et al., 2022). Stress, time pressure and problems with self-study were mentioned by a number of TR pilots and instructors (Moesl et al., 2022). The use of AR was considered to improve learning conditions by enabling remote training and collaboration (Moesl et al., 2022). Thus, the results of Moesl et al. (2022) indicate that the AR technology has the potential to improve the next generation of TR training when used as a complement to traditional instruction methods in the classroom and the full-flight simulators.

Follow-Up Workshops

In further workshops, the data collected with the surveys were reviewed with two senior managers (two men, one with a business degree, and the other with an engineering degree who is also an active airline pilot) and two middle managers (one woman with a degree in business education and one man with an engineering degree who is also an active pilot and instructor; Schaffernak et al., 2022). Due to the complexity and amount of syllabus content, pairwise comparisons were used to establish a priority list, and a set of nine criteria were applied (e.g., benefits for the trainee, frequency of occurrence, quality improvement and decrement, gender diversity, financial benefit and impact on resources; Schaffernak et al., 2022). Based on this priority list and the high-level use cases, in another workshop the final list of prioritised use cases was determined (Schaffernak et al., 2022). In the following section, the evaluation of such a use case (Table 14.1) is described.

Table 14.1. Use case "outside check"

Use case element	Description
Triggering event	Classroom instruction for the aircraft external visual inspection with regard to the location of each item and purpose of its inspection.
Description	The outside check is carried cut on a virtual aircraft model that is augmented with holographic, interactive elements, explanations and guidance for transition between elements.
Actors	TR trainee.
Precondition	The trainee stands in a room that is large enough in front of a virtual model of a business jet.
Flow of activities	A full-scale holographic model of a business jet is displayed. The trainee walks around the model and inspects specific parts following the procedure. AR cues suggest the required actions, such as turning handles or opening doors. For elements that would be out of reach in a real aircraft, the virtual model will be orientated and moved accordingly.

Note. AR = augmented reality; TR = type rating. Adapted from Table 2 in Schaffernak et al., 2022. (Extract changed from the original to explain the content in a more concise way.)

Evaluation of the Use Case Outside Check

Before any flight, a pilot must go through a checklist and inspect various parts of the aircraft to identify potential damage and carry out functional tests. An AR guide was designed for HoloLens 2 (Microsoft, n.d.) that can be used to teach trainees how to conduct the outside check for a business aircraft using a full-size holographic model (Wimmesberger, 2022). An example of the AR guide is illustrated in Figure 14.2, which is part of a sequence (Wimmesberger, 2022). Six trainees, three men and three women, tested the guide independently for about 45 min each time and filled out an evaluation questionnaire afterwards (Wimmesberger, 2022). The ratings provided for voice control, field of view, hologram quality and descriptions for each step were good or very good, but the wearing comfort of the AR device was rated as only medium (Braunstingl et al., 2022; Wimmesberger, 2022). The trainees liked the AR application and found the AR cues to be helpful (e.g., hands indicating certain elements, numbers for the order and arrows showing the next action or item; Wimmesberger, 2022).

Some trainees reported feeling very slight dizziness but did not interrupt the usage, although they could have stopped using it at any time without providing a reason (Wimmesberger, 2022). All women noted that the AR guide was useful and

Figure 14.2. View from the augmented reality device of the outside check instruction. From "Product-Lifecycle-Management Enhanced With Augmented Reality," by S. Wimmesberger, 2022, Unpublished Master's Thesis, Graz University of Technology. © S. Wimmesberger. Reprinted with permission.

said that they would like to use the AR guide and AR in general for future training (Braunstingl et al., 2022; Wimmesberger, 2022). Although the evaluation sample was small, the results indicate that this use case meets the preferences of female and not only of male trainees (Wimmesberger, 2022). Thus, the social sustainability requirement was addressed. This use case also meets economic and environmental sustainability objectives, because it allows the individuals to be trained independently (i.e., without an instructor) and without needing to travel to an airport.

The Ecosystem Analysis

Another main question was how the introduction of AR-based innovations could address the needs and values (e.g., social, ecological, technical, economic, ethical) of various stakeholders involved in the sociotechnical system of the flight training. Which benefits and limitations or which value-generating and value-hampering exchanges (Vorraber, Mueller, et al., 2019) can be identified and addressed at an early stage? Therefore, a pre-conceptual analysis of the ecosystem for flight training in business aviation was conducted using the framework described by Schierlinger-Brandmayr et al. (2022). A workshop with management representatives of an ATO from various departments was conducted (Schaffernak et al., 2022). In this workshop, the economic entities of the prospective ecosystem required for

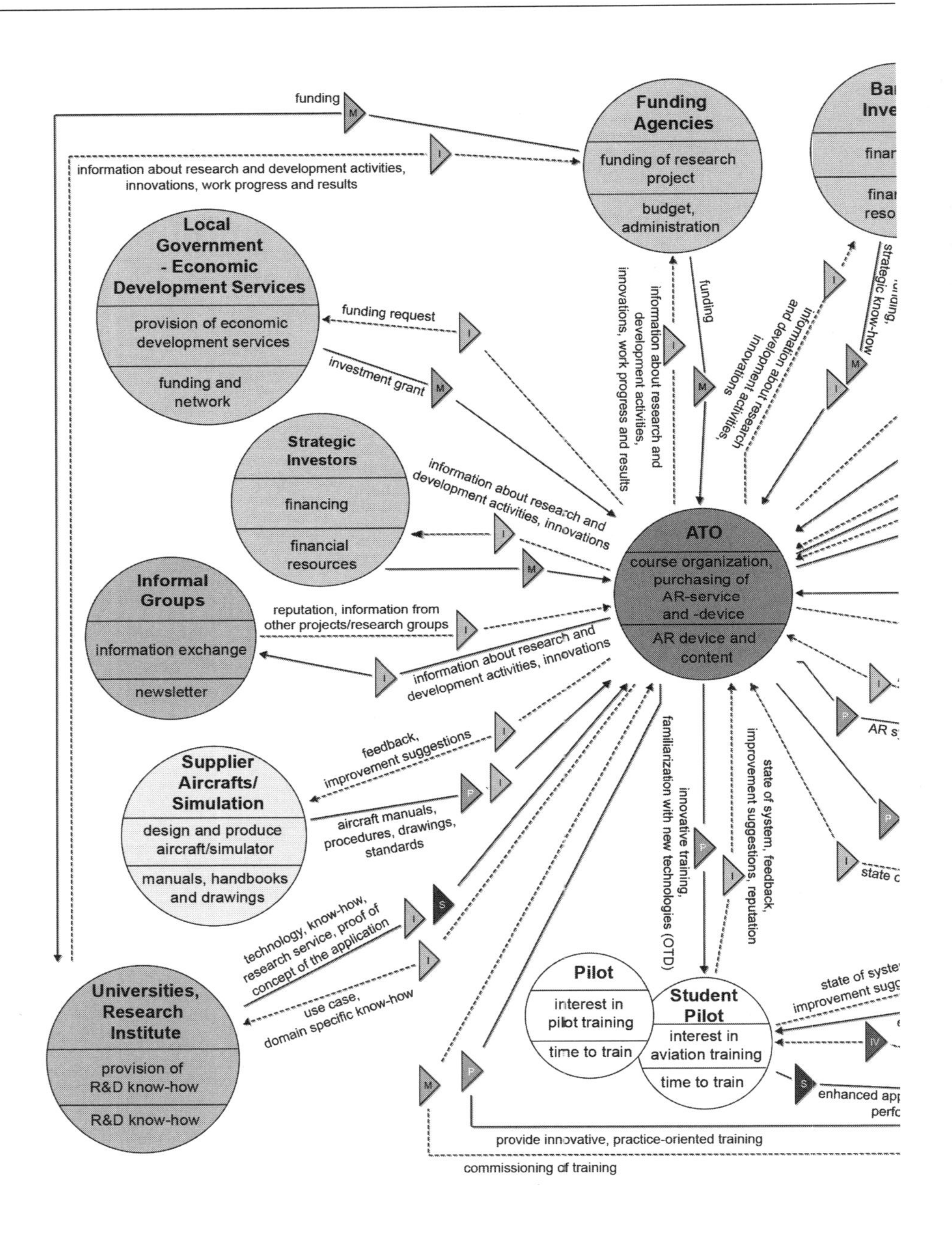

Figure 14.3. Preconceptual ecosystem model of augmented reality services for pilot training. Reprint of Figure 5 in Schaffernak et al., 2022.

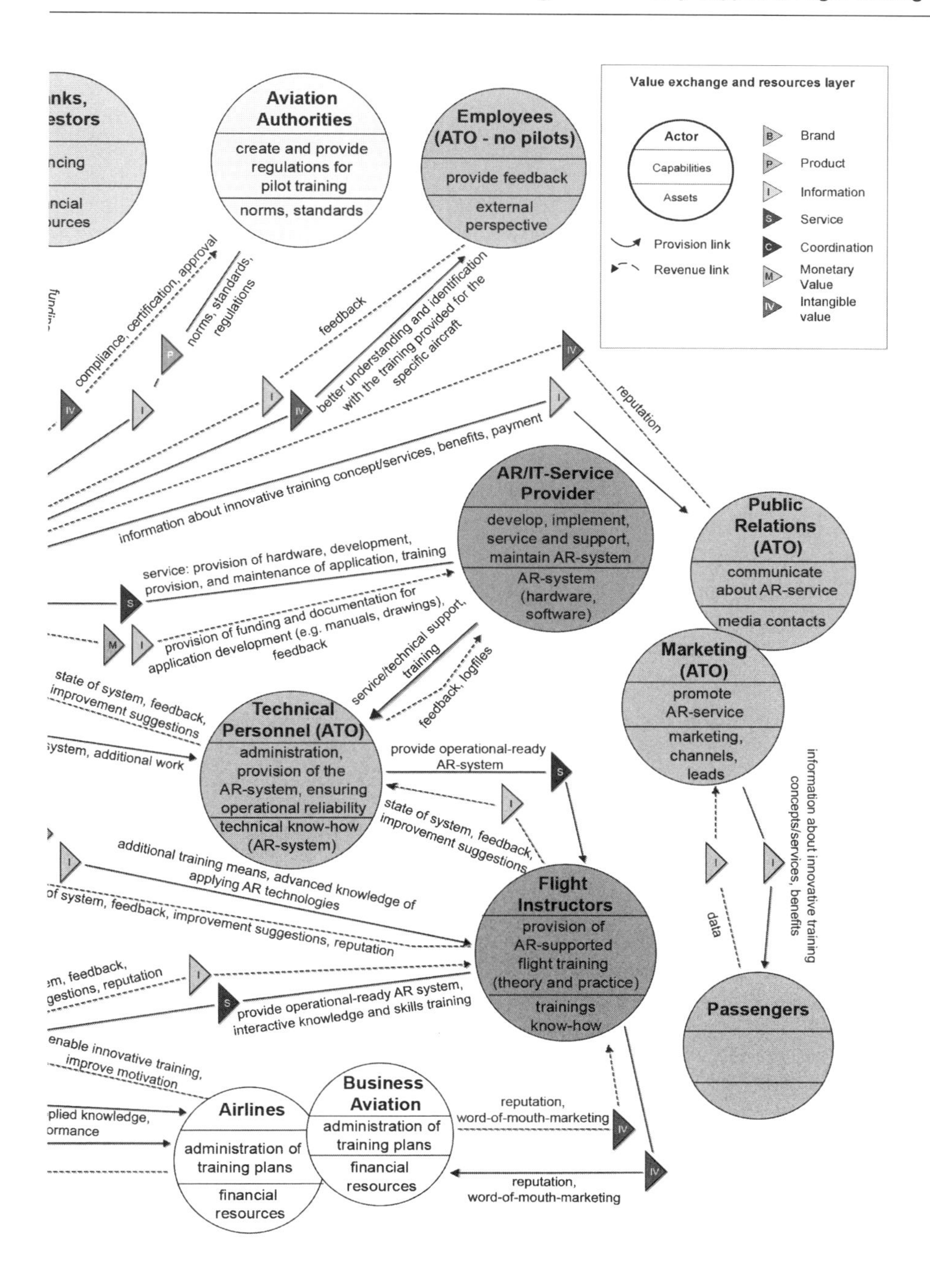

Value exchange and resources layer
Actor
Capabilities
Assets
Provision link
Revenue link
Brand
Product
Information
Service
Coordination
Monetary Value
Intangible value
Aviation Authorities
create and provide regulations for pilot training
norms, standards
Employees (ATO - no pilots)
provide feedback
external perspective
compliance, certification, approval
norms, standards, regulations
feedback
better understanding and identification with the training provided for the specific aircraft
reputation
information about innovative training concept/services, benefits, payment
AR/IT-Service Provider
develop, implement, service and support, maintain AR-system
AR-system (hardware, software)
Public Relations (ATO)
communicate about AR-service
media contacts
service: provision of hardware, development, provision, and maintenance of application, training
provision of funding and documentation for application development (e.g. manuals, drawings), feedback
service/technical support, training
feedback, logfiles
Marketing (ATO)
promote AR-service
marketing, channels, leads
state of system, feedback, improvement suggestions
Technical Personnel (ATO)
administration, provision of the AR-system, ensuring operational reliability
technical know-how (AR-system)
provide operational-ready AR-system
state of system, feedback, improvement suggestions
information about innovative training concepts/services, benefits
data
additional training means, advanced knowledge of applying AR technologies
Flight Instructors
provision of AR-supported flight training (theory and practice)
trainings know-how
provide operational-ready AR system, interactive knowledge and skills training
enable innovative training, improve motivation
Passengers
Airlines
administration of training plans
financial resources
Business Aviation
administration of training plans
financial resources
reputation, word-of-mouth-marketing
reputation, word-of-mouth-marketing

implementing the AR-based use cases were identified, starting with the ATO and adding successive nodes to the network (e.g., ATO, student pilot, flight instructors, technical ATO personnel, AR/IT service provider; see also Schaffernak et al., 2022). The EcoViz tool (Schierlinger-Brandmayr et al., 2022; Vorraber & Müller, 2019) was used to create a visualisation of the ecosystem and the value exchange relations between the economic entities involved in TR training (Figure 14.3; Schaffernak et al., 2022). The circles in Figure 14.3 depict economic entities or actors, and the directed arcs show the value exchange between these (e.g., continuous lines are used for provision links, and dashed lines are used for revenue links; Schierlinger-Brandmayr et al., 2022). Labelled triangles indicate the type of value exchanged (e.g., product, service; Schierlinger-Brandmayr et al., 2022). In a further step, the economic entities were grouped into clusters (e.g., customers, ATO technical staff, ATO financial partners), and colour codes were used to mark the clusters (e.g., pilot, student pilot, airline, business aviation) (Schaffernak et al., 2022).

The results of this analysis show that AR-related services and applications can improve the understanding of the training as well as the identification with the "service" training of ATO employees who were not pilots (Schaffernak et al., 2022). In addition, a role change in various ATO personnel was identified, such as additional tasks for the ATO staff (e.g., technical staff), and the demand for resources and incentives (Schaffernak et al., 2022). Thus, the system design and the incentive design will need to address these aspects (Schaffernak et al., 2022).

Controlled Experiments

After the surveys, controlled experiments (Moesl et al., 2023a, 2023b) were planned and conducted in the laboratory in parallel to the follow-up workshops. The assessed AR training concepts were specified based on the analysis of tasks and on the stages of skill development (Anderson, 1982; Fitts & Posner, 1967). The purpose of performing these experiments was to explore the effects of AR training with larger groups of participants, enabling statistical analyses of AR training and gender effects to be carried out, as shown in the next subsections.

The AR Application for Training the Approach to Landing

An AR-supported training concept for the approach to landing was developed in connection with a fixed-base flight simulator and evaluated in controlled experiments (for more details, see Moesl et al., 2023a). The learning content was augmented by AR cues such as the flight path vector (FPV), altitude, and speed, which were projected into the visual field during approaches to landing in the flight simulator (Moesl et al., 2023a). If these are not provided by these AR cues, the trainee must estimate the FPV, and read the altitude and speed indication in the cockpit (Moesl et al., 2023a). The learning conditions were targeted by practising self-assessment on the basis of several criteria and by receiving feedback and scores for both the self-assessment and the objective performance (Moesl et al., 2023a). The latter feature was AR-supported for the experimental group and in a paper-and-pencil form for the control group (Moesl et al., 2023a). Multimodal AR features were used, such as visual cues, audio cues, gesture, and language interaction (Moesl et al., 2023a). The assessment was conducted with 59 trainees (28 women and 31 men) assigned to an experimental or a control group (Moesl et al., 2023a). Gender groups were nested within the experimental and the control groups (Moesl et al., 2023a). The pre-test and post-test were conducted without AR for both groups (Moesl et al., 2023a). The control group trained with a flight simulator, and the experimental group trained with the AR application connected to the flight simulator (Moesl et al., 2023a). The results of tests conducted without AR do not indicate significant differences between the AR training group and the conventional simulator training group (Moesl et al., 2023a). In addition, the pre-test and post-test comparisons show that a positive transfer of knowledge had occurred for both types of training (Moesl et al., 2023a). Interestingly, the quality of the approach improved after training in both gender groups, but the advance was larger for women than for men (Moesl et al., 2023a). The training significantly decreased the trainees' workload, fear of failure, and negative emotions, and this impact was stronger in women than in men (Moesl et al., 2023a). Furthermore, a significant positive correlation was observed between the performance in the post-test and the self-reported AR presence and comfort with AR reported during training (Moesl et al., 2023a). Although more research is necessary to better understand the effects of AR-supported training, the results of this assessment indicate that such applications based on AR technology would address the desired social sustainability aspects and could bridge the gap between theoretical and practical instruction (Moesl et al., 2023a).

The AR Application for Training of Traffic Procedures

Although not a typical subject for TR training, traffic procedures according to visual flight rules (VFR) are relevant due to the broad operational context in the field of business aviation. Two multimodal AR applications for the training of traffic procedures were developed and assessed with 59 trainees (28 women, 31 men) assigned either to an experimental group (AR training) or a control group (simulator training; for more details, see Moesl et al., 2023b). One AR application in the form of a holographic moving "ball" guided a trainee's eyes in terms of direction and speed required for scanning the outside scenery (Moesl et al., 2023b). This was designed in accordance with the traffic scanning pattern in VFR flight described by the European General Aviation Safety Team (2010; Moesl et al., 2023b). In addition, the AR application provided numeric feedback to the trainee that compared their scanning pattern with the correct one (Moesl et al., 2023b). Another AR application simulated traffic encounters and provided a quiz to assess the collision detection accuracy and the trainees' decisions with respect to the collision avoidance manoeuvre and the right-of-way (Moesl et al., 2023b). This AR application also provided objective feedback (Moesl et al., 2023b). The results of an AR post-test show a significant positive effect of AR training on the performance of the scanning pattern (Moesl et al., 2023b). However, the other performance aspects that were evaluated without AR in the simulator pre- and post-tests for both groups indicate similar levels of performance (Moesl et al., 2023b). The factor gender had a significant effect on the emotions, motivation, and preferences for AR features, but not on the performance (Moesl et al., 2023b). The results indicate that these AR applications would address the social sustainability aspect and could be used in ways that complement theoretical and practical instruction (Moesl et al., 2023b).

Summary and Outlook

This chapter presented an interdisciplinary perspective and a mix of methods and tools applied to inform sustainable innovations in TR training. The sociotechnical view of the UCTM framework (Vorraber, Neubacher, et al., 2019) applied here facilitates the harmonization between the technology-driven and the human-centred and process-driven approaches. The ecosystem analysis results provide an overview of the value exchanges that are necessary to generate a sustainable business network (Schaffernak et al., 2022). The outcome of the process and business model analysis workshops in relation to potential AR innovations was an initial set of high-level use cases for AR-supported training (Schaffernak et al., 2022). These use cases were later refined, and the priorities were synchronized with the results of the human-centred studies conducted with pilots, flight instructors, and trainees (Schaffernak et al., 2022).

The surveys indicated the TR training content items that were most difficult to learn and enabled an initial screening of the content that could be easier to learn if AR was used (Moesl et al., 2022). The analysis of the survey content also revealed a number of training conditions that were experienced as difficult by the TR trainees (due to, e.g., stress, time pressure, high workload; Moesl et al., 2022). In addition, the surveys revealed the pilots' and instructors' preferences for generic AR use cases, and a number of gaming concepts that could be implemented in AR-based training applications (Schaffernak et al., 2020). The human-centred perspective was shaped by the analysis of tasks and skill development stages (Anderson, 1982; Fitts & Posner, 1967) and by taking a gender-sensitive approach to redesign the TR training in a socially sustainable manner (Moesl et al., 2022). Carrying out use case studies with smaller, gender-balanced samples of users allowed us to perform an early assessment and to identify issues that could subsequently be addressed at an early design stage (Wimmesberger, 2022). The results of the controlled experiments in laboratory settings (Moesl et al., 2023a, 2023b) partly confirm the conclusions of Kaplan et al. (2020), who indicated that the AR training was as effective as the conventional simulator training. Furthermore, the AR application for training the traffic scanning pattern led to significantly better performance as compared to training with conventional means (Moesl et al., 2023b). The AR applications described by Moesl et al. (2023a, 2023b) provided objective, data-based performance evaluation and formative feedback, and the trainees could practise self-assessment and integrate objective feedback on the self-evaluations. Thus, the issue of unclear assessment criteria that created difficulties for some trainees, as highlighted by the surveys (Moesl et al., 2022), could be addressed.

We can conclude that the AR applications could facilitate part-task training and learning performance by bridging the gap between theoretical and practical instruction and transforming the ATO processes in socially and economically sustainable ways. These results are encouraging for future work to increase the operational maturity of the AR applications, as well as the assessment of such applications, enabling these to transit from the laboratory to an operationally relevant environment and, later on, to the real training setting.

Notwithstanding the advantages highlighted by the evaluation of the AR applications presented in this study, a number of limitations were also recognised. At this state of knowledge and level of AR technology, it is not recommended that AR training be offered exclusively. Instead, it should be offered in combination with the traditional training means (McCoy-Fisher et al., 2019), if any are available. As noted by Wimmesberger (2022), some users encounter discomfort, cybersickness, or calibration issues when using the current AR devices, and these individuals should have the possibility to opt out of AR training. Furthermore, as the ecosystem analysis results show, both value-generating and value-hampering exchanges are expected as a result of introducing AR technology (Schaffernak et al., 2022). These effects on the ecosystem need to be further examined and addressed once the AR-based processes and business models have been developed.

The methods and tools presented in this chapter can be used for various technologies and domains of application. For instance, the UCTM framework was recently applied to artificial intelligence innovations in aviation training (Nguyen et al., 2023). Also, the ecosystem analysis was recently referenced as a tool for societal readiness assessment of new technologies (Büscher et al., 2023).

Acknowledgments

The research studies presented in this chapter were funded by the Austrian Federal Ministry for Climate Action, Environment, Energy, Mobility, Innovation, and Technology and the Austrian Research Promotion Agency, FEMtech Program "Talente," grant number 866702.

References

Anderson, J. R. (1982). Acquisition of cognitive skill. *Psychological Review, 89*(4), 369–406. https://doi.org/10.1037/0033-295x.89.4.369

Billinghurst, M., Clark, A., & Lee, G. (2015). A survey of augmented reality. *Foundations and Trends in Human-Computer Interaction, 8*(2–3), 73–272. https://doi.org/10.1561/1100000049

Billinghurst, M., & Kato, H. (2002). Collaborative augmented reality. *Communications of the ACM, 45*(7), 64–70. https://doi.org/10.1145/514236.514265

Billinghurst, M., Kato, H., & Poupyrev, I. (2001). The magic book-moving seamlessly between reality and virtuality. *IEEE Computer Graphics and Applications, 21*(3), 6–8. https://doi.org/10.1109/38.920621

Blundell, J., & Harris, D. (2023). Designing augmented reality for future commercial aviation: A user-requirement analysis with commercial aviation pilots. *Virtual Reality, 27*, 2167–2181. https://doi.org/10.1007/s10055-023-00798-9

Braunstingl, R., Vorraber, W., Mösl, B., & Schaffernak, H. (2022). *Gender issues in augmented learning* [Technical report]. Graz University of Technology, Graz, Austria.

Brown, L. (2019, September 20). *Augmented reality in international pilot training to meet training demands* [Poster presentation]. Workshop: Fall Convocation 2019, Western Michigan University, Kalamazoo, MI, USA. https://scholarworks.wmich.edu/instructional-development-grants/1/

Büscher, M., Cronshaw, C., Kirkbride, A., Spurling, N. (2023). Making response-ability: Societal readiness assessment for sustainability governance. *Sustainability, 15*, Article 5140. https://doi.org/10.3390/su15065140

European General Aviation Safety Team. (2010). *Collision avoidance.* European General Aviation Safety Team. Cologne, Germany. https://www.easa.europa.eu/en/document-library/general-publications/egast-leaflet-ga-1-collision-avoidance

European Union Aviation Safety Agency. (August, 2023). *Easy access rules for aircrew (Regulation (EU) No. 1178/2011).* https://www.easa.europa.eu/en/document-library/easy-access-rules/easy-access-rules-aircrew-regulation-eu-no-11782011#group-publications

Feng, S., He, X., He, W., & Billinghurst, M. (2023). Can you hear it? Stereo sound-assisted guidance in augmented reality assembly. *Virtual Reality, 27*, 591–601. https://doi.org/10.1007/s10055-022-00680-0

Ferla, M., & Graham, A. (2019). Women slowly taking off: An investigation into female underrepresentation in commercial aviation. *Research in Transportation Business & Management, 31*, Article 100378. https://doi.org/10.1016/j.rtbm.2019.100378

Fitts, P. M., & Posner, M. I. (1967). *Human performance*. Brooks/Cole.

Hancock, P. A. (2017). On bored to Mars. *Journal of Astro-Sociology, 2*, 103–120.

Huang, W., Wakefield, M., Rasmussen, T.A., Kim, S., & Billinghurst, M. (2022). A review on communication cues for augmented reality based remote guidance. *Journal of Multimodal User Interfaces, 16*(2), 239–256. https://doi.org/10.1007/s12193-022-00387-1

Kaplan, A. D., Cruit, J., Endsley, M., Beers, S. M., Sawyer, B. D., & Hancock, P. A. (2020). The effects of virtual reality, augmented reality, and mixed reality as training enhancement methods: A meta-analysis. *Human Factors, 63*(4), 706–726. https://doi.org/10.1177/0018720820904229

McCoy-Fisher, C., Mishler, A., Bush, D., Severe-Valsaint, G., Natali, M., & Riner, B. (2019). *Student naval aviation extended reality device capability evaluation* [Technical report NAWCTSD-TR-2019-001]. Naval Air Warfare Center Training Systems Division Orlando, FL.

Microsoft. (n.d.). *HoloLens 2* [Equipment]. https://www.microsoft.com/de-de/hololens?rtc=1

Milgram, P., Takemura, H., Utsumi, A., & Kishino, F. (1995). Augmented reality: A class of displays on the reality-virtuality continuum. *Telemanipulator and Telepresence Technologies, 2351*, 282–292. https://doi.org/10.1117/12.197321

Mitchell, J., Kristovics, A., Vermeulen, L., Wilson, J., & Martinussen, M. (2005). How pink is the sky? A cross-national study of the gendered occupation of pilot. *Employment Relations Record, 5*, 43–60.

Moesl, B., Schaffernak, H., Vorraber, W., Braunstingl, R., Herrele, Th., & Koglbauer, I.V. (2021). A research agenda for implementing augmented reality in ab initio pilot training. *Aviation Psychology and Applied Human Factors, 11*(2), 118–126. https://doi.org/10.1027/2192-0923/a000214.

Moesl, B., Schaffernak, H., Vorraber, W., Braunstingl, R., & Koglbauer, I. V. (2023a). Performance, emotion, presence: Investigation of an augmented reality-supported concept for flight training. *Applied Sciences, 13*(20), Article 11346. https://doi.org/10.3390/app132011346

Moesl, B., Schaffernak, H., Vorraber, W., Braunstingl, R., & Koglbauer, I. V. (2023b). Multimodal augmented reality applications for training of traffic procedures in aviation. *Multimodal Technologies and Interaction, 7*(1), Article 3. https://doi.org/10.3390/mti7010003

Moesl, B., Schaffernak, H., Vorraber, W., Holy, M., Herrele, Th., Braunstingl, R., & Koglbauer, I. V. (2022). Towards a more socially sustainable advanced pilot training by integrating wearable augmented reality devices. *Sustainability, 14*(4), Article 2220. https://doi.org/10.3390/su14042220

Nguyen, B., Sonnenfeld, N., Finkelstein, L., Alonso, A., Gomez, C., Duruaku, F., & Jentsch, F. (2023). Using AI tools to develop training materials for aviation: Ethical, technical, and practical concerns. *Proceedings of the Human Factors and Ergonomics Society Annual Meeting, 67*(1), 1343–1349 https://doi.org/10.1177/21695067231192904

Schaffernak, H., Moesl, B., Vorraber, W., Holy, M., Herzog, E.-M., Novak, R., & Koglbauer, I.V. (2022). Novel mixed reality use cases for pilot training. *Education Sciences, 12*, Article 345. https://doi.org/10.3390/educsci12050345

Schaffernak, H., Moesl, B., Vorraber, W, & Koglbauer, I. (2020). Potential augmented reality application areas for pilot education: An exploratory study. *Education Sciences, 10*(4), Article 86. https://doi.org/10.3390/educsci10040086

Schierlinger-Brandmayr, F., Moesl, B., Url, P., Vorraber, W., Vössner, S. (2022). A modeling tool for exploring business ecosystems in a (pre-)conceptual phase. In D. Karagiannis, M. Lee, K. Hinkelmann, & W. Utz (Eds.), *Domain-specific conceptual modeling* (pp. 315–338). Springer https://doi.org/10.1007/978-3-030-93547-4_14

United Nations. (n.d.). *Sustainable development goals. Goal 5: Achieve gender equality and empower all women and girls.* https://www.un.org/sustainabledevelopment/gender-equality/

Vergara, D., Extremera, J., Rubio, M. P., & Dávila, L. P. (2020). The proliferation of virtual laboratories in educational fields. *ADCAIJ Advances in Distributed Computing and Artificial Intelligence Journal, 9*(1), 85–97. https://doi.org/10.14201/ADCAIJ2020918597

Vlasblom, J., van der Pal, J., & Sewnath, G. (2019, May 14–16). *Making the invisible visible – increasing pilot training effectiveness by visualizing scan patterns of trainees through AR* [Paper presentation]. Stockholm, Sweden.

Vorraber, W., Neubacher, D., Moesl, B., Brugger, J., Stadlmeier, S., & Voessner, S. (2019). UCTM-an ambidextrous service innovation framework-a bottom-up approach to combine human- and technology-centered service design. *Systems, 7*(2), Article 23. https://doi.org/10.3390/systems7020023

Vorraber, W., & Müller, M. (2019). A networked analysis and engineering framework for new business models. *Sustainability, 11*(21), Article 6018. https://doi.org/10.3390/su11216018

Vorraber, W., Mueller, M., Voessner, S., & Slany, W. (2019). Analyzing and managing complex software ecosystems: A framework to understand value in information systems. *IEEE Software, 36*(3), 55–60. https://doi.org/10.1109/MS.2018.290100810

Wimmesberger, S. (2022). *Product-lifecycle-management enhanced with augmented reality* [Unpublished master's thesis]. Graz University of Technology.

Woods, B. (2022, November 12). How Airlines plan to create a new generation of pilots amid fears of decade-long cockpit crisis. *CNBC*. https://www.cnbc.com/2022/11/11/how-airlines-plan-to-create-new-generation-of-pilots-at-time-of-crisis.html

Wu, H. K., Lee, S. W. Y., Chang, H. Y., & Liang, J. C. (2013). Current status, opportunities and challenges of augmented reality in education. *Computers & Education, 62,* 41–49. https://doi.org/10.1016/j.compedu.2012.10.024

Chapter 15

How to Conduct and Interpret Meta-Analyses

Monica Martinussen and Sabine Kaiser

Abstract

Research questions may be answered with empirical studies or by aggregating existing research. In contrast to single studies, meta-analyses combine the quantitative findings of multiple primary studies to estimate an overall effect size, for example the mean predictive validity of spatial tests used for pilot selection. This aggregation allows researchers to make more reliable conclusions about the size of a specific relationship, and how much it varies between studies. Moderator analyses can be applied to examine the influence of different factors on the estimated effect size, such as whether the predictive validity varies with the type of aircraft (fixed wing or rotary) or sample (ab initio vs experienced pilots). The aim of this chapter is to present the different steps taken to conduct a meta-analysis and explain different meta-analytic approaches for estimating the mean effect sizes and variation between studies. The chapter also presents different types of software available for both screening studies and meta-analysis calculations, and uses examples from aviation and human factors explaining how the results may be interpreted and used in applied settings.

Keywords

meta-analysis, aviation, synthesising research, effect size, moderator analyses

What Is a Meta-Analysis

Meta-analyses summarise research findings from many studies, for example validation studies in selection research or studies of the effectiveness of different training methods. By pooling and analysing findings from various articles and reports, meta-analyses enable researchers to draw more robust conclusions regarding the magnitude and variability of a specific relationship or group difference. The primary distinction between a systematic review and a meta-analysis lies in their approaches and the type of studies that may be included. While both aim to summarise research, a meta-analysis uses statistical techniques to summarise and compare the results of individual studies, whereas a review does not utilise these statistical techniques. In a meta-analysis, the inclusion of studies with quantitative findings is a prerequisite, whereas a systematic review does not have such a requirement. In some cases, articles may employ both approaches, wherein quantitative findings are subjected to meta-analytic techniques, while studies lacking quantitative data or exhibiting significant heterogeneity are qualitatively described and evaluated.

The popularity of meta-analyses has increased drastically in the past 30 years (Borenstein et al., 2021). In aviation, meta-analyses have been used in selection research to examine the predictive validity of a number of predictors of pilot performance (Hunter & Burke, 1994; Martinussen, 1996), to predict performance based on personality assessments (Campbell et al., 2010), and for predicting military pilot performance (ALMamari & Traynor, 2019) as well as for the selection of air traffic controllers (Dehn & Damitz, 2022; Martinussen et al., 2000). Meta-analysis has also been used to combine other types of studies such as effectiveness studies of crew resource management (CRM) training (O'Connor et al., 2008) or studies estimating the prevalence of neck pain among fighter pilots (Riches et al., 2019). Validation studies typically report correlations between tests and criteria, and effectiveness studies typically examine mean differences between an intervention and a control group, whereas studies examining the prevalence of a condition will report this as a percentage or proportion. These examples represent different study designs where the findings are analysed and reported differently.

When to Do a Meta-Analysis

A meta-analysis is appropriate to conduct when there are "enough" primary studies with quantitative findings that can be summarised in the respective analyses. It is difficult to set the minimum number of studies or rather effect sizes needed to conduct meaningful analyses. Some would argue that three is the absolute minimum, while others would claim that five should be the minimum. Others, again, would argue that a much higher number is needed. Regardless of which rules are used, the number of studies will determine what kind of conclusions can be drawn based on the meta-analysis, especially with respect to what extent the findings can be generalised.

Which Method to Use

There are different meta-analytic approaches available to perform the meta-analyses. Traditionally a distinction has been made between the approach developed by Hunter and Schmidt (Hunter & Schmidt, 1990, 2004; Hunter et al., 1982;

Schmidt & Hunter, 1977, 2015) and Glass (1976) and later Hedges and Olkin (1985). The approach pioneered by Glass (1976) was initially designed to aggregate findings from psychotherapy research, specifically focusing on differences between groups that received treatment and those that did not. By contrast, Schmidt and Hunter (1977) developed a procedure to assess the generalisability of test validity across various settings. For example, they explored whether intelligence tests consistently predicted work performance irrespective of job type or applicant characteristics. Over time, both research traditions have expanded beyond their original scope. However, within the field of work and organisational psychology, the Hunter and Schmidt method remains widely favoured, while Hedges and Olkin's approach is more commonly employed in clinical and medical research.

How to Conduct a Meta-Analysis

Conducting a meta-analysis takes time and includes several steps which are described in the following and depicted in Figure 15.1. The process resembles the ordinary research process found in a primary study, which starts with a research question, data collection and analysis, and ends with a research article.

The Research Question

The initial step in conducting a meta-analysis is to formulate one or more research questions. These questions can range from specific inquiries to broader topics, depending on the researcher's interests and the available studies. Sometimes the authors choose to write a protocol that describes the planned meta-analysis, and also register it in a database such as the International Prospective Register of Systematic Reviews (PROSPERO) for health sciences.

Inclusion and Exclusion Criteria

Inclusion and exclusion criteria define which studies should be included or excluded from the analyses. They can be outlined before the literature search is conducted. Inclusion and exclusion criteria are important in order to be systematic, transparent, and select the right studies. Examples of inclusion criteria are English language or the year when the study was published (e.g., after 2000). More specific criteria are the type of design (e.g., an experiment), or reporting statistics that can be used for the meta-analytic calculations (e.g., frequencies, means and standard deviations, correlations). Exclusion criteria could, for example be that a certain group is excluded (e.g., commercial pilots versus general aviation pilots).

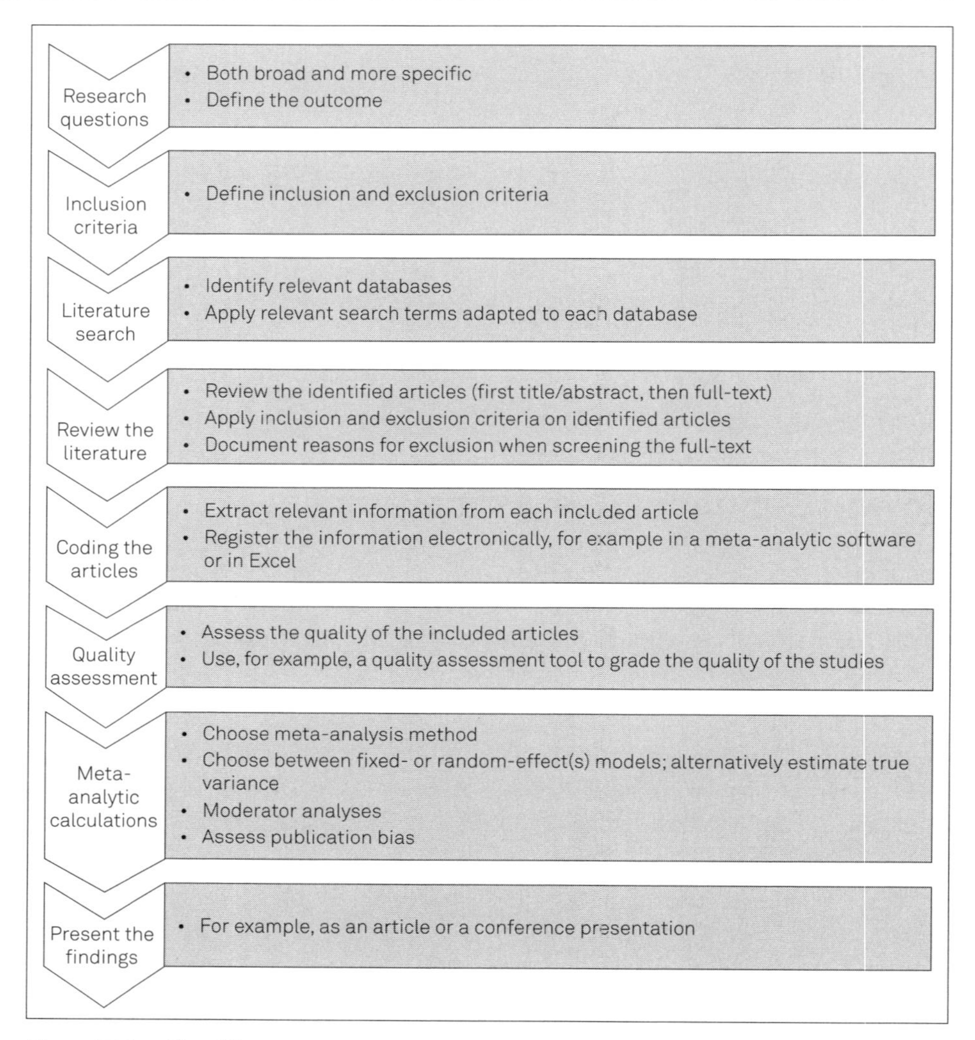

Figure 15.1. The different steps in a meta-analysis.

The Literature Search

The starting point for every meta-analysis is a comprehensive and systematic literature search that covers the most relevant databases for the respective research field and appropriate search key terms to identify relevant primary studies. Also, grey literature (e.g., reports) is of importance and should be searched for by, for example contacting relevant authors in the field of interest to acquire unpublished material. Accordingly, the method section should describe which databases were used (one is usually not enough), the search terms (which might have to be adapted from database to database), and how many articles that were identified in total. The literature search is very complex, and it might take several rounds before finding the right databases and search terms. It is an advantage if a librarian can assist in developing the search strategy or at least review it. After the literature search is conducted, the identified studies are reviewed for their inclusion or exclusion.

The Review Process

The review process involves the screening of the titles and abstracts of the identified studies to determine their potential relevance. Next the selected articles are thoroughly screened in full text to ascertain their eligibility for inclusion in the meta-analysis. Also, there are different tools that can help in conducting the screening of the identified articles (e.g., Rayyan, Covidence). While some meta-analyses solely rely on published articles, others include a wider range of sources, such as reports, conference presentations, and dissertations. Using only published articles ensures a certain level of quality through the peer-review and editorial process, but incorporating grey literature can help mitigate publication bias and provide a more comprehensive overview of the research field. The review process includes a screening of the identified articles from the literature search based on the defined inclusion and exclusion criteria. The review process usually starts with the title and abstract screening to determine whether an article seems relevant for the meta-analysis or not, by applying the pre-defined inclusion and exclusion criteria. If the literature search identified a large number of articles, one can consider starting with a title screening first and continuing with an abstract screening in a second step. Those articles that are not excluded after the title and abstract screening are usually read in full text to inform a final decision about their inclusion or exclusion. In this final step of the review process, it is also important to document the reason for exclusion for every article and report them in the manuscript, for example in a flow diagram.

A flow diagram can be a useful way to depict the different stages of the literature search and review process, such as how many articles were identified through the original search, how many were excluded after title and abstract screening, what were the reasons for exclusion when the full text was examined, and last but not least, what was the final number of included studies. One can either create a new flow diagram or use a template from the Preferred Reporting Items for Systematic Reviews and Meta-Analyses (PRISMA).

Coding of Articles

Once the final sample of studies is determined, the subsequent step entails coding all relevant information, including study characteristics, sample sizes, and findings (effect sizes). This can either be entered in a separate file or directly into the software that is going to be used to conduct the meta-analyses. The extracted information can, for example include the first author (for identification purposes), publication year of the study, sample size, the relevant statistics, and possible moderators (e.g., type of aircraft [fixed wing or rotary] or sample [ab initio vs experienced pilots]). Involving two or more individuals in both the study screening and coding processes can enhance the quality of the literature selection and improve coding accuracy. Furthermore, the consistency between the coders can be calculated using different types of statistics (e.g., intraclass correlation coefficient or Cohen's κ).

Quality Assessment of Included Articles

Often the included studies are assessed based on their quality regarding different criteria such as sampling, study design, method, and statistical analyses. This information can be used to describe the studies in a table or to examine differences in effect sizes based on the individual studies quality score. There are different quality assessment tools (i.e., checklists) depending on the study design you are interested in examining (e.g., RoB 2, which is Cochrane risk of bias tool in randomised trials). Possible quality indicators are, for example questions such as "Is the applied method appropriate to answer the research question?" or "Is the sample size large enough to conduct the respective analysis?" The quality assessment thus indicates the quality of the reporting in the primary study as well as the quality of the included studies (Guyatt et al., 2008).

The Meta-Analytic Calculations

The next step encompasses the actual meta-analysis calculations, where the synthesised effect sizes are computed, and variation between studies is estimated and examined.

What Is an Effect Size?

Research studies investigating the same research question often exhibit variations in their design and measurement approaches. For instance, when examining pilot selection, studies may employ different types of tests and different criteria to assess pilot performance. Similarly, studies investigating the effectiveness of CRM training may adopt either a control-group design or a simple pre-test–post-test design. Moreover, the outcomes measured in these studies can vary, ranging from instructor ratings to self-reported satisfaction. To synthesise findings from studies with diverse measures and designs, it is crucial to identify commonalities that enable the integration of results across studies.

Meta-analyses rely on effect sizes derived from each study as their fundamental building blocks. Effect sizes are statistical measures that quantify the magnitude or strength of a finding. They may encompass standardised measures of co-variation between variables or differences observed between groups. Co-variation is often expressed through correlation coefficients (r), which indicate the strength and direction of the association between two variables. Correlation coefficients range from −1 to +1, where a value close to 0 signifies no association. A correlation coefficient of +1 or −1 indicates a perfect correlation, with a positive correlation implying that higher values on one variable align with higher values on the other, and a negative correlation indicating that higher values on one variable correspond to lower values on the other. The closer the correlation coefficient is to −1 or +1, the stronger the association. To aid the interpretation and communication of findings, Cohen (1988) coined various labels for different effect sizes in his book on statistical power calculations. Table 15.1 presents some guidelines for the interpretation of effect sizes and Figure 15.2 illustrates an example of a large correlation.

Table 15.1. Interpretation of effect sizes

	Effect size	
Effect size labels	**Correlation**	**Hedges' *g***
Small effect	$r = .10$	$g = 0.20$
Medium effect	$r = .30$	$g = 0.50$
Large effect	$r = .50$	$g = 0.80$

Note. Based on Cohen, 1988.

Another frequently used effect size is the standardised mean difference, typically utilised when studies report means for different groups. This effect size quantifies the difference between means (M) divided by a standard deviation (SD), often the pooled standard deviation based on both groups. One example from this family of effect sizes is denoted Hedges' g ($= M_1 - M_2 / SD_{pooled}$). Alternatively, the SD from the control group can be employed to compute the effect size. The standardised mean difference represents the difference between two groups in terms of SDs. In validation studies, where the focus lies on the correlation between test results and a measure of pilot performance, the common effect size is a correlation coefficient, which can be combined across studies. When evaluating CRM training and comparing different groups based on an outcome, the most relevant effect size is the standardised mean difference.

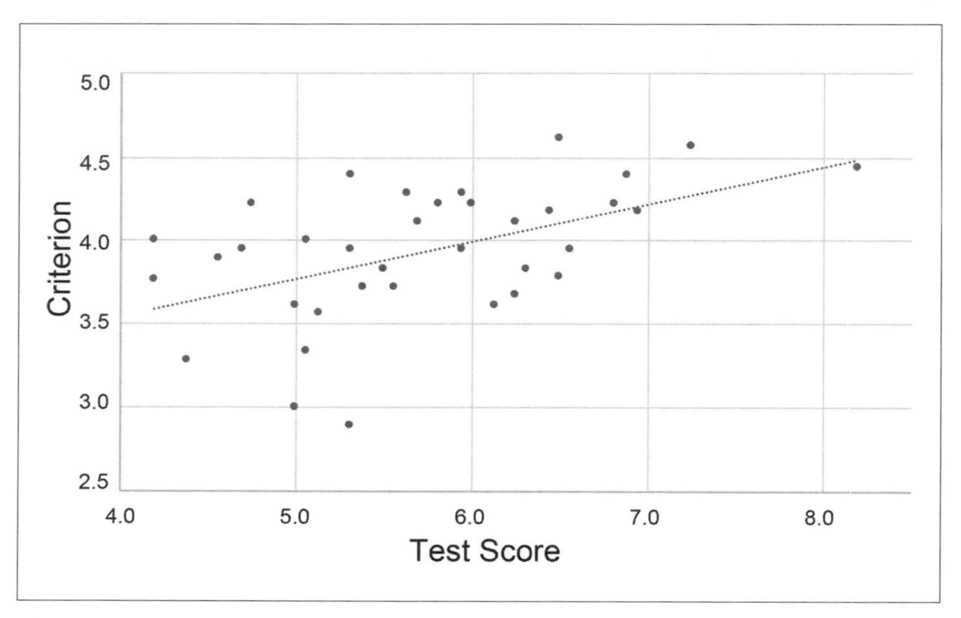

Figure 15.2. Scatterplot of a large correlation.

Statistical Artefacts

There are several potential reasons for variations in results across studies. Firstly, sampling error contributes to random fluctuations due to the examination of samples rather than the entire population. The smaller the sample sizes, the greater the sampling error. Secondly, true variation among studies may arise from specific aspects of the research design. For instance, certain interventions might prove more effective than others, or certain tests may serve as better predictors of pilot performance.

The third factor involves statistical artefacts or methodological issues that can influence the magnitude of the observed effects. In validation studies, a common problem is conducting research within a highly restricted sample (e.g., limited to accepted applicants), which reduces variation in test scores and subsequently lowers the observed correlation between the test and the criterion. Another artefact occurs when the measures used exhibit poor reliability, leading to a decrease in the observed correlation. Additionally, using a dichotomous criterion (e.g., pass/fail) instead of a more continuous measure, such as performance ratings, can introduce another artefact. These artefacts are cumulative, resulting in lower observed validation coefficients compared to a situation without range restriction, with perfectly measured variables and a continuous performance measure.

It is possible to statistically adjust the observed correlations to account for these artefacts, either addressing a specific artefact (e.g., range restriction) or considering the combined effect of many artefacts. If the presence and magnitude of these biases differ between studies, they can contribute to the observed variation. Schmidt and Hunter (2015) describe some of these artefacts specifically in relation to validation studies, while others apply to other types of studies and effect sizes. For instance, poor reliability always diminishes the observed effect. Correcting for range restriction is not a straightforward process and depends on various factors, including the selection ratio and the number of tests used, as well as their intercorrelations. Correction for reliability relies on the reported reliability coefficient, which may or may not be reported in the primary studies. Correction for dichotomisation depends on where the split is, that is how uneven the group sizes are. The further away the groups are from a 50/50 split, the greater the correction.

Mean Effect Size

There are two primary steps involved in conducting a meta-analysis. The first step is to calculate the mean effect size, typically a weighted mean. In the Hunter and Schmidt method (2004), for instance, the effect size is weighted based on the sample size. Ideally, the effect sizes should be adjusted or corrected for any statistical artefacts before computing the mean effect size. In some meta-analyses, both corrected and uncorrected mean effect sizes are reported, and it is important to specify the type of corrections employed. Table 15.2 provides a meta-analysis example for correlations between test and criterion on simulated data where the data include seven studies with individual correlations ranging from .06 to .20, as well as correlations corrected for criterion reliability. The mean sample-size weighted correlation is .14, and corrected for lack of criterion reliability, slightly higher (.16) when using the software Metados (Martinussen & Bjørnstad, 1999). See Appendix for formulas and calculations.

Table 15.2. Meta-analysis example for correlations (r between test and criterion)

Study	N	r	r_{yy}	r_c
1	200	.20	.60	.26
2	150	.20	.70	.24
3	1,500	.10	.80	.11
4	200	.15	.60	.19
5	100	.13	.70	.16
6	1,200	.20	.80	.22
7	400	.06	.70	.07

Note. N = sample size; r = correlation; r_{yy} = criterion reliability; r_c = correlation corrected for criterion reliability.

Alternative meta-analytic approaches may utilise weighting methods slightly different than direct sample size (as suggested by Hedges and Olkin and implemented in the software "Comprehensive Meta-Analysis version 4"; see for example Borenstein et al., 2022). The approach taken in a meta-analysis depends on the specific research question. In some cases, it may be adequate to calculate an overall effect size. However, it is more common to compute mean effect sizes for different sub-groups in addition to the overall effect.

How to Study Variation Between Studies

When synthesising findings from numerous studies, the process typically involves two steps. Firstly, calculating the mean correlation provides an overall estimate of the effect. Secondly, the focus shifts to examining the variation among the studies. This entails determining whether the observed effect such as correlation coefficients exhibit greater variation than expected and assessing the extent to which the observed variation is influenced by sampling error.

In the approach proposed by Schmidt and Hunter (2015), the variation is estimated by comparing the observed variation with the variation attributed to sampling error. If the observed variation is small or predominantly caused by sampling error, then the mean effect size serves as a reliable estimate of the true effect. However, if substantial variation exists, the subsequent step involves investigating potential factors, known as "moderators", that might explain the differences in variation, such as variations among sub-groups. In the example in Table 15.2, the estimated population variance was 0.001 and the corresponding standard deviation was 0.032. This estimated standard deviation may be used to calculate an interval (credibility interval) where the true effect is likely to be found (Whitener, 1990). Moreover, the percentage of explained variance was 64% indicating that there is some variance left to be explained. Hunter and Schmidt (2004) have suggested a 75% rule, meaning that if 75% or more of the variance is explained by sampling error, there is no need to examine moderators.

The most basic form of meta-analysis is referred to as a "bare-bones" meta-analysis (Schmidt & Hunter, 2015), where effect sizes are not adjusted for statistical artefacts such as reliability or range restriction. Alternatively, corrections for statistical artefacts can be performed individually or by using compiled information from studies included in the meta-analysis (using artefact distributions) to improve the estimates of both the mean effect sizes as well as how the true variation between studies is estimated.

An alternative approach to estimating the variation between studies is to conduct a significance test. In this case, the null hypothesis posits that there is no significant variation among the studies. If the test yields a significant result, the null hypothesis is rejected, indicating the presence of true variation between studies. Consequently, the subsequent step involves investigating potential moderators in a similar manner as previously outlined. The examination of variation through significance testing was initially described by Hedges and Olkin, and later expanded upon by Borenstein et al. (2009, 2021). The weighting scheme employed in these calculations is contingent upon the chosen model, namely the fixed effect model or the random effects model.

In the fixed effect model, it is assumed that the included studies can be regarded as replications. However, this assumption may not hold true in most meta-analyses, as there are typically variations across studies in terms of study design, participant characteristics, and measurement methodology. On the other hand, the random effects model considers that effect sizes in the population are not constant and that factors beyond sampling error can contribute to the observed variation in effect sizes.

In the random effects model, studies are weighted by the inverse of the variance components, which encompass both random variation (sampling error) and variation between studies. Consequently, this results in more equal weights being assigned to the studies compared to the fixed effect model where the weighting is more similar to the one employed by Schmidt and Hunter (2015).

Publication Bias

Publication bias refers to the possible overrepresentation of significant findings in published studies, and that non-significant findings are less likely to be published. If a meta-analysis is based on published studies only, it may overestimate the true effects. It is of interest to examine the extent to which the findings from unpublished studies may influence the meta-analytic results and therewith the overall conclusion (Lipsey & Wilson, 2001). This can, for example, be done by comparing the mean effect sizes between published and unpublished studies but there are also other methods that can be used (for more information, see Cooper et al., 2009).

Presenting the Findings

In the final step, the results of the meta-analysis are presented to a broader audience, and hopefully provide an answer to the initial research question. This is typically done in the form of a research article or conference presentation.

Which Software to Use

There are different types of software that can be used to conduct meta-analyses; the more general ones are R, SAS, or the newer versions of SPSS. The R software offers a package for performing meta-analysis calculations according to the Hunter and Schmidt method (1990). There is also more specialised software for conducting meta-analyses such as Comprehensive Meta-Analysis (CMA; Borenstein et al., 2022) or Review Manager 5 (RevMan 5; The Cochrane Collaboration, 2020). CMA uses the meta-analytic approach by Borenstein et al. (2021), while RevMan 5 is Cochrane's software and widely used.

How to Interpret Meta-Analyses

Interpreting meta-analyses includes an assessment of the different steps that are described in this chapter and that need to be completed in order to conduct a meta-analysis. It includes assessing the quality of the meta-analysis, possible biases, and interpreting effect sizes and other findings such as from the moderator analyses. To assess a meta-analysis quality, it is important to use a systematic and transparent approach throughout the study. The literature search and the review process should be described in detail and inclusion and exclusion criteria that were applied to select the studies should be stated explicitly.

Summary and Outlook

Well conducted meta-analyses can be an important contribution to the research field. They do not just summarise statistical findings that are more reliable than the results from single studies, but also generate new knowledge beyond this. Meta-analyses provide an insight into the status quo of a research field, can identify the need for further research into a specific direction, and provide answers to research questions that cannot be answered by conducting primary studies alone. The number and quality of the primary studies will determine the overall quality of the meta-analysis, as well as the conclusions that can be drawn. As always, the more data the better, or rather the more studies the better the meta-analysis.

References

ALMamari, K., & Traynor, A. (2019). Multiple test batteries as predictors for pilot performance: A meta-analytic investigation. *International Journal of Selection and Assessment, 27*(4), 337–356. https://doi.org/10.1111/ijsa.12258

Borenstein, M., Hedges, L. V., Higgins, J. P., & Rothstein, H. R. (2009). *Introduction to meta-analysis*. John Wiley & Sons. https://doi.org/10.1002/9780470743386

Borenstein, M., Hedges, L. V., Higgins, J. P., & Rothstein, H. R. (2021). *Introduction to meta-analysis* (Vol. 2). John Wiley & Sons. https://doi.org/10.1002/9781119558378

Borenstein, M., Hedges, L. V., Higgins, J. P., & Rothstein, H. R. (2022). *Comprehensive Meta-Analysis* (Version 4) [Computer software]. Biostat. https://www.meta-analysis.com/

Campbell, J. S., Castaneda, M., & Pulos, S. (2010). Meta-analysis of personality assessments as predictors of military aviation training success. *The International Journal of Aviation Psychology, 20*(1), 92–109. https://doi.org/10.1080/10508410903415872

Cohen, J. (1988). *Statistical power analysis for the behavioral sciences* (Vol. 2). Lawrence Erlbaum Associates.

Cooper, H., Hedges, L. V., & Valentine, J. C. (2009). *Handbook of research synthesis and meta-analysis*. Russell Sage Foundation. http://www.jstor.org/stable/10.7758/9781610441384

Dehn, D. M., & Damitz, M. (2022). Do personality traits predict training performance? *Aviation Psychology and Applied Human Factors, 12*(2), 72–83. https://doi.org/10.1027/2192-0923/a000222

Glass, G. V. (1976). Primary, secondary, and meta-analysis of research. *Educational Researcher, 5*(10), 3–8. https://doi.org/10.2307/1174772

Guyatt, G. H., Oxman, A. D., Vist, G. E., Kunz, R., Falck-Ytter, Y., Alonso-Coello, P., Schünemann, H. J., & GRADE Working Group (2008). GRADE: An emerging consensus on rating quality of evidence and strength of recommendations. *BMJ (Clinical Research* Ed.*), 336*(7650), 924–926. https://doi.org/10.1136/bmj.39489.470347.AD

Hedges, L. V., & Olkin, I. (1985). *Statistical methods for meta-analysis*. Academic Press.

Hunter, D. R., & Burke, E. F. (1994). Predicting aircraft pilot-training success: A meta-analysis of published research. *The International Journal of Aviation Psychology, 4*(4), 297–313. https://doi.org/10.1207/s15327108ijap0404_1

Hunter, J. E., & Schmidt, F. L. (1990). *Methods of meta-analysis: Correcting error and bias in research findings*. Sage Publications.

Hunter, J. E., & Schmidt, F. L. (2004). *Methods of meta-analysis: Correcting error and bias in research findings* (2nd ed.). Sage Publications.

Hunter, J. E., Schmidt, F. L., & Jackson, G. B. (1982). *Meta-analysis: Cumulating research findings across studies*. Sage Publications.

Lipsey, M. W., & Wilson, D. B. (2001). *Practical meta-analysis*. SAGE Publications.

Martinussen, M. (1996). Psychological measures as predictors of pilot performance: A meta-analysis. *The International Journal of Aviation Psychology, 6*(1), 1–20. https://doi.org/10.1207/s15327108ijap0601_1

Martinussen, M., & Bjørnstad, J. F. (1999). Meta-analysis calculations based on independent and nonindependent cases. *Educational and Psychological Measurement, 59*(6), 928–950. https://doi.org/10.1177/00131649921970260

Martinussen, M., Jenssen, M., & Joner, A. (2000, September). *Selection of air traffic controllers: Some preliminary findings from a meta-analysis of validation studies* [Presentation]. 24th European Association for Aviation Psychology (EAAP) conference, Crieff, Scotland.

O'Connor, P., Campbell, J., Newon, J., Melton, J., Salas, E., & Wilson, K. A. (2008). Crew resource management training effectiveness: A meta-analysis and some critical needs. *The International Journal of Aviation Psychology, 18*(4), 353–368. https://doi.org/10.1080/10508410802347044

Riches, A., Spratford, W. A., Witchalls, J., & Newman, P. (2019). A systematic review and meta-analysis about the prevalence of neck pain in fast jet pilots. *Aerospace Medicine and Human Performance, 90*(10), 882–890. https://doi.org/10.3357/AMHP.5360.2019

Schmidt, F. L., & Hunter, J. E. (1977). Development of a general solution to the problem of validity generalization. *Journal of Applied Psychology, 62*(5), 529. https://doi.org/10.1037/0021-9010.62.5.529

Schmidt, F. L., & Hunter, J. E. (2015). *Methods of meta-analysis: Correcting error and bias in research findings* (3rd ed.). Sage Publications. https://doi.org/10.4135/9781483398105

The Cochrane Collaboration. (2020). *Review Manager* (Version 5.4) [Computer software]. Cochrane. https://training.cochrane.org/online-learning/core-software/revman

Whitener, E. M. (1990). Confusion of confidence intervals and credibility intervals in meta-analysis. *Journal of Applied Psychology, 75*(3), 315–321. https://doi.org/10.1037/0021-9010.75.3.315

Appendix

Formulas

The mean sample-size weighted correlation: $\bar{r} = \frac{\Sigma\,[N_i r_i]}{\Sigma\,N_i}$

The population variance: $\sigma_\rho^2 = \sigma_o^2 - \sigma_e^2$

Results

Meta-analysis calculations (bare-bones) using the Hunter and Schmidt method:

The number of studies = 7

The total number of participants = 3,750 (mean number per study = 535)

The mean correlation (sample size weighted) = .14

The mean correlation unweighted = .15

The estimated population variance = 0.001 and standard deviation = 0.032

Lower credibility value (90%): .10

Percentage of observed variance accounted for by sampling error = 64%

Corrected and *N*-weighted mean correlation (for criterion reliability) = .16

Chapter 16

Single European Sky Air Traffic Management Research Case Study

Where Have We Been and Where Are We Going?

Adriana-Dana Schmitz and Rubén Rodríguez Rodríguez

Abstract

More often than not, human performance (HP) experts share a common denominator in their work: How can we increase visibility of HP and how can we ensure a better integration of the human elements, from early research to industrial development and deployment? In the context of the Single European Sky Air Traffic Management (ATM) Research (SESAR) 3 Joint Undertaking Programme, a recognised leader in advancing global interoperability in aviation, there was a recent shift in the way maturity evaluation of projects is classified. With the introduction of technology readiness levels (TRLs), widely recognised in various industries, the categorisation of HP maturity assessments had to be reclassified as well. As a result, the SESAR 3 JU HP Assessment Methodology has been updated to fit all stages of research from TRL0-TRL2 (exploratory research) to TRL3-TRL6 (industrial research) and TRL7-TRL8 (digital sky demonstrators). The traceability between HP and TRLs is not novel; however, it is the first time that it has been formalised in a research approach at such a large scale in ATM research in Europe. Moving a step forward, could a future integration of human readiness levels (HRLs) finally set the scene for HP to become more substantiated in the evaluation of the research maturity levels and make it more visible and clearer to non-HP experts? The hypothesis is that such an approach would ensure a better consideration given to the impact on the human, from innovation to implementation, by making HP more tangible and structured for multidisciplinary researchers.

Keywords

human performance, maturity evaluations, technological readiness levels, human readiness levels, human centric design

Introduction

Programmatic guidance for the assessment of human performance (HP) design requirements has been provided to NextGen (USA) and Single European Sky ATM Research (SESAR) 3 Joint Undertaking (hereafter referred to as "SESAR 3 JU") research and development (R&D) program managers. Their major aim was to strengthen the business case of innovative concepts and emerging technologies along the R&D stages. It was recognised that a failure to fully understand, communicate and consider HP during the R&D phases would increase the risk of cost-overrun, deployment delay or even stop the projects. Whether we talk about exploratory research or implementation phases, ensuring safety remains one of the utmost priorities in aviation. Our goal is to minimise the potential for occurrences and constantly improve performance and efficiency, all the while striving to uphold or even enhance HP. This goal, nevertheless, is more intricate than ever amidst the growing prevalence of artificial intelligence (AI) and machine learning techniques, together with higher levels of automation and the consequent new and innovative ways of interacting between the human and the machine. Technological advancements are soaring, with the long-term goal being to have them compensate for the human limitations, including to significantly reduce "human error." *But are we sure that we are adequately integrating HP knowledge into industrial developments?*

Repeatedly, we come across staggering statistics in relation to human error (HE) in high-risk industries. To use a few examples, DeMott (2014) has cited HE as being responsible for 60%–80% of failures, accidents and incidents. Over 90 % of today's road accidents can be put down to HE (National Highway Traffic Safety Administration [NHTSA], 2019). During 2004 in the USA, pilot error was listed as the primary cause of 78.6% of fatal general aviation accidents and the primary cause of 75.5% of general aviation accidents overall (Shively, 2013), while in cybersecurity as many as 95% of data breaches are identified to be caused by HE (World Economic Forum [WEF], 2022). However, the reiteration of HE-related statistics might have long lost their meaning without helping us move further. First, there is a lack of consensus on what HE means (Shorrock, 2014) and second, we all know the famous saying: "To err is human." So, what next? There is no denying that it is in our human nature to make mistakes that are linked to our limitations. An approach typical to Safety I, based on a "causality credo" where "humans are viewed predominantly as a liability or hazard" (EUROCONTROL, 2013), becomes less and less tangible (unless we talk about intentional and malicious acts). Aviation has become an increasingly complex socio-technical system, where we cannot afford to look at its components independently. And steadily, we are moving towards finding proactive ways in finding answers, in order to look beyond the concept of error and to find ways to understand success, as well. From this perspective, the human is seen more and more as the "resource necessary for system flexibility and resilience" (Hollnagel, 2014) with considerable strengths that support the provision of several flexible solutions to many different problems. The field of

HP requires us to look beyond human limitations, recognising that individuals do not operate in isolation but within the larger system. It creates a need to analyse and address contributions to performance in the context of dynamic and interconnected systems, where changes in one component can affect the entire system. *But do we fully incorporate this knowledge in research and are we disseminating it properly outside the HP community?*

The multifaceted nature of the field of HP should be seen as an enabler for attaining success and hence safety. However, we see that HP is often misunderstood. Thus, its integration in research activities is susceptible to being poorly integrated into the development and validation of new concepts and technologies, and even when it is done, it is often very late in the process. This can have significant cost implications, especially if substantial modifications are required from a technological perspective to fit the human capabilities, or worse in the case that developments are not fit for purpose. Until we reach the era of full automation, where the roles of the human will be reshaped, we need to (re)acknowledge that the human is still the binder of it all. We must recognise and appreciate the crucial role of humans as the foundation of everything, from conceptualisation to coordination at all levels. The recurring challenge we face is ensuring the effective integration of HP expert knowledge into future developments and maintaining a balanced focus on both humans and the rapidly advancing technologies.

SESAR and Human Performance

SESAR is a recognised leader in advancing global interoperability in aviation that has acknowledged HP as a distinct transversal area, from the very beginning. In the definition process of SESAR, the European Commission and EUROCONTROL co-funded the development of the ATM Master Plan (SESAR Joint Undertaking [SJU], 2020) in order to link R&D activities with deployment scenarios, to achieve the performance objectives of the Single European Sky. Despite not having any defined performance "ambitions" at the level of the ATM Master Plan, due to its qualitative nature, HP has been seen as a fundamental driver. As such, "failure to manage human performance issues properly" in the development and implementation of the target concept, has been identified as one of the most significant of the seven risks in the ATM Master Plan (SJU, 2020). Therefore, one of the largest ATM research endeavours globally has explicitly considered HP as a pillar for future developments. SESAR acknowledged and promoted the crucial role the human actors have in achieving success and at the same time the need to continuously assess technological developments against human capabilities. The SESAR HP Assessment Methodology (HPAP)[1] built on the existing HF Case (EUROCONTROL,

1 Documentation openly available only to the SESAR community or upon request. Further enquiries should be made with SESAR 3 JU.

2011), was consequently developed to ensure a standardised approach to HP integration into research activities. It established a clear expectation that as soon as a potential impact on the human actor is foreseen, following the introduction of a new concept or technology, an HP assessment shall be performed.

The SESAR HPAP uses an argument and evidence approach, that allows the HP expert to explore potential HP considerations (i.e., issues and benefits) for the following four HP areas: human roles and operating methods, human and the system, teams and communication, and transition factors. These are further broken down at more granular levels, during the assessment process. A change assessment is first performed with the scope of identifying the potential areas of impact from an HP perspective and, further on, potential issues and benefits are identified. On this basis, a list of HP activities is defined to address all the impacts previously mentioned. These are either included in project validation activities such a prototyping session, human-in-the-loop simulations etc., or performed additionally (and preferably together with safety experts) as specific HP activities (i.e., task analyses, cognitive walkthroughs etc.). These steps are systematically documented in an HP assessment plan. The final goal of the HP assessment is to create and formulate an operational concept by identifying essential recommendations and requirements to proactively prevent or minimise any potential adverse effects of the proposed conceptual development. These findings are encompassed in an HP assessment report (HPAR).

Human Performance Maturity Assessments

To establish the level of progress and viability of the projects, SESAR outputs are subject to regular maturity evaluations, relying on content and evidence captured in project deliverables. Rather than simply indicating whether the criteria for a given maturity level is met or not, the primary aim of the maturity evaluations is to ensure the existence of documented evidence that substantiates the assessment. Therefore, to ensure the provision of robust SESAR solutions[2], the level of maturity attained throughout the development life cycle will be evaluated at relevant exit maturity gates. SESAR I (2009–2016) and SESAR 2020 (2017–2023) required an evaluation of the maturity criteria of the projects from the European Operational Concept Validation Methodology (EOCVM) point of view, as developed by EUROCONTROL in 2010 and the criteria applicable stemmed from V1 (feasibility) to V3 (pre-industrialisation). An integral part of the HPAR is the assessment

2 SESAR solutions refer to new or improved operational procedures or technologies that aim to contribute to the modernisation of the European and global ATM system. Each solution is described in a range of documentation, including: operational services and environment descriptions; safety, performance and interoperability requirements; technical specifications; regulatory recommendations; safety and security assessments; human and environmental performance reports.

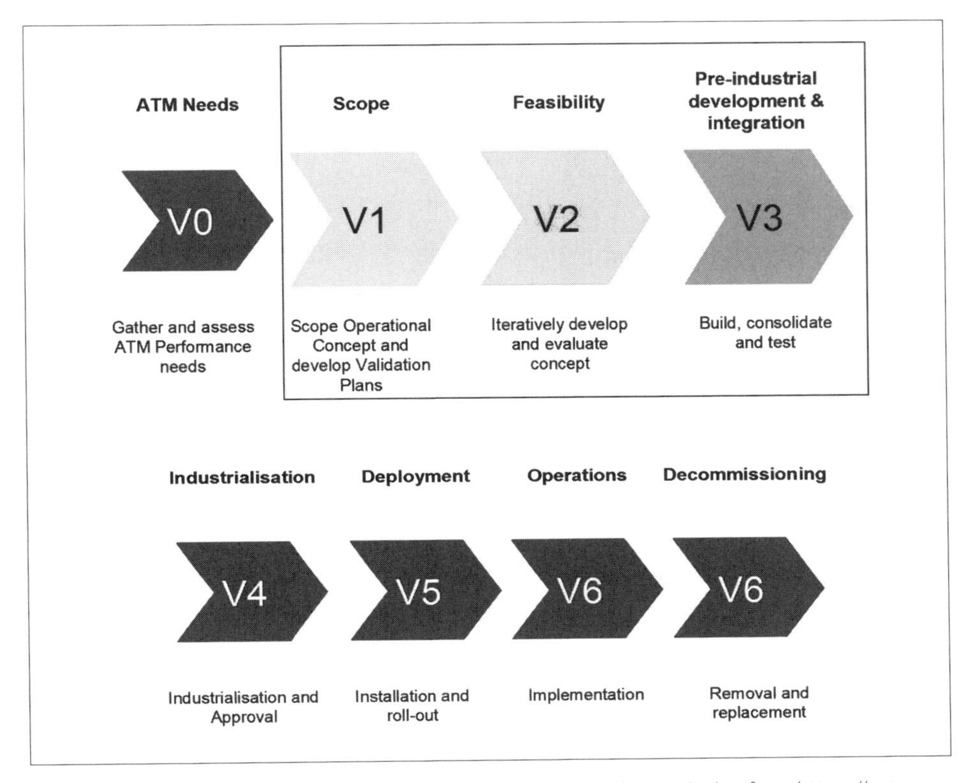

Figure 16.1. E-OCVM concept life cycle model. Reprinted with permission from https://www.eurocontrol.int/sites/default/files/2019-08/e-ocvm3-vol1-figures.pdf (© EUROCONTROL, 2010).

of the maturity level achieved by the solution, from an HP perspective. The HP maturity criteria were tailored therefore for V1–V3 maturity levels, in accordance with E-OCVM principles (EUROCONTROL, 2010), on the basis of the concept life cycle model (Figure 16.1).

The E-OCVM methodology was introduced as relevant to anyone involved in the ATM R&D of new operating procedures, human–machine interactions and supporting technologies. However, despite the fact that the E-OCVM created a solid baseline for the evaluation of maturity criteria from the perspective of the various transversal areas, the risk of having the "separate viewpoints operate in parallel throughout the process of concept development and evaluation" (EUROCONTROL, 2010, p. 53), was identified as a potential risk of "fragmentation, which can lead to incoherent/ inconsistent results" (EUROCONTROL, 2010, p. 53).

SESAR 3 JU

With SESAR 3 JU, a shift from an evaluation of the maturity criteria from a V-level perspective has taken place and now SESAR has moved on to a maturity categorisation from a TRL standpoint (SJU, 2022). Various sectors of the manufacturing industry were already accustomed to using TRLs, as initially proposed by National Aeronautics and Space Administration (NASA; Mankins, 1995), which is perceived to provide a finer structure. As illustrated in Figure 16.2, in order to align with the new approach, the SESAR 3 JU HPAP has been updated to match the maturity criteria to TRL levels and expanded its coverage from exploratory research (TRL0) all the way to digital sky demonstrators (TRL7–8).[3]

Consequently, the spectrum of applicability of the HPAP was officially enlarged. In the context of SESAR 3 all solutions are required to perform an HPAP as soon as a potential impact on the human is foreseen, even in early phases of exploratory research. As a result, the HPAP has aligned its approach and redefined the characteristics of the HP expectations in terms of involvement and expected output for each of the three constituent pillars of research in SESAR (exploratory research, industrial research, and digital sky demonstrators).

For *exploratory research (TRL0-TRL2),* the HPAP aims to allow for an incipient identification of potential impacts on HP, through feasibility studies and exploration of the transition from scientific to applied research. The main focus is on defining the initial operational concept, followed up by a necessary exploration of different alternatives and the identification of potential show-stoppers. The HP assessment process is expected to already be guided by the HP specialist. It involves all the stakeholders and requires a strong cooperation with other actors of the process, in particular safety experts, operational concept designers and systems designers.

3 Fast-track projects are excluded from the description in this chapter, as fundamentally they do not cover a different process from an HP perspective. The main difference is that fast-track projects target a rapid delivery of new products and services towards TRL7.

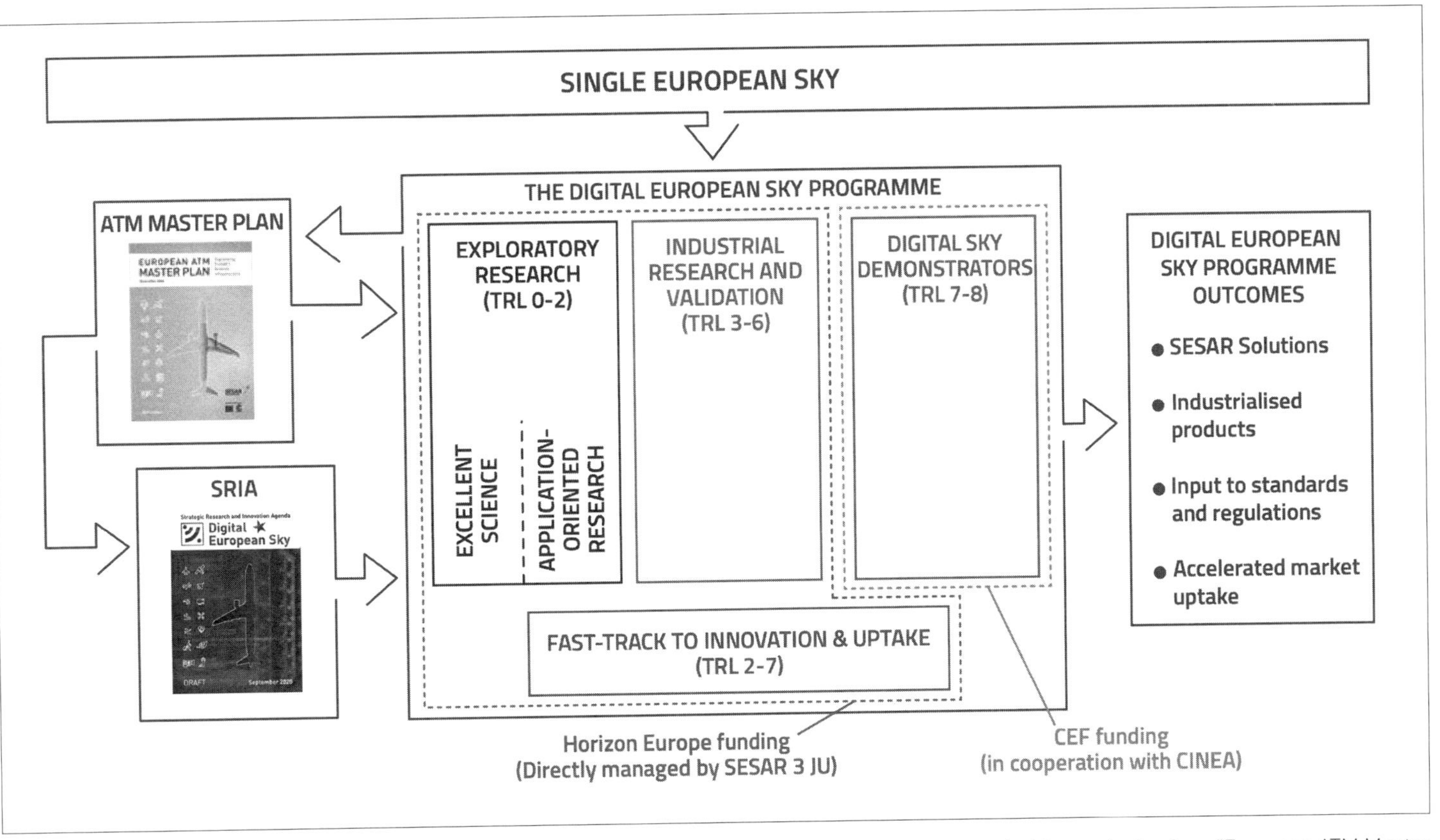

Figure 16.2. TRL levels in the context of SESAR 3 (SJU, 2022). TRL = technology readiness levels. Reprinted with permission from "European ATM Master Plan 2020 Executive View," by SESAR Joint Undertaking, 2020. © SESAR Joint Undertaking.

For *industrial research (TRL3-TRL6),* the HPAP focuses on determining the initial operational/technical concept from an HP point of view, resulting in the definition of a concept that is mature enough to be prototyped and tested. The process at this stage necessarily includes the identification of HP needs to be covered, as well as the identification of both potential benefits and aspects that may become impediments or issues for the transition at a later stage. The HP activities at this stage can cover a wide range of activities from task analyses to real-time simulations.

For *digital sky demonstrators (TRL7-TRL8),* the HPAP follows through the exposure of all relevant HP issues and benefits identified in industrial research in an operational environment. The final goal is to validate the feasibility of the concept/technology for further authorisation/certification. In TRL8 the validation and verification process is expected to be finalised, capitalising on all HP findings.

This transition from V levels to TRL levels is considered by the SESAR HP community as a great opportunity to increase the visibility and understanding of the HP work by non-HP experts as it paves the way for the future introduction of the human readiness levels (HRL) framework that will be discussed in the section Human Readiness Levels.

Human Performance "As Done" and Human Performance "As Imagined"

It cannot be argued that the baseline for an accurate integration of HP in research is missing in SESAR. On the contrary, throughout its lifecycle, SESAR has continued to prioritise HP and bring visibility to the importance of carrying adequate assessments and embedding the findings in key deliverables. The final goal of SESAR is to "optimise the overall performance of the socio-technical ATM system and maximise human performance" (SJU, 2020). There is a well-understood and accepted idea that in order to further progress and to adapt to the ever-increasing complexity and capacity demands, there is a need to continue to develop the role of the human in parallel with ATM concepts and technological developments (SJU, 2020). Whether we talk about cybersecurity threats or human–AI teaming concepts where collaborative, co-adaptive and joint intelligence models of decision-making are used, the balance between technology and human capabilities becomes more crucial than ever.

The industry has developed at a remarkable pace in the past decade, but the demands from the industry are at times barely kept up with by the human components (at system and individual levels). That is why we consider it is more important than ever to re-evaluate our understanding of human capabilities, to ensure a balance between technological and human components. One of the pioneers of systems thinking, Russel Ackoff pointed very pertinently that "it is possible to improve the performance of each part or aspect of a system taken separately and simultaneously reduce the performance of the whole" (Ackoff, 1999, p. 36). This is probably more accurate than ever when dealing with fast technological maturations. *Are we keeping up with the HP component?*

Despite the strong fundamentals of HP described in the SESAR documentation, in practice there seems to still be a discrepancy between *work as done* and *work as imagined* when it pertains to HP. Biede and Pelchen-Medwed (2017) emphasised the effectiveness of the HPAP in SESAR projects as well as the "major achievement" of allowing for a standardised approach to be applied at such a high scale, involving multinational, multi-organisational and multidisciplinary teams. However, they additionally acknowledged that "further evaluation of HP is vital" to ensure a "more homogeneous application of HPAP for the next phases" (Biede & Pelchen-Medwed, 2017, p. 8). The overview of the discussions held in the context of the HP Community of Practice of SESAR has often indicated that HP is still not fully understood outside the HP community. There seems to be a perception that HP evaluates performance instead of also actively contributing with knowledge in supporting the design of technological prototypes or associated procedures. Occasionally, this resulted in a delayed integration of HP in exchanges with operational experts or an inadequate allocation of HP resources. Such conclusions are not intended under any circumstances to condescend any of the experts that have collaborated with HP experts in SESAR activities, but rather to show that there needs to be more (and more than ever) of an emphasis on HP. Or better said, we need to find better ways to *walk the talk* in non-HP expert groups. To overcome such instances, progress has been made in the SESAR 3 JU related documentation. To date, there was no document outlining the expected interactions between experts at transversal areas, but this has been addressed in the context of SESAR 3. A roadmap has been created to provide a clear overview of the HP work process and the expected contributions by an HP expert. It also outlines the anticipated interactions the HP expert should have with the solution team throughout all stages of the HPAP.

Technology Readiness Levels

In the mid-1970s, NASA's researchers proposed the first TRL scale with the aim of having a measurement system that would allow for a standardised maturity assessment of technologies (Sadin et al., 1989). First, it was composed of seven levels, until the 1990s when NASA adopted a scale with nine levels (TRL1-TRL9), which gained widespread acceptance across industry and remains in use today in various industries from aerospace to medicine. The nine-scale maturity evaluation (Table 16.1) comprises the technological conception (TRL1-TRL3) and the development and demonstration phase (TRL4-TRL6). After TRL6, the standard engineering development cycle for new designs continues all the way up to the successful operation of ground systems supporting launch and in flight (TRL9 completed; NASA, 2020).

Table 16.1. Technology readiness levels (TRL)

Level	Definition
TRL1	Basic principles observed and reported
TRL2	Technology concept and/or application formulation
TRL3	Analytical and/or experimental performance/function proof of concept
TRL4	Assembly/component breadboard validation in a laboratory environment
TRL5	Assembly/component brassboard validation in a relevant environment
TRL6	System/subsystem prototype demonstration in a relevant environment
TRL7	System prototype demonstration in a space environment
TRL8	Actual system completed and "flight qualified" through test and demonstration (ground or flight)
TRL9	Actual system "flight proven" through successful mission operations

Note. Adapted from NASA, 2020, Figure 2-1.

By having a standardised approach to interpreting maturity levels, technological developments have been better framed and easier to relate to by multidisciplinary teams, creating a common denominator. However, despite the widely used TRL scale in high-risk industries, the approach has met with criticism in relation to the fact that by being technologically ready it cannot be assumed that a system is automatically a "successful product" (Héder, 2017). But perhaps the most relevant gap addressed by this book chapter refers to the inability of TRLs to inform about the maturity in relation to the human-related components of technological developments (Boring, 2022; Salazar et al., 2020; Salazar & Russi-Vigoya, 2021).

Human Readiness Levels

To address this limitation, the development of HRLs has emerged as a solution to incorporate an early and systematic consideration of the human element into design and development processes. Salazar et al. (2020) gave a detailed overview of the development of HRLs, from the first mention of an HRL approach by Dr Hector Acosta in 2010, to the refinement made by the working group led by Sandia

National Laboratories during 2019–2020. The Human Factors and Ergonomics Society (HFES) published in 2021 a draft paper meant to detail the applicability of HRLs throughout the design and development phases, synthesising the HRLs scale (Table 16.2).

Table 16.2. Human readiness levels (HRLs) as described in the Human Factors and Ergonomic Society draft paper

Phase	HRL Level
Basic Research and Development Scientific research, analysis, and preliminary development on paper and in the laboratory occur. This phase culminates in a validated proof of concept that addresses human needs, capabilities, limitations and characteristics.	**HRL1:** Basic principles for human characteristics, performance and behaviour observed and reported. **HRL2:** Human-centered concepts, applications, and guidelines defined. **HRL3:** Human-centered requirements to support human performance and human–technology interactions established.
Technology Demonstrations The technology is demonstrated at increasing levels of fidelity, first in the laboratory and later in relevant environments. This phase concludes with demonstration of a representative system in a high-fidelity simulation or actual environment, with evaluation of human systems designs provided by representative users.	**HRL4:** Modelling, part-task testing and trade studies of human systems design concepts and applications completed. **HRL5:** Human-centered evaluation of prototypes in mission-relevant part-task simulations completed to inform design. **HRL6:** Human systems design fully matured as influenced by human performance analyses, metrics, prototyping and high-fidelity simulations.
Full-Scale Testing, Production and Deployment Final testing, verification, validation and qualification occur, with human performance evaluations based on representative users. This phase concludes with operational use of the system and continued systematic monitoring of human-system performance.	**HRL7:** Human systems design fully tested and verified in operational environment with system hardware and software and representative users. **HRL8:** Total human–system performance fully tested, validated and approved in mission operations, using completed system hardware and software and representative users. **HRL9:** System successfully used in operations across the operational envelope with systematic monitoring of human–system performance.

Note. Reprinted with permission from "Human Readiness Level Scale in the System Development Process," by the Human Factors and Ergonomics Society, 2021.

Following the same approach as the TRLs, the purpose of the HRLs was to create a standardised approach for the maturity assessment of the human dimension in relation to the usability and effective use of a novel technology or system. By having the human dimension integrated as early as possible in the design phase, the reduction of cost implications as well as a minimisation of *human error* are expected. By integrating the human dimension next to the technological one, the aim is to identify and address human-related system issues at the earliest possible stage, thereby mitigating their impact. To mirror the TRL approach, the HRL approach covers the same levels of maturity, from exploratory research to the implementation phase. The nine HRLs can be categorised in three main phases: basic research and development (HRL1–HRL3), technology demonstrations (HRL4–HR6) and finally the full-scale testing, production and deployment phase (HRL7–HRL9).

In order to gain the full benefit of the two scales, it is highly desirable that progression through the two scales is done in parallel and in alignment (HFES, 2021). The misalignment of the two is associated with risks particularly if HRLs lag behind TRL ratings. The risk of having insufficient time to address the human implications or the increase in costs associated with such evaluations at this stage increases significantly. On the other hand, if TRL ratings lag behind HRLs, the risk increases in higher maturity levels if the HRL ratings reach the highest value in the context of an insufficiently mature technological improvement (HFES, 2021). Perhaps the most critical practical application of the HRL is that it would allow multidisciplinary teams to verify the progression of the concept from its two most crucial dimensions, more fluently.

SESAR and HRLs

To refer specifically to SESAR undertakings, the hypothesis is that by having an equal emphasis put on both the technological and human level from the very beginning of a project, there will be more synchronicity in maturity progressions. If we recap the risk identified by E-OCVM about a potential fragmentation of findings between transversal areas, we believe that by having a direct link between TRLs and HRLs, a smoother integration of findings from both directions could be facilitated. The current evaluation approach of the HP maturity criteria is incorporated in the overall criteria for the maturity assessment. However, it is not visible enough for the multidisciplinary experts as it is performed in silo by the HP expert allocated to the project. By incorporating both TRLs and HRLs as two pillars of maturity assessments, the expectation is that all individuals involved in the project, irrespective of their background or specific role, naturally prioritise both dimensions. Standardising a human-centred approach of project maturity assessments, in multidisciplinary teams and not just in the framework of HP assessments, would formalise the crucial role of a human design-oriented approach. Instead of ensuring HP is covered by the individual work of HP experts, HRLs would create a common framework for the multidisciplinary teams to collaborate and prioritise the centrality of HP in their work. The SESAR programme is a step closer to creating this bridge between the disciplines by having already transitioned to TRL levels. The next iteration of the SESAR documentation (post-SESAR 3) is highly recommended to take into account the integration/adaptation of the HRL scale in addition to the TRL scale.

Acknowledgements

The baseline of this work has been started by the ongoing collaboration between EUROCONTROL and the Federal Aviation Administration (FAA), under the CP 3.2 project. The authors of this chapter remain available for consultation on the developments of the project in the European arena and would like to sincerely acknowledge the FAA counterparts for a very fruitful collaboration.

References

Ackoff, R. L. (1999). *Ackoff's best: His classic writings on management*. John Wiley & Sons. https://doi.org/10.1093/oso/9780195123876.003.0003

Biede, S., & Pelchen-Medwed, R. (2017). *Effectiveness of the application of the Human Performance Assessment Process in SESAR* 1 – *sharing lessons learned.* Air Traffic Management Research and Development Seminar, Seattle, Washington, USA. https://www.atmseminar.org/past-seminars/12th-seminar/papers/

Boring, R. (2022). Implications of human reliability analysis for human readiness levels. In R. Boring (Ed.), *Human error, reliability, resilience, and performance. 13th international conference on applied human factors and ergonomics (AHFE 2022).* https://doi.org/10.54941/ahfe1001573

DeMott, D. L. (2014, July, 7). *Human reliability and the cost of doing business* [Paper presentation]. 2018 Science Applications International Corp., Houston, TX, USA.

EUROCONTROL. (2010). *European operational concept validation methodology, Vol. I, Version 3.0.* https://www.eurocontrol.int/publication/european-operational-concept-validation-methodology-eocvm

EUROCONTROL. (2011, August, 23). *Support material for human factors case application (*Ed. *3.0).* https://skybrary.aero/sites/default/files/bookshelf/4559.pdf

EUROCONTROL. (2013). *From Safety I to Safety II: A white paper* https://skybrary.aero/sites/default/files/bookshelf/2437.pdf

Héder, M. (2017). From NASA to EU: The evolution of the TRL scale in Public Sector Innovation. *The Innovation Journal: The Public Sector Innovation Journal, 22*(2), 1–23, Article 3. https://www.innovation.cc/discussion-papers/2017_22_2_3_heder_nasa-to-eu-trl-scale.pdf

Hollnagel, E. (2014). *Safety I and Safety II. The past and future of safety management*. Ashgate Publishing.

Human Factors and Ergonomics Society. (2021). *Human Readiness Level Scale in the system development process.*

Mankins, J. (1995). *Technology readiness level – a white paper*. National Aeronautics and Space Administration.

National Aeronautics and Space Administration (NASA). (2020). *Technology readiness assessment: Best practices guide.* https://ntrs.nasa.gov/api/citations/20205003605/downloads/%20SP-20205003605%20TRA%20BP%20Guide%20FINAL.pdf

National Highway Traffic Safety Administration. (2019, September). *Traffic safety facts 2017. A compilation of motor vehicle crash data* (Report No. DOT HS 812 806). https://crashstats.nhtsa.dot.gov/Api/Public/Publication/812806

Sadin, S. R., Povinelli, F. P., & Rosen, R. (1989). The NASA technology push towards future space mission systems. *Acta Astronautica, 20*, 73–77. https://doi.org/10.1016/0094-5765(89)90054-4

Salazar, G., See, J. E., Handley, H. A. H., & Craft, R. (2020). *Understanding human readiness levels.* https://www.osti.gov/servlets/purl/1807329

Salazar, G., & Russi-Vigoya, M. N. (2021). *Technology readiness level (TRL) as the foundation of human readiness level (HRL) understanding the TRLs as the foundation of human readiness level (HRL).* https://ntrs.nasa.gov/api/citations/20210000183/downloads/EID%20Abstract%20Summary%20092920_Technology%20Readiness%20Level%20(TRL)%20as%20the%20foundation%20of%20Human%20Readiness%20Level%20(HRL).docx.pdf https://doi.org/10.1177/10648046211020527

SESAR Joint Undertaking. (2020). *European ATM Master Plan 2020 executive view.* Publication Office of the European Union. https://www.sesarju.eu/sites/default/files/documents/reports/European%20ATM%20Master%20Plan%202020%20Exec%20View.pdf

SESAR 3 Joint Undertaking. (2022). *Multiannual work programme* 2022-2031. Publications Office of the European Union. https://data.europa.eu/doi/10.2829/60154

Shively, J. (2013, August, 12). *If human error is the cause of most aviation accidents, then shouldn't we remove the human?* [Paper presentation]. Meeting of the Association of Unmanned Vehicle Systems, Washington, DC, USA.

Shorrock, S. (2014, December, 2). *Human error still undefined after all these years.* https://humanisticsystems.com/2014/12/02/human-error-still-undefined-after-all-these-years/

World Economic Forum. (2022). *The global risks report 2022* (17th ed.). https://www3.weforum.org/docs/WEF_The_Global_Risks_Report_2022.pdf

Chapter 17

Human and Organisational Factors Integration in the Aviation Industry

Maturity Analysis and Methodological Recommendations to Answer Society Trends

Florence Reuzeau

Abstract

The integration of human and organisational factors (HOF) is well established in the aviation industry. HOF contribute to the design of aircraft, production tools, workstations, procedures and organisations. Significant efforts have been made in this domain since the 1990s with the objective of reinforcing aviation safety and more broadly strengthening the performance of flight operations. This chapter addresses a number of questions. Do HOF bring the full potential of added value to the aeronautical industry today? How are HOF integrated in the end-to-end processes of the aircraft life cycle? Are human-centred approaches as developed today sufficient with regard to the evolution of safety requirements, the emerging technological trends and the demands of society? Reflections on these questions and a way forward to bring more value for human operators and organisations alike are described.

Keywords

human factors, organisational, maturity analysis, safety, performance

Introduction

Mobilising humans to design, produce, fly or maintain an aircraft is what characterises the aeronautical industry. Humans are an asset for aviation, but also a real challenge. Aircraft are operated or produced worldwide by a very diverse population of operators. Thus, it is required to define and implement standards of system design, operational use, and industrial means to adapt the technical and operational systems to this impressive diversity. The integration of human and organisational factors (HOF) is implemented in the processes of aircraft manufacturers such as Airbus. HOF contribute to the design, development and in-service support of aircraft (cockpit, cabin, maintenance), production tools, workstations, procedures and organisations. This chapter questions how HOF is applied and integrated, and explores how to pursue the effort to bring even more value to human operators and to the business.

Historically, significant investment efforts have been made since the 1990s, contributing towards enhancing air safety, protecting the health and safety of humans and improving operational performance. The issues targeted by HOF evolved over time to embrace a global approach of human considerations. In Airbus, in 1984, the first "ergonomics" department was set up within the final assembly line (FAL). It was part of the human resources organisation. The first challenge for the initial team was to demonstrate that ergonomics can bring real value to the company in order to guarantee a safe and healthy working environment. The current team encompasses a systemic approach of the human at work where the interface between the human operator, the system and the organisation is adjusted to the physical and psychological capacity and limitations of the human. In the 1990s, Airbus Training Centre (1988) and Airbus Engineering (1995) hired a few human factors (HF) specialists to integrate HF into the different aircraft development processes as cockpit, cabin, maintenance design as well as pilots' training programs. Aviation authorities were very proactive to push HF development in the airlines, manufacturers, suppliers, research centres and regulatory bodies. This step was favoured by the ability of HOF specialists to pragmatically bring knowledge about humans at work into the industry and by the concrete regulatory materials brought by the airworthiness authorities. The report by Abbott et al. (1996) put forward 50 recommendations aimed at improving HF considerations and boosting flight safety. In this era, the question was how to improve the interaction between human operators and technologies to create an appropriate coupling between humans and new automated systems. How to make the machine understandable and usable by humans? The first question is about human/machine authority sharing and what information is needed by the human operators to keep the proper level of control of the situation. The development of automation has raised many ergonomic issues: How to engage and disengage the autopilot, how to be prepared to take over in any circumstances. Research and preliminary studies identified the key HF issues to be covered, and provided preliminary guidelines to integrate automated systems in a cockpit as "automation should be easy to use and to understand". This development was pursued during a second era (2000–2015) and was not Airbus specific. Most high-risk and complex systems industries integrated HF into their own major processes. The first specific HF certification requirement, CS 25-1302, was applied in 2007 (European Union Aviation Safety Agency [EASA], 2007).

The third era (since 2012) can be characterised by two main developments. First the growing role of the "organisational" part of HOF thanks to the introduction of the safety management system (International Civil Aviation Organization [ICAO], 2013; Stroeve et al., 2022). The means to achieve an efficient human–machine interaction today is still based on the definition of automation design requirements (cockpit design, pilot training and operational procedures) that have been formalised in key references (Federal Aviation Administration [FAA], 2022; Yeh et al., 2013). Crew resources management (CRM) training was developed also to adapt pilot training to the more automated aircraft (Jimenez et al., 2015).

The second major development is the growing contribution to social sciences for a better integration of the psychosocial considerations (EASA, 2021). These three eras of HOF development were based on HOF academic competencies such as psychology, sociology, physiology, linguistics, aeromedicine, and neuroscience. These competencies were introduced to provide the knowledge on human behaviour that is necessary to understand human performance and limitations at work from a broader perspective.

This chapter provides a critical review of how HOF currently support the development of the aviation industry from an aircraft manufacturer's point of view. The analysis is driven by three main questions:

- Do HOF bring their full potential of added value to the aeronautical industry today (Section 1)?
- How are the new technologies that are powerfully pushing industries considered by HOF (Section 2)?
- How are the societal considerations taken into account by HOF (Section 3)?

Section 1: Do HOF Offer Their Full Potential of Added Value to the Aeronautical Industry?

The questions addressed in this section include: Where do aircraft manufacturers stand in terms of taking HF into account? Are HOF integrated into the end-to-end aircraft life cycle? At what level of maturity?

The main focus of HF and ergonomics or HOF can be roughly summarised as finding the reasonable adaptation between human needs, strengths and limitations and the working situation requirements including how reality of operations can make hazards emerge. HF scientists and professionals dedicated several publications to formalise how HOF structured their knowledge, process, methods and tools to be directly applicable to high-risk industries (Pelchen-Medwed et al., 2021; Wolf & Mollard, 2020). These processes have been partly streamlined and traced in reference documents (ICAO, 2021; International Organization for Standardization [ISO] 11228-1, 2021; ISO 9241-210, 2010) as well as in the Airbus repositories. It could be said that the theoretical and methodological background of HOF is defined and sufficiently advanced for large-scale deployment. This is partially true, but HOF still often struggle to find their place in companies. This can stem from a lack of visibility or recognition of the value produced by HOF for the business. It can also be due to the difficulty for very technological companies to know how to embed HOF into their strategic objectives. To clarify this point, the key HOF impacts are summarised here.

The Key HOF Impacts for an Aircraft Design and Manufacturing Company

The HOF approach must contribute to maintaining or going beyond the ultra-safe level of aviation safety.

The HOF approach must be part of a modern vision of quality, safety and security management. It offers a systemic approach where the human operators, the working situation and the organisational contexts go hand in hand to produce performance while ensuring people and product safety. To do that, the interactions between the different structuring elements of the working situation are designed with respect to the needs, capacities and limitations of the human operators. One of the most difficult challenges is to articulate safety requirements with the operational needs and to bring the evidence that safety is never compromised. High-level engagement of HF before entry into service to understand and test the systems and the implementation into in-service operational experience are paramount.

The HOF approach must support design and innovation.

HOF should contribute to the design and demonstration of product, service and industrial means. In particular, HOF should support innovation by guaranteeing the health, safety, comfort and well-being of passengers, crew, workers and all ground personnel involved or potentially impacted by air transport. Both designer and manufacturer companies such as Airbus must respond to strong productivity challenges in a highly competitive market. This is why HOF are part of the modernisation of production tools to accelerate the industrial performance of factories without degrading working conditions or even by finding opportunities to improve them. HOF should be involved in the definition of an increased level of automation and digitalisation by proposing a philosophy of human–automation and human–robot interaction, including how to use digitalisation in the service of human beings.

The HOF approach must contribute to understanding and responding to the needs and aspirations of civil society and of passengers.

The need for travel is manifold. People seek travel for the purpose of visiting their relatives, for business needs, for leisure, because it is part of the cultural background or for other reasons best known to them. Transport of goods is also essential for society. The HOF approach helps in understanding why and how people would want to fly. How will the willingness to fly match with climate change consciousness? Since the commercial aviation fleet is expected to increase by 33%

between 2023 and 2033 (Prentice et al., 2023), HOF should be involved to provide a deep understanding of people's behaviour and guide the company in the change management.

The HOF approach helps to analyse in-service events and production events.

HOF should be part of root cause analysis when human performance is a potential contributor to the event. HF specialists (together with design teams, quality teams or other relevant organisations) shall be fully competent to lead an in-depth analysis of human behaviours and to identify the actual root cause. For example, HOF specialists perform a systemic analysis of the working situation and explain how technical or organisational dysfunctions can be prone to human error. For example, a worker interrupted several times during a 15-min task is likely to make a mistake in the sequence of the sub-tasks because short-term memory has been disturbed.

The HOF approach must contribute to designing and developing a sustainable and responsible industry.

Air transport is a specific and complex industry; extremely expensive aircraft are designed to be operated for 30–40 years by a wide variety of human operators worldwide, who are not yet born in most cases. It requires a high capacity of projection of the human operators of the future. What will their competencies and arousal be, and how can the attractiveness of aeronautical professions be maintained? HOF can greatly contribute to anticipating the generation and cultural effects on human performance. The training of end users and human operators is paramount to pursuing a continuous match between work and humans. Training should not be considered as a first means to correct a design issue, but as the way to mutually expose the human being and working situations in order to allow for a smooth evolution of both parties.

The Airbus HOF Approach

To meet these aforementioned requirements, Airbus has developed a HOF approach, adapted to different applications, products or industrial means. These are capabilities implemented in the company that are based on a few major fundamentals (Reuzeau, 2013).

An "integrative" approach in which humans are at the heart of operations.

And it is not just words: The pilot of an aeroplane and blue-collar workers are fully in the loop and "in control" of their mission. The HOF approach is based on a non-dissociative combination of methods from ergonomics to the more recent human and social sciences. For example, Airbus has integrated the full basement of the French-Speaking Ergonomics Society for participatory design, user-centred design, and human–computer and human–system interaction methods, with a robust and mandatory use of work analysis.

A systemic and systematic HOF approach focusing on the analysis of working situations.

Whether end user or human operator, the human at work is caught in an operational and contextual situation that influences his/her behaviour. Technical tools, information systems, procedures or instructions, light or sound environments, working conditions, schedules, internal company policies – all these factors influence, positively or not, human behaviour and produce either job performance or unreliability. “Systemic analysis of real situations” is perhaps the mantra of the HOF approach, whatever the application – an aircraft, an industrial design or the analysis of in-service events.

An approach based on deep knowledge and skills in the human and social sciences.

The theoretical knowledge of people at work or of the processes, methods and tools is provided by human and social sciences. High levels of skills, based on strong academic education, are paramount for rationalising and arguing about human behaviour while avoiding personal opinions and stereotypes. This is key to influencing the decision about the design of a working organisation or a working station. Airbus group governs HOF competency in the expert career path as one of the critical competencies.

A “dynamic, open and learning” approach that feeds on the different applications in Airbus, military and commercial.

While the fundamental values and principles of the HOF approach apply to any activities of the Airbus group, the methods and degree of application are adapted to the different industrial contexts. The HOF network brings together HOF experts and specialists from the various divisions, whether they work on products or in production sites. As a vector for disseminating the HOF best practices, each division has customised a HOF approach to the specificities of its applications and products (such as regulatory requirements for commercial aviation certification, military contracts with government, etc.), but the core and the pillars of HOF approach are standardised.

The Maturity of the HOF Approach

Several papers describe how to define the level of maturity of embedded processes (Edmonds & Gray, 2019; Kirwan, 2022; Stroeve et al., 2022; Van Veenendaal & Wells, 2012). Inspired by these references, Airbus defined five levels of maturity for the integration of HOF into Airbus processes (Reuzeau, 2020). In a nutshell, in the initial level, the HOF studies are opportunistic and not repeatable. In Level 2, the HOF process is documented and the HOF competencies are not always appropriate. In Level 3, the HOF process is defined, in place and visible but there is no measurement of its effectiveness. Levels 4 and 5 represent a mature HOF process with effectiveness measurement and proper governance.

At group level, the current level of maturity is roughly 3+ as some elements are at Level 4 or 5 but there is a lack of consistency between the various divisions and organisations (Reuzeau, 2020). The sustainability of HF approaches in the company is a permanent issue. It mainly depends on two key factors: (1) the company's competency in the fields of HF and (2) the robust adoption of HOF by the company, which means that the HOF processes, methodologies, skills, and organisations are fully part of the foundations of the company. For example, is HOF part of the decision-making process in the same way as aerodynamics, fluid mechanics or safety? The route to Level 5 needs to progress on several key points, as described next.

HOF Competencies

The effective and lasting installation of HOF competencies, both in terms of the level of expertise and the number of resources, is still fragile (Biede et al., 2023).

- The recruitment of HF graduates is constantly increasing but the skills are neither easy to recruit nor easy to integrate into new organisations when the latter do not already have an HF heritage.
- The HOF specialists are a minority. The aeronautical field is a field requiring high technological skills. Consequently, it traditionally employs around 90% highly qualified engineers. Specialists in the human and social sciences are therefore a very small minority, traditionally less valued than the engineering sciences. "As a minority, Human Factors specialists have to constantly demonstrate their added value in front of a 'monocultural engineering world'" (Reuzeau, 2020, p. 36).
- Engineers used to claim doing "HOF like Mr Jourdain speaks prose without knowing". It happens that engineers take on HOF issues without having the complete appropriate skills, applying general knowledge about human behaviours, capacities and limitations such as (a) more or less true stereotypes, (b) biased knowledge due to cognitive bias or culture background, or (c) too simple human behaviour assumptions. For example, an assumption could be that "the worker will know how to do the task because he/she will strictly apply the instruction manual" or when the procedure is complicated "the worker knows her job" or "he will see the information because it is written in front of him" – even if this information is buried in the middle of about 20 other important information. Task interruptions, management of unexpected situations , or other factors affecting the performance of the human operator can be sometimes underestimated if HOF specialists are not part of the team. Proper HOF skills allow for consideration of human capacities, limitations and aspirations and address

the HOF issues properly. For example touch screens are more than recommended on the ground, but should be used very carefully in a cockpit (Barbé, Mollard, & Wolf, 2014). The location of the screen, the type of interaction, and the duration of the task should drive the installation of touch screens in the cockpit considering the vibratory conditions and the vertical or side access to the displays. Without in-depth studies, the design choice could be technology driven and fail to offer the expected benefits.

- The engineering world likes figures and statistics. While the field of human physiology has produced standards of human capacities and limitations (e.g. limitation in carrying heavy loads), the fields of psychology, sociology and work analysis are more inclined towards a complementarity of descriptive, qualitative and quantitative analysis. An exclusively quantified approach to human work dehumanises these operators from their personality, their culture, their anchoring in society – the key parameters of the actual situation which of course influence human behaviour. A kind of checklist that would verify that the design respects the capacities and limitations of percentiles of the population is unsatisfactory to guarantee human and industrial performance. Facing real-situation characteristics by prototyping, simulation and observation is paramount to achieving human performance at work.

A conclusion on this point is that the training of engineers in the HOF profession is an interesting solution for disseminating the HOF mindset, but is insufficient for dealing with complex HOF issues that require deep knowledge of the human being at work.

A Permanent Adoption of the HOF by the Company

The deployment of HOF approaches requires the acceptance and commitment of engineers and management. The wealth and motivation of engineers to push technological innovations is an indisputable benefit for the company. Engineers are naturally a techno-driven force of innovation, motivated by a real pleasure in innovation. From a HOF point of view, it is not easy to support numerous new ideas, no matter how brilliant they may be. The more promising innovations must pass an extensive HOF analysis to attempt to project that innovation into the next 40 years – the life cycle of an aircraft from first inspiration to end of life. The case of artificial intelligence (AI) is an example of scientific debate. What applications, and how to integrate AI algorithms for the benefit of operational performance and the benefit of users, for what use, how to regulate these uses? The only way to answer these questions is to explore future operational situations (Barcellini, 2022; Daniellou, 2005; Nelson et al., 2009) to understand the issues and anticipate what the reality of operations could be.

Without sufficient penetration of the HOF discipline, without direct intervention in project progress milestones, without sponsorship from decision-making bodies, the best HOF fail to produce significant impacts on the design of a product or the modification of an industrial means. This is the challenge of establishing solid governance. For example, Airbus has set up a co-design activity for defining how to embed a user-centred design (UCD) process in Airbus engineering by involving engineers, HOF specialists, pilots and managers in the project. This cooperation was beneficial in getting design stakeholders on board in an approach that made sense to the engineering and operational testing community.

Section 2: How Are the New Technologies That Are Powerfully Pushing Industries Considered by HOF?

The following are the main questions addressed in this section: What are the new questions raised by some new technologies such as digitalisation and artificial intelligence? How does HOF understand the complexity of the penetration of new technologies in the working environment?

This section questions the relevance of human-centred approaches as developed today with regard to technological transformations and the evolution of safety and security requirements (i.e. safety management system). New technologies are supposed to assist the human operators, increase their capacities, replace them in painful tasks or even replace expensive environments such as flight simulators in the context of human operator training (Khenak, Bach, Drouot, et al., 2023). Sometimes, new technologies are introduced by technophiles. Technology pushers sell their immediate merits in terms of productivity or an enriched user experience (i.e. immersive glasses). Thus, the question is not only how these technologies will be efficient to achieve the expected goals but how these technologies will or could drastically affect humans' health and their performance at work or even change the role of human beings at work (Laroche & Reuzeau, 2022) when using these technologies.

The Push of Digital and Immersive Technologies

Aeronautics is slowly and cautiously taking hold of these new technologies by challenging their real ability to assist human operators to achieve the mission for which they are responsible, without degrading their perception of the level of control over the machine (Bernard & Arnold, 2019). A very high-level critical analysis of the potential impacts and HOF issues of promising technologies is used to decide whether these technologies have a future for aeronautical products, services, manufacturing and production. This is illustrated in a few examples; for instance using virtual reality (VR) at the service of pilot training. Despite the wealth of research on VR, Airbus carried out a study (Khenak, Bach, & Buratto, 2023) to understand the relationship between cybersickness and usability through the HF dimensions. They investigated the potential effect on pilots' sickness during use and the after-effects. Indeed, the authors showed that previous research studies are not always transferable to the population of pilots or to their types of tasks, since neither the population sample tested (young people in general) nor the time duration, type of tasks or type of working station was representative of the pilot population.

The Push of a New Generation of Artificial Intelligence Technologies

Recently, there has been a surge in new-generation AI-based systems that can be promising for developing human assistance such as machine learning, deep learning, image-based vision and natural language processing. The announcement to great fanfare of ChatGPT's market in 2022 is a striking example: Millions of users have taken hold of this tool, as a discovery, a game, a boon, even a thought-creation tool. Technological developments happen very quickly and one of the major questions raised by the human and social sciences is the question of the time necessary for humans to analyse, test and evaluate whether a technology offers benefits that outweigh the negatives or whether there are even harmful side effects for humans.

The use of smart technologies raises the questions of how they can reduce people's ability to produce knowledge by themselves or to reduce one's ability to think critically by using without hindsight results that have the appearance of truth. Aeronautics is fully conscious of the risk of pilots' complacency and the erosion of competencies that have been identified since the introduction of more automated systems. This is the reason why the aviation industry is conducting studies to understand the HOF issues and benefits of AI bricks according to their different applications such as decision-making, supporting human visual sense or others. EASA (2023) summarised the first categorisation of the role of AI in five levels depending on the function AI is endorsing in the human–machine interaction. In Levels 1A and 1B, AI supports information acquisition and analysis, whereas in Levels 3A and 3B, AI can make and apply decisions, without being overridable by human operators.

- For example, the "speech to text" technology promises to help the pilot and the air traffic controller to streamline and secure understanding of ground/on-board communications, which can be very noisy and unclear due to radiofrequency technology that is highly sensitive to environmental disturbances. A priori, this technology should bring a great benefit to pilots, providing a robust and clear air traffic control message. Level 1A in terms of HOF does not replace a task valued by the pilot community and does not change their mission and role.

- A second example is image-based vision technology. It can help to analyse a human face and recognise the signs of an operator losing vigilance in order to warn him/her when these signs are sufficiently robust and unambiguous, whoever the individual. This is a key difficulty that must be pointed out when technology is intended to capture information about humans in order to analyse human behaviours or states. Each human is different owing to their physical traits (a unique face, type of skin, etc.), singular attitudes, facial expressions, and accessories worn such as glasses or jewellery. These issues have to be solved. The success of a system technology in this area will require the provision of outputs that are highly representative of the reality and are fair and reliable. At Level 1A the use of these technologies is limited to providing alertness systems. The humanities and social sciences support these initiatives to guide how the design should be made for the pilots to use these devices, perceive them as useful and then accept to use them.
- A third example is decision-making support systems. The use of massive in-service experience data and contextual actual flight data can allow the machine to offer a crew a pre-selection of the two or three best options to manage a diversion (Bernard & Arnold, 2019). This kind of technology is more Level 1B. We know that pushing a decision option on a human tends to limit their ability to develop critical thinking by themselves and consequently their competency to make robust decisions. Consequently, the question for HOF specialists is, "Does this specific decision-making competency support a need or a task valued by the pilots? If yes, how can this information be displayed? How can it support pilots to maintain their competencies?"

Each technology will take time to get its stripes, because it must be considered in relation to a global need of the human operator at work. Coming back to HOF, new questions should be addressed and traced during the integration of the new technologies in the aircraft life cycle. How will this technology and its associated usage change the human operators' role? How does this system make operators grow? How should these technologies be used? What are the limits of these new technologies for users and operators? Is the information true, false, complete, gendered or culturally biased? What is the impact on one's free will? What are the social benefits or costs for human operators?

Section 3: How Are the Societal Considerations Taken Into Account by HOF?

Society is evolving and the diffusion of emerging technologies is incredibly fast. HOF should consider how to accompany these evolutions to adapt the work, product and services to the humans' expectations and aspirations for social purpose.

What are the major challenges of tomorrow? They are numerous and diverse, some are unpredictable, but some others will change our lives. It takes time to understand and respond to the aspirations of civil society, considering the impact of psychosocial factors on the behaviours and attitudes of civil society as well as on human operators such as aeronautical personnel. Some of the following challenges deserve more investigation and specific attention.

Fighting Climate Change

Fighting climate change is at the top of the list of social aspirations of society, at least in Europe, Australia, and some other parts of the world.

Aeronautics will face global challenges such as the recent pandemics or climate change. Aviation is battling to reduce the impacts of flights in terms of CO_2 emissions or any other emissions. Aviation is also engaged in reducing noise nuisance and improving the overall efficiency of air transport. Aviation needs to face many challenges related to the planet in order to enable passengers to continue to visit and take care of their family, to discover the world, or to do their job and earn money.

Improving Working Conditions

Professionals aspire to see their profession and working conditions improving.

The meaning and essence of professions, salaries, the balance between professional and private life, and well-being at work occupy a large place in the social debate in Western countries. Professionals such as pilots are very proud of their job and value their profession by their engagement to protect the safety of the flight and of the passengers. HOF should evolve and integrate a very broad scope of all the factors that can impact human behaviours.

Dealing With Demographics

Demographic characteristics (age, level of education, birth rate, etc.) are not to be neglected and are sometimes reduced to stereotypes of little value.

How do pilots approach the question of automation, or a more automated aircraft? Is automation a question of a generational gap? The question is not so simple: Generation Z, despite being accustomed to the use of automated technologies (automation information), still wants to understand what automation does because it is one of the conditions for remaining in "control" of the machine. It is therefore important to rationalise the integration of new technologies as presented in Section 2, from a psychosocial point of view. There are demographic changes, but they are not sufficient to explain or project future human behaviours. Young generations (alpha, beta) are born in a time of digital toolsets and social media. Social networks influence more and more how society perceives the world. Data, both true and false, are circulated with high speed. This is why HOF should understand what the effects of demography at work are, and what can be expected from the operator of the future (Pons et al., 2022) who will work together with the operators of today (Tose & Tazi, 2022).

Considering the Emotional Dimension

The emotional dimension has taken a more valued place in the social realm. HOF considers the way end users and human operators perceive how a change makes sense for them and how they value it.

HOF considers emotional dimensions when designing or assessing a working situation together with user experience specialists. Beyond utility, usability, comfort, health protection and task performance, the emotional factors are considered more systematically. The emotional factors are considered from the perspective of improving working conditions, improving the performance at work, well-being, and the engagement of human operators. It questions how a technological change or a new way of working can impact the attractiveness of the job (e.g. AI, robotisation, supporting tools for remote work). The consideration of emotional affect goes beyond the way it is used to assess the customer's preferences. Developed initially by web marketing, it was used for profiling each consumer to provide an ever more incredibly personalised user experience and to develop their consumerist appetite. For example, a pilot's emotion is captured to assess how new information is perceived in the cockpit, but it is not used to multiply the options in the presentation of information in a cockpit, to stick to any pilot's individual preferences. Diversification of information could even go against safety principles, such as the standardisation of cockpits, which guarantees that (a) each pilot can build benchmarks (stable organisation of the command post, type of information and commands, organisation of work established by the Standard Operating Procedure [SOP]); (b) design and training items are limited and then can be safely demonstrated; and (c) each pilot can refresh their skills by recurrent training sessions. On the other hand, any potential development in terms of crew composition is studied in depth to understand the impact on pilots or cabin crew. Gaining a robust understanding of what is valued today by human operators in their profession and tasks, and why these jobs are so attractive to the different generations with varying professional profiles, contributes greatly towards adapting the developments accordingly.

Balancing the Private and the Personal

There is an increasingly blurred boundary between the works sphere and the private sphere, between the human operator at work and the consumer.

This confusion can lead to methodological biases that need to be formally considered. For example, it is not appropriate to use only a large-scale questionnaire (based on individual opinions collection) to conduct a needs analysis of human operators for the development of a high-risk mission. An opinion collected with time constraints is instantaneous and very biased (superficial thought, context-dependent, unconscious cognitive bias). Individuals and humans at work cannot be reduced to a set of opinions, preferences or statistical rules. We have gone from the old AI "expert system" in the 1980s extracting knowledge from a single expert (lasting 100 hr) to the brief and expeditious interaction of 10 min by theme, or even multiple-choice questions without elaborating on the data. "Did you like the movie? 5/5 I slept well, the armchair is great"'. HOF should go beyond the data flash dictatorship and the illusion of the quantitative. HOF are there to understand the rationale behind the opinions or verbalisations. This requires individual or collective interviews with solid methodologies and know-how to enable the interlocutor to express their rationale, their deep thoughts buried in reasoning or emotional layers. HOF should be used to develop observations of the field or simulations of reality to also access non-conscious behaviours or thoughts that cannot be captured by opinion surveys. This is why HOF continues to promote and apply the work analysis to define the bases, invariants and sociocultural variables of human behaviours in their diversity.

These findings suggest that companies and HOF specialists must equip themselves with the skills and means to carry out in-depth studies to better understand the society of today and tomorrow from a psychosocial and anthropological perspective. Sorting out the short-lived effects of profound changes and identifying the invariant behaviours of certain social groups can be critical, also because some behaviours have their roots in ancient history. Corporate culture, professional culture of certain sectors (such as airline pilots), and regional cultures (South East Asia, North America, etc.) are important for understanding the behaviours of certain social groups. In fact, part of the HOF mission is being able to guide the company in the design choices and in how to accompany the introduction of new technologies or a new operational concept in an aeroplane cockpit.

The underlying question is how to reconcile technological developments as previously described with societal expectations. Can they be the public or human operators at work? If an innovation does not make sense, it falls into disuse on its own. The technology is more likely to be rejected or misused when the users do not benefit from it. If an innovation seems to make sense, then it must be studied in a design process integrating the HOF approach, scrutinising the real contexts of use and carrying out typical HF and psychosocial studies for valid and reliable understanding and reporting of the social impacts. The appropriation by society or professional groups of a major change, whether technological, climatic or social, must be studied upstream by anthropologists and psychosocial specialists in order to better adapt these changes to the affected populations and support these changes.

Summary and Outlook

HOF were originally focused on the human operator at work. Studies were mainly focused on the means to successfully achieve the tasks structured by constraints and requirements. Today, these approaches bring a global approach to humans at work and encompass the job, the human tasks, and the environment in which the activity takes place as well as the individual and social characteristics. Nevertheless, there remains the need to manage a series of HOF points of vigilance in order to maintain the sustainability of HOF processes and therefore guarantee that HOF can deliver the proper added values. HOF specialists should be involved in major decisions where human performance, health and safety, motivation and engagement are at risk. This requires reinforcing the identification of HOF risks across the company and reinforcing the visibility of what HOF can provide to the business to enhance people and product safety, operational efficiency, industrial performance and well-being at work. There is a risk of HOF knowledge erosion due to the loss of HOF specialists (e.g. attrition, retirement) and resource scarcity. In addition, the definition of a HOF core competency is key for its sustainability. HOF competencies should be adapted to the high-speed evolution of technologies or ways of working. HOF specialists should be permanently exposed to new technologies to anticipate HOF-related issues.

HOF competencies should be reinforced to accommodate more psychosocial and anthropological competencies, to get a global understanding and continue to develop appropriate approaches encompassing both society's and human operators' expectations for social purposes. The aeronautical system has shown that it has evolved considerably over the past two decades and that it will continue to change. Society evolves. Designers, users, organisational decision-makers will always be human. Adaptation mechanisms must be found that satisfy global diversity and enable adaptation across cultures. Training, simplifying the use of technical systems, and the active participation of front-line actors in change are crucial for adapting to this changing sector.

References

Abbott, K., Slotte, S. M., & Stimson, D. K. (1996). *The interfaces between flightcrews and modern flight deck systems.* Federal Aviation Administration. https://www.tc.faa.gov/its/worldpac/techrpt/hffaces.pdf

Barbé, J., Mollard, R., & Wolf, M. (2014). Ergonomic approaches to integrate touch screens in future aircraft cockpits. *Journal européen des systèmes automatisés, 4–6,* 303–318. https://doi.org/10.3166/JESA.48.303-318

Barcellini, F. (2022). The design of "future work" in industrial contexts. In H. Laroche, C. Bieder, & J. Villena-López (Eds.), *Managing future challenges for safety: Demographic change, digitalisation and complexity in the 2030s* (pp. 75–83). Springer International Publishing. https://doi.org/10.1007/978-3-031-07805-7_10

Bernard, D., & Arnold, A. (2019). Cognitive interaction with virtual assistants: From philosophical foundations to illustrative examples in aeronautics. *Computers in Industry, 107,* 33–49. https://doi.org/10.1016/j.compind.2019.01.010

Biede, S., Detaille, F., Narotra (Balachandran), T., Petrovic, K., Rathje, H., Rea, A., Schwarz, M., & Vereker, A. (2023). A Competency framework for aviation psychologists and human factors specialists in aviation. *Aviation Psychology and Applied Human Factors, 13*(1), 58–68. https://doi.org/10.1027/2192-0923/a000245

Daniellou, F. (2005). The French-speaking ergonomists' approach to work activity: Cross-influences of field intervention and conceptual models. *Theoretical Issues in Ergonomics Science, 6*(5), 409–427. https://doi.org/10.1080/14639220500078252

Edmonds, J., & Gray, K. (2019). Assessing human factors maturity. *Chemical Engineering Transactions, 77,* 481–486. https://doi.org/10.3303/CET1977081

European Union Aviation Safety Agency. (2007). *Certification specifications for large aeroplanes CS-25.* https://www.easa.europa.eu/sites/default/files/dfu/CS-25_Amdt%203_19.09.0_Consolidated%20version.pdf

European Union Aviation Safety Agency. (2021). *Study on the societal acceptance of urban air mobility in Europe.* https://www.easa.europa.eu/sites/default/files/dfu/uam-full-report.pdf

European Union Aviation Safety Agency. (2023). *Concept paper: Guidance for level 1 & 2 machine learning applications* (Proposed Issue 02). https://www.easa.europa.eu/en/easa-concept-paper-first-usable-guidance-level-1-machine-learning-applications-proposed-issue-01pdf

Federal Aviation Administration. (2022). *Flightpath management.* https://www.faa.gov/regulations_policies/advisory_circulars/index.cfm/go/document.information/documentID/1041433

International Civil Aviation Organization. (2021). *Doc* 10151-*manual on human performance (HP) for regulators.* https://www.icao.int/safety/OPS/OPS-Normal/Pages/HP.aspx

International Civil Aviation Organization. (2013). *Annex 19-safety management: International standards and recommended practices.* https://www.icao.int/safety/SafetyManagement/Documents/Annex%2019%20-%20ICAO%20presentation%20-%20self%20instruction%2024September2013.pdf

International Organization for Standardization. (2010). *Ergonomics of human-system interaction – Part 210: Human-centred design for interactive systems (ISO Standard No.* 9241-*210:2010).* https://www.iso.org/standard/52075.html

International Organization for Standardization. (2021). *Ergonomics – manual handling – part 1: Lifting, lowering and carrying (ISO Standard No.* 11228-*1:2021).* https://www.iso.org/standard/76820.html

Jimenez, C., Kasper, K., Rivera, J., Talone, A. B., & Jentsch, F. (2015). Crew resource management (CRM): What aviation can learn from the application of CRM in other domains.

Proceedings of the Human Factors and Ergonomics Society Annual Meeting, 59(1), 946–950. https://doi.org/10.1177/1541931215591274

Khenak, N., Bach, C., & Buratto, F. (2023). Understanding the relationship between cybersickness and usability through the human factors dimensions. In *Proceedings of ergonomie et informatique avancée* (pp. 1–10). ACM. https://doi.org/10.1145/3624323.3624342

Khenak, N., Bach, C., Drouot, S., & Buratto, F. (2023). *Evaluation of virtual reality training: Effectiveness on pilots' learning.* IHM'23 – 34e Conférence Internationale Francophone sur l'Interaction Humain-Machine, AFIHM; Université de Technologie de Troyes, France.

Kirwan, B. (2022). Towards an action-oriented safety culture maturity scale. In P. Arezes & A. Garcia (Eds.) *Safety management and human factors. AHFE international conference* (Vol. 64). AHFE International. https://doi.org/10.54941/ahfe1002631

Laroche, H., & Reuzeau, F. (2022). Learning from the military. In H. Laroche, C. Bieder, & J. Villena-López (Eds.), *Managing future challenges for safety: Demographic change, digitalisation and complexity in the 2030s* (pp. 25–31). Springer International Publishing. https://doi.org/10.1007/978-3-031-07805-7_3

Nelson, J., Buisine, S., & Aoussat, A. (2009). Assisting designers in the anticipation of future product use. *Asian International Journal of Science and Technology in Production and Manufacturing Engineering, 2*(3), 25–39. https://hal.science/hal-00787618/document

Pelchen-Medwed, R., Save, L., Heintz, A., Reuzeau, F., & Biede-Straussberger, S. (2021). The evolution toward a common air/ground framework for human performance assessments in Europe. In I. V. Koglbauer & S. Biede-Straussberger (Eds.), *Aviation psychology: Applied methods and techniques* (pp. 1–16). Hogrefe Publishing.

Pons, B., Rodriguez, J. H., & Reuzeau, F. (2022). Airbus global workforce forecast (GWF). In H. Laroche, C. Bieder, & J. Villena-López (Eds.), *Managing future challenges for safety: Demographic change, digitalisation and complexity in the 2030s* (pp. 63–66). Springer International Publishing. https://doi.org/10.1007/978-3-031-07805-7_8

Prentice, B., DiNota, A., Sargent, S., Hayes, L., Franzoni, C., Mishra, U., & Stelle, M. (2023). *Global fleet and MRO market forecast* 2023–2033. Oliver Wyman. https://www.oliverwyman.com/our-expertise/insights/2023/feb/global-fleet-and-mro-market-forecast-2023-2033.html

Reuzeau, F. (2020). The key drivers to setting up a valuable and sustainable HOF approach in a high-risk company such as Airbus. In B. Journé, H. Laroche, C. Bieder, & C. Gilbert (Eds.), *Human and organisational factors: Practices and strategies for a changing world* (pp. 31–39). Springer International Publishing. https://doi.org/10.1007/978-3-030-25639-5_5

Reuzeau, F. (2013). *Human factors design process: Benefits and success factors.* 3rd Human Dependability Workshop (HUDEP), ESA, Munich, Germany.

Stroeve, S., Smeltink, J., & Kirwan, B. (2022). Assessing and advancing safety management in aviation. *Safety, 8*(2), Article 20. https://doi.org/10.3390/safety8020020

Tose, A., & Tazi, D. (2022). Careers surpassing a half-century: A look at Japan and France. In H. Laroche, C. Bieder, & J. Villena-López (Eds.), *Managing future challenges for safety: Demographic change, digitalisation and complexity in the 2030s* (pp. 51–57). Springer International Publishing. https://doi.org/10.1007/978-3-031-07805-7_6

Van Veenendaal, E., & Wells, B. (2012). *Test maturity model integration TMMi: Guidelines for test process improvement.* UTN Publishers.

Wolf, M., & Mollard, R. (2020). *Pratiques de l'ergonomie: de la méthode aux applications* [Ergonomics practices: From methods to applications]. Octarès.

Yeh, M., Jo, Y. J., Donovan, C., & Gabree, S. (2013). *Human factors considerations in the design and evaluation of flight deck displays and controls* (Version 1.0, No. DOT-VNTSC-FAA-13-09). John A. Volpe National Transportation Systems Center (US). https://rosap.ntl.bts.gov/view/dot/11986

Contributors

Editors

Ioana V. Koglbauer received her PhD and *Habilitation* in psychology from the University of Graz, Austria, and is an experienced aviation psychologist accredited by the European Association for Aviation Psychology. Ioana is affiliated with Airbus Defence and Space GmbH in Germany, and is also a lecturer at Graz University of Technology in Austria. She has led multiple research projects between academia and the aviation industry, and has authored numerous publications. In addition, Ioana served as an expert for various international organisations: The Civil Air Navigation Services Organization (CANSO) and the EASA Human Factors CAG. She was Editor-in-Chief of the peer-reviewed journal *Aviation Psychology and Applied Human Factors* and Member of the Board of the Directors of the European Association for Aviation Psychology for 4 years (2014–2018). Ioana is also a passionate pilot and instructor with experience in teaching human factors and safety for "Airline Transport Pilot" and "Type Rating Instructor" courses.

Sonja Biede-Straussberger is an expert for cognitive psychology in the Human Factors and Ergonomics Department of Airbus Operations. She received her PhD in psychology in 2006, and holds a master's degree in educational science, both from the Karl Franzens University of Graz, Austria. Between 2003 and 2006, her studies with EUROCONTROL were focused on monotony in air traffic controllers, and then extended to the human integration in the global aeronautical system. Between 2009 and 2018 she led the Airbus human factors contribution to SESAR, endeavouring to give human operators a central role throughout the engineering cycle. Beyond that, she is actively involved in promoting human factors integration in the industry, such as the French Association for System Engineering or the European Association of Aviation Psychology. She holds a private pilot licence and taught human factors for student pilots as well as engineers. Today, she applies her expertise to continuously improving or developing new products by optimising the human factors contribution along all phases of cockpit design.

Authors

Cheryl Agyei is a scientist with a research interest in the physiology of extreme environments, particularly within human spaceflight and high-altitude physiology. She has a background in biomedical science and completed her MSc in human & applied physiology at King's College London. She has worked for the European Space Agency (ESA), where her research focused on evaluating key findings from two decades of life science research on the International Space Station (ISS) and assessing their operational impact on space medicine.

Cédric Bach works as a human factors expert at Human Design Group in France. He has a PhD in psychology and is specialised in usability engineering for immersive systems.

Julia Behrend is a psychologist with a PhD in cognitive neuroscience affiliated with Ecole Normale Supérieure, Paris. Dr. Behrend is the Head of Safety Innovation and Human Performance at Air France's Flight Safety Department. Her current role is to ensure an up-to-date understanding of human factors implied in flight safety.

Leonie Bensch holds degrees in psychology and human–computer interaction. She is a PhD candidate at the German Aerospace Center (DLR) investigating augmented reality applications for future lunar extravehicular activity (EVA) missions in collaboration with the European Space Agency and the Institute of Industrial Engineering and Ergonomics at RWTH Aachen, Germany.

Denys Bernard is affiliated with Airbus Commercial. He is an aeronautical engineer (ISAE-SupAero) and holds a PhD in computer science (UPS Toulouse). He is an expert in artificial intelligence (AI) for human–system interaction. He contributed to decision support systems initially for aircraft maintenance and now for pilot decisions in operations. He also participates in regulatory initiatives for the certification of AI.

Brittany Bishop is a First Lieutenant in the United States Air Force. She earned her bachelor degree from the United States Air Force Academy and her masters degree from the Massachusetts Institute of Technology in aeronautical/astronautical engineering. At MIT, Brittany spent her time within Lincoln Laboratory Group 66 and within the Engineering Systems Laboratory focusing on the analysis of defense and communication systems.

Reinhard Braunstingl received his PhD in mechanical engineering from Graz University of Technology, Austria. He is a professor with the Graz University of Technology, specialised in mechanics, flight mechanics and flight simulation. He has conducted numerous research projects in cooperation with the aviation industry. Reinhard is also an experienced and passionate pilot, flight instructor and examiner.

Judith Irina Buchheim is a clinician scientist trained in anaesthesiology and working at the Laboratory for Translational Research "Stress and Immunity", Department of Anaesthesiology, LMU University Hospital, Munich, Germany. She has over 10 years of experience in the management and coordination of space flight (analogue) and clinical research studies. Dr. Buchheim is the Principal Investigator of the CIMON Pilot Study.

Florence Buratto works as a human factors expert for Airbus aircraft division, in particular for the Flight Operations and Training Standard Department. She has a PhD in cognitive psychology and is specialised in human-machine interaction and human performance, with a particular focus on the disturbing factors of the performance.

Andrea E. M. Casini holds a PhD in aerospace engineering. He currently works for the German Aerospace Center (DLR) as Science and Operations Manager for the LUNA analogue facility, a technology and training centre to prepare the forthcoming missions to the Moon.

Mickaël Causse is Professor of Neuroergonomics at the Human Factors Laboratory of ISAE-SUPAERO in Toulouse, France, where he obtained his PhD in neuroscience in 2010. Prof. Causse's research focuses on analysing the activity of civil and military pilots, in particular through the use of eye-tracking and neurological measurements.

Ralf Christe is a passionate UX/UI designer and brand expert affiliated with APOGAEUM Design, Durmersheim, Germany. He designed the animated face and the digital surface of the CIMON robot. Ralf studied design at the State Academy of Fine Arts Stuttgart and worked as a managing partner in a UX design company based in Karlsruhe, Germany.

Anne-Claire Collet works as an R&D engineer at Human Design Group in France and is specialised in human physiology and cognition. She has a PhD in cognitive neuroscience and an engineering degree.

Maxime Cordeil is a senior lecturer at the School of Electrical Engineering and Computer Science, specialising in human-centred computing, located in Brisbane, Australia. He earned his PhD in 2013. Dr. Cordeil's research focuses on investigating how virtual and augmented reality technologies empower users to gain a deeper understanding of and interact more effectively with complex data.

Aidan Cowley holds a PhD in electronic systems. After 3 years as a researcher and lecturer at the National Centre for Plasma Science and Technology (NCPST), in 2014 Dr. Cowley joined the European Space Agency as a scientist, stationed at the European Astronaut Centre in Cologne, Germany.

Paul de Medeiros is an industrial design engineer at NASA's Center for Design and Space Architecture. His research focuses on the adaptation of strategic human-centred design methods to early-stage development of human space systems.

Daniel Dreyer is affiliated with Airbus Defence and Space GmbH. After studying aeronautical engineering at the Technische Universität München, Daniel joined Airbus as a research engineer and received his PhD. He worked on European research projects focusing on virtual reality, human factors and cockpit design for civil and military applications, where he acted as Human Factors Architect and Human Factors Integration Manager.

Stéphane Drouot works as an instructor for Airbus aircraft division, especially on the specifications of pilot training programs and training media for the Flight Operations and Training Standard Department. He is a former military and instructor pilot on fighter and maritime patrol aircraft, and is currently simulator instructor on A320/A330 and A380 aircraft.

Florian Dufresne is an aerospace engineer and a PhD candidate at the Arts et Métiers Institute of Technology in Laval, France. Following his motto "Ad astra, per virtualis", he aims at studying the avatar–user relationship in extended reality (AR, VR) and its effects on the user experience in cross-platform applications.

Till Eisenberg is the project manager for the project CIMON at Airbus Defence and Space GmbH. He studied mechanical engineering at RWTH Aachen, focussing on human spaceflight topics. He started his career in the Department of Life Science at Airbus, formerly EADS Space Transportation, in 2005.

Anna Fogtman (PhD) is a senior scientist at the European Astronaut Centre, European Space Agency (ESA). Professionally involved in spaceflight for seven years, she leads the radiation protection operations to safeguard European crewed missions to space. She manages ESA operations on the International Space Station (ISS) and plans for Moon missions. She works with science teams to coordinate ESA space research

Hans-Gerhard Giesa is the Airbus Senior Expert in the field of human factors in cabin and cargo, affiliated with Airbus Operations GmbH. He holds a doctorate in human–machine systems from the Technische Universität Berlin, and has an interdisciplinary background in engineering and psychology. After joining Airbus in

2003, he progressed through a combined management and expert position in Cabin and Cargo Human Factors and Security. In 2016, he was nominated as the Senior Expert and provides advice and consultancy for technical decisions on ergonomics, comfort, health, usability and operational safety and efficiency.

Adrianos Golemis (MD, MSc) is a medical doctor with a master's degree in space studies from the International Space University (ISU). He has worked for the European Space Agency (ESA) to implement medical experiments at Concordia Station, Antarctica, and supported bed-rest and dry immersion physiology studies for the French Institute MEDES. Since 2018 he is supporting European astronauts before, during, and after their spaceflight while in 2022 he went through ESA's selection of new astronauts himself.

Enrico Guerra is a researcher and enthusiast in the fields of extended reality, robotics and human spaceflight. He holds a bachelor's and a master's degree in computer science and is currently pursuing a PhD in the field of human–robot interaction at the University of Duisburg-Essen in Germany.

Pauline Harrington is a master's candidate at the Massachusetts Institute of Technology in the Engineering Systems Laboratory, MA, USA. She graduated from Tufts with a BS in engineering psychology in 2021. Her research interest is in using human factors techniques to enhance safety analysis on complex sociotechnical systems.

Don Harris is Professor of Human Factors in the Faculty of Engineering, Environment and Computing at Coventry University, UK. He is a fellow of the Chartered Institute of Human Factors and Ergonomics and a chartered psychologist. Don is the human factors technical advisor to Flimax Ltd.

Christophe Hurter is Professor of Computer Science at the French Civil Aviation University in Toulouse, France, where he earned his PhD in 2010. Professor Hurter's research focuses on explainable artificial intelligence, big data, information visualisation, immersive analytics, and human–computer interaction.

Sylwia Kaduk (PhD) is a scientist working in the Space Medicine Team (ESA HRE-OM) at the European Astronaut Centre (EAC) in Cologne, Germany. She is also a human factors engineer at the Austrian Space Forum. In 2021 she obtained a PhD in human factors engineering from the University of Southampton. Previously she worked as a research technician at Surrey Sleep Research Centre and as a research assistant in the behavioural medicine group at the University of Maastricht.

Sabine Kaiser (PhD) is professor at the Regional Centre for Child and Youth Mental Health and Child Welfare at UiT The Arctic University of Norway in Tromsø.

Her PhD work was related to collaboration, burnout, engagement, and service quality in the healthcare sector. Her research interests also include research methods, meta-analysis, and youth mental health and welfare.

Christian Karrasch is Program Manager at the German Aerospace Center. He led the innovative and successful CIMON project for 4 years from the initial phase in early 2017 until the beginning of 2021. Christian studied physics and earned his PhD in material physics with experiments conducted on the International Space Station (ISS).

Nawel Khenak works as an R&D engineer at Human Design Group in France. She has a PhD in computer science and is a specialist in spatial presence and cybersickness in immersive systems.

Nancy Leveson is a professor in the Department of Aeronautics and Astronautics, Massachusetts Institute of Technology, USA. Professor Leveson works on safety in complex systems. Her research areas include safety engineering, system engineering, human factors, software engineering, and the social and managerial aspects of safety. The emphasis is on how to integrate all of these factors when considering safety.

Heikki Mansikka (LtCol, ret.) is a former F/A-18 fighter pilot and an adjunct professor with the Department of Military Technology, National Defence University in Finland.

Monica Martinussen (PhD) is Professor of Psychology and Department Head at the Regional Center for Child and Adolescent Mental Health at UiT The Arctic University of Norway. She is also Professor II at the Norwegian Defence University College. She conducted her doctoral research in the area of pilot selection, and has been engaged in research on that topic for many years. She is currently Associate Editor of the journal *Aviation Psychology and Applied Human Factors*.

Birgit Moesl is a PhD candidate at the Institute of Engineering and Business Informatics at the Graz University of Technology in Austria. Her research interest is in sociotechnical systems, responsible research and innovation, and gender studies. In addition, she is an active pilot.

Randall J. Mumaw (PhD) is employed by San Jose State University in California and works on-site at the NASA Ames Research Center. Dr. Mumaw's research career has focused on supporting operator performance in complex, safety-critical systems. He has studied air traffic controllers, commercial pilots (NASA, Boeing), and nuclear power plant operators (Westinghouse).

Thomas Müller holds a doctorate in engineering and a *Habilitation* in ergonomics. He is a lecturer at TU Berlin and was previously at the Hochschule für Technik und Wirtschaft in Berlin. After several years as a senior scientist at TU Berlin he owned an engineering/ergonomics consultancy. He worked for over 10 years at the Cabin and Cargo Human Factors department at Airbus (Hamburg) and recently as an external employee on a project basis for the Airbus Cabin and Cargo Human Factors Department. He actively supports standardisation (DIN, CEN, ASD-STAN) in the field of aircraft cabin design.

Federico Nemmi works as a human monitoring systems, neuroscience and R&D engineer at Human Design Group in France. He has a PhD in cognitive neuroscience with a particular focus on human monitoring systems and data science applied to human behaviour and physiology.

Tommy Nilsson is affiliated with the European Space Agency (ESA). He is a postdoctoral research fellow at the European Astronaut Center in Cologne, Germany. His work is primarily situated within the field of human–computer interaction and focuses on investigating the potential applications of virtual reality as a preparatory tool for humanity's upcoming missions to the Moon.

Alexander Rabl studied psychology at Regensburg University and Charles University in Prague and received his PhD in a collaboration project with Airbus. He worked at Airbus as a human factors researcher, interface designer and project manager for national and international programs regarding multimodal displays and user assistance for cockpits in highly agile aircraft. He is affiliated with Airbus Defence and Space GmbH.

Gerhard Reichert is a renowned product designer with a strong engineering background. He has worked at prestigious companies such as Studio De Lucchi Milano. In 2000, he founded his own studio, where office furniture, medical technology, lighting technology, among other things, are developed. Mostly he is leading the supervision for the whole process from concept to realization. He was responsible for the industrial design and together with Ralf Christe for the digital surface of the CIMON robot. Since 2007 he has been teaching as a professor for product design and has conducted international research projects.

Florence Reuzeau is an aeronautical engineer and doctor in cognitive ergonomics. She worked as a safety specialist on complex avionics systems for 5 years before setting up the Human Factors Organisation in Airbus Engineering. She integrated human sciences into engineering for an optimised human-centred design applied to A380, A400M, A350 and ATR aircraft. She contributed to the definition of human factors regulations. Since 2014, she has been Human Factors Executive Expert for Airbus overall, defining the Airbus Group human factors strategy for research and development. She is involved in international standardisation groups and is very active in the organisation of worldwide human factors conferences. She is a European expert for European project evaluation.

Rubén Rodríguez Rodríguez is an aeronautical engineer at CRIDA A.I.E and an associate lecturer at University Carlos III of Madrid. He has worked in air traffic management in the definition and validation of operational concepts and systems, and he is currently acting as the human performance lead in SESAR3.

Flavie Rometsch is affiliated with the European Space Agency (ESA). Flavie is a space systems engineer working at the Concurrent Design Facility (CDF) in the European Space Research and Technology Centre, in the Netherlands, where she actively contributes to the Argonaut lander project. With a strong interest in extended reality technologies, Flavie envisions their potential for future human space exploration applications.

Rodrigo Rose is a system safety engineer and a recent master's graduate from the Department of Aeronautics and Astronautics at the Massachusetts Institute of Technology, USA. He obtained his bachelor's degree in aerospace engineering from the Georgia Institute of Technology, focusing on commercial aviation safety. At MIT, his research focused on applications of systems theory for the improvement of safety in aviation and healthcare.

Harald Schaffernak is a PhD student at the Institute of Engineering and Business Informatics at Graz University of Technology. Beyond his academic endeavors, he is a co-founder of Guid.New GmbH, a private company in the software engineering industry. His research focuses on software quality and augmented reality.

Adriana-Dana Schmitz is a certified aviation psychologist and human performance (HP) expert at EUROCONTROL. Dana was the HP Lead in the Single European Sky ATM Research (SESAR) between 2020 and 2023 and a co-lead of the Coordination Plan CP3.2, between SESAR and NextGen (Federal Aviation Administration). She holds an MSc in clinical neuropsychology and is currently working in the Safety Unit of the Network Manager Directorate.

Matthew J. W. Thomas (PhD) is an associate professor at Appleton Institute at Central Queensland University in Adelaide, Australia, and Director of Westwood-Thomas Associates. His research focuses on error management and non-technical skills across high-risk industries.

Andreas Treuer holds an MSc in aerospace engineering from RWTH Aachen University, Germany. Having contributed to several European Space Agency (ESA) projects, including the Argonaut Lunar Lander, his current position at the

European Astronaut Center involves supporting the LUNA analogue facility project and the XR Lab as a visualisation, haptics and extended reality engineer. He is affiliated with the ESA.

Christoph Vernaleken, currently Senior Expert Flight Deck and Human Factors at Deutsche Aircraft GmbH and lecturer at TU München, Germany, has been working in flight deck design and research since 2001, starting with a doctorate in aerospace engineering (TU Darmstadt). Previously, he served almost 15 years with Airbus Defence and Space.

Kai Virtanen is a professor in the joint professorship of Operations Research with Systems Analysis Laboratory, Department of Mathematics and Systems Analysis, Aalto University, Finland, and the Department of Military Technology, National Defence University, Finland.

Anna Vock is currently completing her MSc studies in design and engineering at the Politecnico di Milano, Italy. Leveraging her internship experience at the European Space Agency, she founded and leads the "Space Architecture" project group within the PoliSpace Student Association, dedicated to exploring human-centred design approaches for future lunar outposts.

Wolfgang Vorraber is an associate professor at Graz University of Technology, Austria. His research interests include business information systems engineering and sustainable service design with a special focus on production, public safety, health care, and aviation. Wolfgang received a PhD in engineering economics and management.

Tobias Weber (PhD) is a space medicine scientist at KBR GmbH for the European Space Agency (ESA). Since 2013, he has been working in the space medicine team at the European Astronaut Centre in Cologne, Germany. His research in space physiology aims to optimize astronaut healthcare and apply findings to the general population. He is also a licensed strength and conditioning coach.

Peer Commentaries

In today's rapidly changing technological and societal landscape, this book is an absolute must-read. It takes a deep dive into the psychological and human factors that are crucial for maintaining safety in aerospace operations and development. Penned by top experts from academia and the aerospace industry, it unveils cutting-edge techniques in system safety engineering, groundbreaking design concepts for cockpits and spacecraft, and savvy AI implementation strategies. It tackles hazard analysis for space, air, and ground operations, and explores the thrilling potential of virtual, augmented, and mixed reality to enhance operator perception and boldly go toward new frontiers. The book also puts forward innovative proposals to elevate industry standards and seamlessly integrate human and organizational factors.

Rosa Maria Arnaldo Valdés, Professor of Aviation Safety and Human Factors, Faculty of Aerospace Engineering, Universidad Polytechnical de Madrid, Spain; Former President of the Spanish Civil Aviation Accident Investigation Authority

From my perspective, the book has the potential to become an essential resource for addressing the challenges posed by the expected increase of automation levels in aviation in the near future. The gradual integration of artificial intelligence into current operations will deeply transform aviation, altering existing paradigms while aiming to maintain a human-centric approach. This book offers an overarching, interdisciplinary approach to understanding the state-of-the-art in human factors and safety in aviation and anticipates questions, trends, and strategies for integrating human factors in a rapidly evolving technological landscape.

I appreciate how the volume systematically progresses, in a well-structured manner, from the system design table to actual deployment, considering crucial factors such as system explainability and the necessary trade-offs between situational awareness, workload, and performance. The specific use cases are illustrative and pertinent, notably the eye-tracking example for pilot attention and the detailed chapter on air traffic management research. This particular chapter explains how the Single European Sky Advanced Research (SESAR) program has analyzed human factors and capabilities in designing new operational and technological solutions intended to transform future operations in Europe. The concept of human readiness levels (HRL) is highlighted as a fundamental indicator.

In a nutshell, this book provides a comprehensive overview of the growing importance of human factors in air traffic management with a future-oriented outlook on the needs and challenges. Practitioners and aviation specialists will find it a valuable guide for anticipating potential questions and strategies, acting as a compass to navigate the challenges related to emerging technologies and higher levels of automation in the aviation domain.

Jose M. Cordero, Leader of the Digital European Sky Performance Management in SESAR; Principal Researcher at CRIDA-ENAIRE

The book Aerospace Psychology and Human Factors, co-edited by Ioana Koglbauer and Sonja Biede-Straussberger, combines basic principles of human factors and scientific research with applied aerospace topics. It spans the arc from Gestalt psychology and its significance for cockpit design to modern concepts of explainable AI. Aspects of design and assistance systems focus on the user and their expectations in the field of aerospace, air traffic control, and human space flight. Essential concepts of cognitive psychology are placed into context and their measurement is discussed, as are the possible consequences of virtual, augmented, and mixed reality. In a technically dominated environment such as aviation, this book clearly emphasizes the human factor and its manifold interdependencies in this socio-technical system and thus represents a valuable link to the neighboring disciplines of aviation psychology.

Oliver Daum, Human Factors Researcher, Psychological Service of the German Bundeswehr; Lecturer, Department of General Experimental Psychology, Johannes Gutenberg University of Mainz, Germany

Are you involved in the design, training, operation, or safety monitoring of future airplanes and space travel, planning to use STPA, AI, VR, or AR? Then this excellent selection of the latest research papers by leading human factors experts from both industry and academia is for you. I highly recommend it as reading for anyone committed to advancing human-machine collaboration in the aviation and space technology field to the next level!

Mirjana Fritsche, Pilot, Aviation Expert, Liaison Member of S7 SAE, committee for Flight Deck and Handling Qualities Standards for Transport Aircraft

Aerospace Psychology and Human Factors is an excellent resource that integrates foundational concepts with contemporary and emerging issues in the field. It covers a wide range of topics, from traditional human factors principles to cutting-edge technologies, and offers invaluable insights into the psychological and human elements critical for safety and efficiency in aerospace systems. The book is an essential read for anyone dedicated to human factors, including practitioners, researchers, and students alike, providing practical techniques and solutions for today's challenges and future developments.

Dr. Patrizia Knabl-Schmitz, Human Factors Specialist, Emirates

This book offers a unique, comprehensive, and insightful exploration of the fundamental connection between human potential and the operational dynamics of aerospace systems. Rich in examples, scientific methods, regulations, and standards, it is an indispensable source for those passionate about putting the people first in the design!

Dr.-Ing. Gernot Konrad, Flight Deck Human Factors Expert

This publication demonstrates that aerospace psychology is not solely concerned with the selection and/or counseling of personnel, but also encompasses forward-thinking concepts in the realms of system design and optimization of human–machine interface adaptation. Incorporating contemporary technologies, it provides guidance on the implementation of statistical methods to ensure an evidence-based approach to methodology development.

Michael Mikas, MSc, Head of the Department for Aviation Psychology, Austrian Armed Forces Vienna

The human factors methods, techniques, and tools outlined in this book focus on the human element of the complex sociotechnical aerospace system. It strives to address the challenges of rapid change while keeping safety firmly as the priority and acknowledging the human as the glue for these systems! The chapters approach the current and emerging challenges to our industry in a multi- and interdisciplinary way, keeping the importance of the end-user in mind at all stages with the knowledge that a holistic/collaborative approach yields the best results.

Nicola Ni Riada, BA, MA, MSc, Operational Air Traffic Controller with over 30 years of experience in ATC; Human Factors Actor; IFATCA Communications Coordinator

Technical and methodological competence is half the battle in the design and operation of aerospace systems. The other half is – as correctly emphasized in the preface – passion for people and society as a whole. In their usual highly professional manner, the two renowned female editors took great care to offer a diverse and exquisite line-up of authors and current topics. The chapters are of practical relevance and make you want to try the methods and techniques yourself. This book is a long-awaited sequel of a series of unique and comprehensive guidelines on how to properly apply aerospace psychology and human factors. Whether you are new to the subject or at expert level – become methodically more competent by immersing yourself in this book!

Dr. Michaela Schwarz, Co-Founder and Deputy Chairwoman of the Austrian Aviation Psychology Association (AAPA), Graz, Austria

From cockpit design to space exploration, this book takes an in-depth look at the world of aeronautical/aerospace human factors. Human factors became an area of interest nearly 80 years ago. Despite eight decades of experience, as we move from simpler, hardware-based systems to complex software-intensive and automation supported systems, practitioners still struggle to integrate properly the human element into overall system design and safety assurance in a simple and practical manner that generates enough value to the aviation business. This book addresses some aspects of this need: It is an easy to navigate, one-stop shop for advanced practices in human factors that every aviation practitioner should read and try to apply.

Dr. Branka Subotic, Head of Human Factors and Transformation, Skyguide, Switzerland